Neurophysiology - Networks, Plasticity, Pathophysiology and Behavior

Edited by Thomas Heinbockel

Published in London, United Kingdom

Neurophysiology - Networks, Plasticity, Pathophysiology and Behavior
http://dx.doi.org/10.5772/intechopen.101295
Edited by Thomas Heinbockel

Contributors
Pipat Chiewvit, Yunxiang Ge, Weibei Dou, Gabriela Alvares-Pereira, Carolina Maruta, Maria Vânia Silva-Nunes, Aynur Özge, Derya Uludüz, Füsun Mayda Domaç, Nikolaos Tsarouchas, Jose Rodrigo Carrillo-Marquez, Jose Damian Carrillo-Ruiz, Jitender Jakhar, Manish Sarkar, Nand Kumar, Maurizio Oggiano, Atif Amin Baig, Farwa Munir, Zainb Anis, Nida Islam, Sehar Bano, Shahzaib Naeem, Muhammad Hassan Nasir, Zaineb Sohail, Mohammad Shabani, Mohammad Banazadeh, Kiarash Saleki, Banafshe Abadi, Zeynab Pirmoradi, Zahra Esmaili, Shiva Amiri, Ramtin Pourahmad, Kristi A. Kohlmeier, Richard M. Millis, Merin Chandanathil, Ayoola Olajuwon Awosika, Fidelis Nwachukwu, Ravindrasingh Rajput, Sheetal Naik, Kishan Kadur, Thomas Heinbockel

First published in London, United Kingdom, 2022 by IntechOpen
IntechOpen is the global imprint of INTECHOPEN LIMITED, registered in England and Wales, registration number: 11086078, 5 Princes Gate Court, London, SW7 2QJ, United Kingdom

British Library Cataloguing-in-Publication Data
A catalogue record for this book is available from the British Library

Additional hard and PDF copies can be obtained from orders@intechopen.com

Neurophysiology - Networks, Plasticity, Pathophysiology and Behavior
Edited by Thomas Heinbockel
p. cm.
Print ISBN 978-1-80356-068-7
Online ISBN 978-1-80356-069-4
eBook (PDF) ISBN 978-1-80356-070-0

For EU product safety concerns:
IN TECH d.o.o., Prolaz Marije Krucifikse Kozulić 3, 51000 Rijeka, Croatia,
info@intechopen.com or visit our website at intechopen.com.

Meet the editor

Thomas Heinbockel, Ph.D., is a professor in the Department of Anatomy, at Howard University College of Medicine, Washington, DC. He holds an adjunct faculty position in both the Department of Anatomy & Neurobiology and the Department of Physiology, at the University of Maryland School of Medicine. Dr. Heinbockel studied biology at Philipps University, Germany. His studies of the brain started during his MS thesis work at the Max-Planck-Institute for Behavioral Physiology, Germany. Dr. Heinbockel earned a Ph.D. in Neuroscience at the University of Arizona, USA. After graduating, he worked as a research associate at the Institute of Physiology, Otto-von-Guericke University, Germany. Dr. Heinbockel's research is focused on understanding how the brain processes information as it relates to neurological and psychiatric disorders. His laboratory at Howard University concentrates on foundational and translational topics such as drug development, organization of the olfactory and limbic systems, and neural signaling and synaptic transmission in the central nervous system.

Contents

Preface

Neurophysiology is the study of the nervous system in terms of its function. The focus is on nerve cells and glial cells as the building blocks of the nervous system and the networks that are formed by them. Individual neurons and their networks are subject to various forms of plasticity and undergo short- or longer-lasting modifications. The outcome of neural network action can be the behavior of specific brain areas and body regions or the behavior of an individual. Finally, pathophysiological processes can affect all organizational levels of the nervous system from individual cells to networks, neural systems, and behavior. *Neurophysiology - Networks, Plasticity, Pathophysiology and Behavior* addresses these various scales in health and disease.

The book reviews novel findings related to neurophysiology, such as neural plasticity, neurological disorders, sensory systems, cognition, and behavior. It provides an overview of the current state of the art of neurophysiology research and focuses on the most important evidence-based developments in this area. Chapters focus on recent advances in specific areas of neurophysiology in different brain regions and experimental models.

The book is divided into two sections. Section 1, "Neural Networks, Neuroplasticity, Behavior," includes Chapters 1–6. Section 2, "Neuropathophysiology," includes Chapters 7–12.

Chapter 1, "Clinical Neurophysiology of Epileptogenic Networks", by Nick Tsarouchas, describes foundational elements of neural networks and how they can give rise to aberrant functions such as epileptic seizures. More specifically, the author demonstrates with artificial neuronal network simulations how physiological brain oscillations (delta, theta, alpha, beta, and gamma range, and transients thereof, including sleep spindles and larger sleep waves) are generated. In addition, the author shows how epileptiform phenomena can develop and be observed at different levels of analysis.

In Chapter 2, "Neurophysiology Involved in Neuroplasticity: Mechanisms of Forgetting", Jose Rodrigo Carrillo-Marquez and Jose Damian Carrillo-Ruiz explain new neurophysiological mechanisms that result in processes of brain plasticity. Furthermore, the authors clarify that neuroplasticity can take place in different complex tasks and discuss the evolutionary advantage of forgetting.

Chapter 3, "Neurophysiology of Emotions", by Maurizio Oggiano, reviews emotions as automatic and primary patterns of purposeful cognitive and behavioral organizations. The author outlines the neurophysiological basis of elements of emotion and their function, namely, coordination, signaling, and information purposes.

Chapter 4, "Neuronal Architecture and Functional Organization of Olfactory Glomeruli", by Thomas Heinbockel, addresses the modular organization of the brain in a sensory system, our sense of smell, and olfaction. The chapter focuses primarily on an insect animal model that has contributed tremendously to our understanding of questions regarding how olfactory information is processed in the brain and how olfactory glomeruli contribute to encoding qualitative and quantitative features of a sensory stimulus.

In Chapter 5, "Quantitative Electroencephalography for Probing Cognitive and Behavioral Functions of the Human Brain", Richard M. Millis, Merin Chandanathil, Awosika Ayoola, Fidelis Nwachukwu, Ravindrasingh Rajput, Sheetal Naik, and Kishan Kadur use exam scores and an individualized self-inventory of psychosocial interactions to provide data for probing behavioral and cognitive performance of medical students. Their study of quantitative electroencephalography could be predictive of academic performance, and, specifically, for improvements in attentional control, cognitive performance, and psychosocial skills. It could also serve as a proxy indicator for neurofeedback training-related changes in neuroplasticity.

In Chapter 6, "Resting-State Brain Network Analysis Methods and Applications", Yunxiang Ge and Weibei Dou address how functional Magnetic Resonance Imaging (fMRI) can be used to construct brain networks. The authors also discuss challenges and pitfalls when analyzing groups of brain networks as well as review the clinical application of resting-state fMRI in the neurorehabilitation of spinal cord injury patients and stroke patients.

In Chapter 7, "Neuroimaging in Common Neurological Diseases Treated by Anticoagulants", Pipat Chiewvit discusses the use of anticoagulant drugs in treating symptomatic patients and as prophylactic therapy in asymptomatic patients. More specifically, this chapter reviews neuroimaging in common neurological conditions such as ischemic stroke, cerebral venous sinus thrombosis, and arterial dissecting disease of head and neck arteries and how anticoagulant drugs contribute to treatment therapy.

In Chapter 8, "Sleep Patterns Changes Depending on Headache Subtype and Covariates of Primary Headache Disorders," Füsun Mayda Domaç, Derya Uludüz and Aynur Özge reviews the bidirectional effects that sleep and headache have on each other. On the one hand, diminished and poor quality of sleep can be a trigger factor for headaches. On the other hand, patients with headaches may have poor sleep quality. The author indicates that the pathophysiology of headaches and sleep disorders share the same brain structures and pathways and suggests that clinicians consider sleep complaints for effective management of headaches.

In Chapter 9, "The Role of Cognitive Reserve in Executive Functioning and Its Relationship to Cognitive Decline and Dementia", Gabriela Álvares-Pereira, Carolina Maruta, and Maria Vânia Silva-Nunes explore how cognitive reserve is implicated in coping with the negative consequences of brain pathology and age-related cognitive decline. The authors focus on compensation mechanisms related to the frontal lobe and its role in maintaining cognitive performance at an advanced age and in dementia.

In Chapter 10, "Understanding the Neuropathophysiology of Psychiatry Disorder Using Transcranial Magnetic Stimulation", Jitender Jakhar, Manish Sarkar, and Nand Kumar present Transcranial Magnetic Stimulation (TMS) as a safe and non-invasive tool to investigate and modulate intracortical circuits. TMS allows for direct stimulation of cortical neurons and generation of action potentials which can provide insight into the pathophysiology of various neuropsychiatric disorders.

In Chapter 11, "Impact of Hypoxia on Astrocyte Induced Pathogenesis", Farwa Munir, Nida Islam, Muhammad Hassan Nasir, Zainab Anis, Shahar Bano, Shahzaib Naeem, Atif Amin Baig, and Zaineb Sohail review the importance of the most abundant cell type in the central nervous system, namely, astrocytes. These glial cells are critical during neuro-inflammation since they activate pro-inflammatory pathways which can lead to neurodegenerative disorders. This in turn impairs neural circulation and blood flow by affecting the blood-brain barrier which results in a lower oxygen concentration, subsequent brain hypoxia, and additional astrocyte-induced pathogenesis.

In Chapter 12, "Astrocytic Abnormalities in Schizophrenia," Kiarash Saleki, Mohammad Banazadeh, Banafshe Abadi, Zeynab Pirmoradi, Zahra Esmaili, Shiva Amiri, Ramtin Pourahmad, Kristi A. Kohlmeier, and Mohammad Shabani discuss the role of astrocytes for an important neuropsychiatric disorder, schizophrenia. The authors focus on astrocytes in schizophrenia at the molecular and behavioral level, including immune system function, changes in white matter, neuroplasticity, and therapeutic implications of targeting astrocytes in schizophrenia.

I am grateful to IntechOpen for initiating this book project and for asking me to serve as its editor. Many thanks go to Sara Tikel at IntechOpen for guiding me through the publication process and for moving the book ahead in a timely fashion. Thanks are due to all contributing authors for their excellent chapters. Hopefully, all contributors will continue their research with many intellectual challenges and exciting new directions. I would like to thank my wife Dr. Vonnie D.C. Shields, Associate Dean and Professor, Towson University, MD, and our son Torben Heinbockel for the time that I was able to spend working on this book project during the past year. Finally, I am grateful to my parents Erich and Renate Heinbockel for their continuous support and interest in my work over many years.

Thomas Heinbockel, Ph.D.
Department of Anatomy,
Howard University College of Medicine,
Washington, DC, USA

Section 1

Neural Networks, Neuroplasticity, Behavior

Chapter 1

Clinical Neurophysiology of Epileptogenic Networks

Nick Tsarouchas

Abstract

Current theories and models of brain rhythm generation are based on (1) the excitability of individual neurons and whole networks, (2) the structural and functional connectivity of neuronal ensembles, (3) the dynamic interaction of excitatory and inhibitory network components, and (4) the importance of transient local and global states. From the interplay of the above, systemic network properties arise which account for activity overdrive or suppression, and critical-level synchronization. Under certain conditions or states, small-to-large scale neuronal networks can be entrained into excessive and/or hypersynchronous electrical brain activity (epileptogenesis). In this chapter we demonstrate with artificial neuronal network simulations how physiological brain oscillations (delta, theta, alpha, beta and gamma range, and transients thereof, including sleep spindles and larger sleep waves) are generated and how epileptiform phenomena can potentially emerge, as observed at a macroscopic scale on scalp and intracranial EEG recordings or manifested with focal and generalized, aware and unaware, motor and nonmotor or absence seizures in man. Fast oscillations, ripples and sharp waves, spike and slow wave discharges, sharp and rhythmical slow waves, paroxysmal depolarization and DC shifts or attenuation and electrodecremental responses seem to underlie key mechanisms of epileptogenesis across different scales of neural organization and bear clinical implications for the pharmacological and surgical treatment of the various types of epilepsy.

Keywords: epilepsy, seizure, epileptogenic networks, epileptogenesis, focal and generalized epilepsies, epileptic syndromes, cortico-thalamo/ganglio-cortical networks (primary generalized tonic-clonic seizures, myoclonic jerks, photoparoxysmal responses, typical and atypical absences), focal neocortical and allocortical or limbocortical networks (focal [auto]motor, aware or unaware seizures with secondary propagation, bilateral spreading and/or generalization), nerve action potential, depolarization and repolarization, excitatory and inhibitory postsynaptic potentials, neuronal and network excitability, structural and functional network connectivity, physiological brain oscillations (delta, theta, alpha, beta and gamma oscillations, sleep spindles, sleep waves), fast oscillations and synchronization, ripples and sharp waves, paroxysmal depolarization and DC shifts, spike and slow wave discharges, sharp and rhythmical slow waves, electrodecremental responses and desynchronization, stochastic resonance, phasic and tonic inhibition, critical-level synchronization, depolarization block, critical global and local transient brain states, decreased network inhibition or defective activation of GABAergic transmission,

IntechOpen

increased network excitation (glutamatergic, cholinergic and monoaminergic transmission) or excitability and synchronization, biological and artificial neuronal networks, excitatory and inhibitory network components, recurrent neuronal networks with inhibitory feedback, pulse-coupled neural networks and neuronal spiking models, cerebral cortex, neocortex and allocortex, thalamus and basal ganglia, hippocampus, entorhinal cortex and limbic system, brainstem, ascending reticular activation system, intracellular and extracellular recordings, local-field potentials, intracranial and scalp-surface electroencephalography, antiseizure or antiepileptic medications, epileptogenesis-modifying medications, pharmacological, neurostimulation, neurosurgical treatments of epilepsy

1. Introduction

A seizure is the clinical manifestation of an abnormal, excessive or hypersynchronous discharge of a population of neurons [1]. Epileptogenesis is the sequence of underlying processes and/or events that can turn a neuronal network into an epileptogenic (hyperexcited or hypersynchronous) one [1–6].

Generalized epileptic seizures are considered to originate at some point within, and rapidly engage bilaterally distributed networks, including cortical (not necessarily the entire cortex) and subcortical structures (diencephalon/thalamus, basal ganglia, limbic system). Even if individual seizure onsets appear localised or asymmetric, the location and lateralisation may not be consistent from one seizure to another [1–6].

Focal epileptic seizures are considered to originate primarily within networks limited to one cerebral hemisphere. These are more discretely localised or distributed, and can originate or involve cortical and subcortical structures independently in either hemisphere. Ictal onset is consistent from one seizure to another with preferential propagation patterns, usually slower when compared to generalized epilepsies, which can potentially evolve and spread to the contralateral hemisphere or eventually engage bilateral hemispheres (bilateral spreading or secondary generalisation). In cases where there are more than one local epileptogenic networks involved corresponding to more than one seizure types, each individual seizure type has a consistent site of onset [1–6].

The fundamental principle of causality implies that both processes, 'focal' and 'generalized', start somewhere locally in the brain. The particular propagation pathways, how rapidly they spread and engage bilateral cortical networks are crucial for the distinction of 'focal' and 'generalized' epileptogenic networks, which may be more of an operational rather than a pragmatic dichotomy.

All diverse clinical patterns of seizures with either focal or generalized underlying pathomechanisms can be classified into a few categorical types of stereotypical epileptic features: seizures with preserved, impaired or lost *awareness* or *consciousness* and with predominantly *motor* (clonic/myoclonic, tonic/myotonic/dystonic, hyperkinetic/paretic or spasms), *limited-motor* (subtle automatisms, negative myoclonus, atonia, behavioural changes) or *non-motor* (sensory, autonomic, perceptual, behavioural arrest or absences) manifestations (**Table 1**) [1–6].

Seizure propagation takes preferential faster or slower pathways through the same neural/cerebral substrate in terms of neocortical structural connections (short-range and long-range association fibers: arcuate fasciculus, uncinate fasciculus, superior and inferior fronto-occipital fasciculi, etc. and interhemispheric association fibers: corpus

I. Generalized onset - usually compromised consciousness/awareness with variable degrees of motor manifestations, as a result of rapid bilateral hemispheric spread from the very beginning of the seizure and involvement of key (not necessarily all) *neocortical and subcortical structures (diencephalon, basal ganglia and limbic system, brainstem and cerebellum)*
A. Seizures with tonic and/or clonic manifestations (tonic-clonic, clonic or tonic seizures) **B. Absences** (typical, atypical or myoclonic absences) **C. Myoclonic seizure types** (myoclonic seizures, myoclonic-astatic seizures or eyelid-myoclonia) **D. Epileptic spasms** (myotonic seizures) **E. Atonic seizures**
II. Focal onset – may or may not (to a variable extent) compromise consciousness/awareness, and show variable degrees of motor and sensory manifestations implying more focal involvement, at least initially confined only to one cerebral hemisphere, of key *neocortical [frontal, temporal, insular, parietal, occipital] and/or subcortical structures (diencephalon, basal ganglia and limbic system, brainstem and cerebellum)*, with potential for ipsilateral, contralateral and/or bilateral hemispheric spreading and/or secondary generalisation
A. Localised to: 1. ***Neocortical*** **- without local spread** (focal clonic, myoclonic or inhibitory-motor seizures, focal sensory seizures with elementary symptoms) or **with local spread** (jacksonian march-seizures, focal tonic [asymmetric] seizures, dysphasic/aphasic seizures or focal sensory seizures with experiential symptoms) 2. ***Limbic-system*** **predominantly** (hippocampal, parahippocampal)
B. With ipsilateral propagation to: 1. ***Neocortical areas*** (includes hemi-tonic, hemi-clonic or hemi-atonic seizures) 2. ***Limbic areas*** (insula, amygdala, hypothalamus, including gelastic seizures)
C. With contralateral spreading to: 1. ***Neocortical areas*** (hyperkinetic seizures) 2. ***Limbic areas*** (dyscognitive seizures with or without automatisms [psychomotor])
D. With bilateral spreading or secondarily generalized: **1. Tonic-clonic seizures** **2. Absence seizures** **3. Epileptic spasms (from focal lesions)**

Table 1.
Basic seizure categorization scheme [1–6].

callosum, anterior and posterior commissures) and functional network connectivities (sensorimotor, central-executive, default-mode, salience, visuospatial attention, language, visual networks, etc.), as well as subcortical structures (thalamus, limbic system and ascending reticular activating system [ARAS]) or subcortical network connections and functional connectivities (thalamocortical, limbic system fibers [cingulum, fornix, medial forebrain bundle, etc.]) [7].

Across diverse seizure patterns the following fundamental seizure types emerge with fairly distinct pathomechanisms in the underlying epileptogenic networks (**Table 2**) [1–5].

Based on further patient and epilepsy characteristics, in particular age at onset and remission (where applicable), seizure triggers, diurnal variation, distinctive co-morbidities such as intellectual, neurological and psychiatric abnormalities, evolution and progression of the condition or not, correlated with the underlying brain pathology, aetiology and pathophysiology, electroclinical, neuroimaging and genetic investigations, epilepsies can be organized into more complex clinical diagnostic entities, so-called epilepsy syndromes. Such syndromes have a typical age of seizure onset,

Focal motor aware or unaware (more extended local networks involved)	Focal motor with ipsilateral or contralateral propagation, bilateral spreading or secondary generalized	Primarily generalized (originate within and rapidly engage bilaterally distributed networks), involving cortical and subcortical structures	Absence seizures	Myoclonic jerks	Epileptic Spasms

Table 2.
Fundamental seizure types emerging from seizure semiology and distinct pathophysiological processes.

specific seizure types and EEG characteristics and other electroclinical and neuroimaging features which, when taken together, allow the specific syndromic epilepsy diagnosis [8, 9]. The identification of an epilepsy syndrome is useful as it provides information on which underlying aetiologies should be considered, what is the current and future prognosis, which pharmacological anti-seizure medications and/or neurosurgical or neurostimulation interventions might be most useful. Certain epilepsy syndromes may manifest seizure exacerbation, modification or ineffective control with particular anti-seizure medications, which can be avoided and seizure control outcomes can be optimized through early syndromic diagnosis [2–6, 8–11]

The differential effectiveness of antiepileptic drugs across seizure types highlights likely distinct seizure pathomechanisms and the underlying pathophysiological processes of different epileptic syndromes (**Table 3**) [8–11].

Antiseizure medications in *italics* are generally avoided or contraindicated for the treatment of idiopathic (genetic) generalized epilepsies (**Table 3**). **Carbamazepine, Oxcarbazepine, (Fos)Phenytoin** (mainly voltage-dependent sodium channel blockers binding in the inactivated sodium channels and preventing high-frequency action potentials) may be used in the rare pure forms of primarily generalized tonic-clonic seizures (GTCS) but are not indicated as first-line for idiopathic generalized epilepsies, either because they are ineffective or may exaggerate/exacerbate certain types of seizures. Carbamazepine may treat manic and depressive symptoms in bipolar disorder by increasing dopamine turnover and GABA transmission.
Eslicarbazepine has lower affinity for inactive voltage-gated sodium channels in the resting state compared to Carbamazepine and Oxcarbazepine, thereby selectively inhibits repeated neuronal firing in the epileptic focus, as well as T-type calcium channels in vitro. **Lacosamide** may be selective for inhibiting depolarized neurons (slow inactivation gating of sodium channels), affecting only those neurons (at the epileptic focus) which are depolarized or active for long periods of time. **Lamotrigine** (acting as voltage-gated inactivated sodium channel and R-type calcium channel blocker, suppressing glutamate release and stabilising membranes) may exaggerate myoclonic jerks in juvenile myoclonic epilepsy and some progressive myoclonic epilepsies. **Ethosuximide** (T-type calcium channel blocker) is only effective for absences and may be effective in negative myoclonus. **Levetiracetam** is an inhibitor of synaptic vesicle protein 2A (SVP2A) and presynaptic neurotransmitter release in high-frequency firing neurons and inhibitor of N-type calcium channels. It may indirectly enhance GABAergic neurotransmission via GABA-A receptors and decrease glutaminergic excitation via modulation of NMDA and AMPA receptors or upregulation of glial glutamate transporters. **Brivaracetam** is the racetam derivative of Levetiracetam with 20 times higher affinity for binding SVP2A, while also inhibiting sodium channels and impairing epileptogenesis through modulation of

AEDs	Focal seizure motor or not, aware or unaware	Focal seizure secondarily generalized GTCS	Primarily generalized GTCS	Myoclonic jerks	Absence seizures
Valproate	Effective	Effective	Effective	Effective	Effective
Ethosuximide	Ineffective	Exaggerates?	Exaggerates?	Effective in negative myoclonus	Effective
Zonisamide	Effective	Effective	Effective	Effective	Effective?
Topiramate	Effective	Effective	Effective	Effective	Effective?
Levetiracetam Brivaracetam	Effective Effective (adjunct)	Effective Effective (adjunct)	Effective Unknown	Effective Unknown	Effective Unknown
Lamotrigine	Effective	Effective	Effective	Exaggerates	Effective
Phenytoin	*Effective*	*Effective*	*Effective*	*Ineffective*	*Exaggerates*
Carbamazepine	*Effective*	*Effective*	*Effective*	*Ineffective*	*Ineffective, Exac of atypical absences?*
Oxcarbazepine	*Effective*	*Effective*	*Effective*	*Exacerbates*	*Exacerbates*
Eslicarbazepine	Effective	Effective	Effective	Ineffective	Ineffective
Lacosamide	Effective	Effective	Effective (adjunct)	Unknown	Unknown
Cenobamate	Effective (adjunct)	Effective (adjunct)	Unknown	Unknown	Unknown
Perampanel	Effective (adjunct)	Effective (adjunct)	Effective (adjunct)	Unknown	Unknown
Phenobarbital	Effective	Effective	Effective	Effective	Exaggerates?
Clobazam	Effective	Effective	Effective?	Effective?	Effective?
Clonazepam	Effective?	Effective?	Ineffective?	Effective	Effective
Tiagabine	*Effective (adjunct)*	*Effective (adjunct)*	*Ineffective*	*Exaggerates*	*Exaggerates*
Vigabatrin	*Effective (adjunct)*	*Effective (adjunct)*	*Ineffective Effect for Epl.Spasms*	*Exaggerates*	*Exaggerates*
Gabapentin *Pregabalin*	*Effective* *Effective (adjunct)*	*Effective* *Effective (adjunct)*	*Ineffective* *Ineffective*	*Exaggerates* *Exaggerates*	*Exaggerates* *Exaggerates*

Table 3.
Adapted from Panayiotopoulos [5] and adjusted based on updated publicly available information from https://www.medicines.org.uk/emc and https://www.fda.gov/drugs and https://go.drugbank.com/drugs and https://bnf.nice.org.uk/

synaptic GABA. **Zonisamide** (blocking repetitive firing of voltage-gated sodium channels, reducing T-type calcium channel currents or binding allosterically to GABA receptors inhibits the uptake of GABA and enhances the uptake of glutamate) and **Topiramate** (voltage-dependent sodium channel blocker, allosteric stimulator of GABA-A receptors and inhibitor of AMPA and Kainate glutamate receptors) are effective in all types of epilepsy. **Cenobamate** reduces repetitive neuronal firing in the epileptic focus by enhancing the inactivation of sodium channels and inhibiting the persistent component of the sodium current and acts as a positive allosteric modulator of GABA-A ion channel subtypes. **Gabapentin** (presynaptic voltage-dependent calcium channel inhibitor and dose-dependent inducer of L-glutamic acid decarboxylase that enhances GABA synthesis) and **Pregabalin** used as adjunctive for focal seizures may have pro-myoclonic effects. **Clonazepam** (1,4-benzodiazepine, full agonist of GABA-A receptors resulting in increase in the frequency of chloride-channel opening) is mainly used for myoclonic jerks, but it may not suppress GTCS of juvenile myoclonic epilepsy. **Clobazam** (1,5-benzodiazepine, partial agonist of GABA-A receptors) licensed as adjunctive therapy may be more efficacious in focal than generalized epilepsies. **Phenobarbital** is a potentiator agonist of GABA-A receptors resulting in increased duration of chloride-channel opening and may also act on Glutamate receptors. **Perampanel** (non-competitive AMPA glutamate receptor antagonist) is mostly used for focal epilepsies and only as adjunctive for primary generalized ones. **Valproate** (directly suppresses voltage-gated sodium channel activities and influences many other channels and neurotransmitters and indirectly enhances GABAergic neurotransmission as inhibitor of succinic semialdehyde dehydrogenase [GABA transaminase]) is effective against all types of epilepsy [5, 8–11]. **Fenfluramine** (a serotonin-releasing agent that stimulates multiple 5-HT receptor subtypes) is used as an adjunctive treatment in Dravet syndrome. **Rufinamide** (prolonging the inactive state of voltage-gated sodium channels and inhibiting mGluR5 subtype receptors at high concentration) is used as an adjunctive treatment of seizures in Lennox-Gastaut syndrome. **Stiripentol** potentiates GABAergic transmission by elevating the levels of GABA and acting as a positive allosteric modulator of GABA-A receptors and is used as an adjunctive treatment in Dravet syndrome. **Cannabidiol (CBD Oil)**, the major component of the resin of *Cannabis sativa* plant (marijuana), is devoid of the psychoactive, euphoric or intrusive effects and abuse liability of the tetrahydrocannabinol (THC) component. Endocannabinoid receptors regulate cognition, pain sensation, appetite, memory, sleep, immune function, fear, emotion or mood and are mostly localized in the hippocampus and amygdala. Cannabidiol may have low affinity for endocannabinoid receptors but may indirectly modulate these receptors by blocking the breakdown of Anandamide. It could also activate the transient receptor potential of Vanilloid type-1 (TRPV1), antagonise the G protein-coupled receptor 55 (GPR55), target abnormal sodium channels, block T-type calcium channels, modulate adenosine receptors or adenosine reuptake, voltage-dependent anion selective channel protein (VDAC1) or tumor necrosis factor alpha (TNFa) release. It has been licenced as adjunctive treatment in Tuberous Sclerosis and (together with Clobazam) in Lennox-Gastaut and Dravet syndromes. (publicly available information at: https://go.drugbank.com/drugs and https://bnf.nice.org.uk/).

2. Physiological brain networks

From individual neurons to dynamic neuronal networks

In the following, embarking from the Hodgkin-Huxley neuronal membrane model we endeavour to create biologically realistic and computationally efficient models of spiking neurons and further on to generate a local spiking neuronal network of a 1000 excitatory and inhibitory neurons. Our aim is to study the behaviour of this simple network under different structural and functional constraints, and critical network parameters, in order to understand the rich network dynamics that emerge, and gain insight into the physiology of cortical neuronal networks and pathophysiology of seizures [12].

2.1 Modeling the neuron

The Hodgkin-Huxley biological neuron model

The Hodgkin-Huxley type models [13] represent the biophysical properties of cell membranes and ionic conductances (current flows) that help determine at any time the neuronal resting and action membrane potentials (for mathematical details [14]). The lipid bilayer is represented as a capacitance (Cm). Voltage-gated and leak ion channels are represented by nonlinear (gn) and linear (gL) conductances, respectively. The electrochemical gradients driving the flow of ions are represented by batteries (E), and ion pumps and exchangers are represented by current sources (Ip) (**Figure 1**) [12, 15].

If the integration of Excitatory Postsynaptic Potentials (EPSP) and Inhibitory Postsynaptic Potentials (IPSP) at the long somatodendritic processes of pyramidal neurons (thousands of synaptic contacts) [16] is sufficient to shift the resting membrane potential at the axon hillock closer to threshold (around −55 mV, inside negative), voltage-gated fast Na-channels open up allowing an influx of Na^+ and depolarization current sufficient to turn the inside of the membrane positive, resulting in the generation of an action potential (up to +40 mV, the inside positive). The local reversal of the membrane potential during the upstroke makes the Na^+ channels rapidly turn into an inactivated (non-conducting absolute refractory) state, while different voltage-gated channels open up allowing together with leaky K^+ channels for the early repolarisation and late after-hyperpolarisation phases of the membrane potential (prolonged relative refractory state). This is a very simplified *integrate-and-fire model* of a neuron and accounts for the action potential generated in a neuron (**Figures 2** and **3a** and **b**) [19].

The processing of post-synaptic potentials is much more than a simple algebraic summation, most likely a geometrical (vectorial) spatiotemporal integration with very

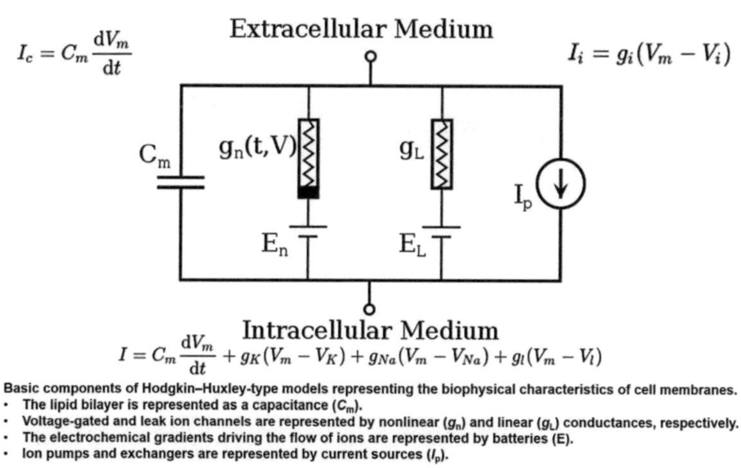

Figure 1.
The electronic circuit equivalent of the Hodgkin-Huxley biological neuronal model [14] from https://commons.wikimedia.org/wiki/File:Hodgkin-Huxley.svg By Krishnavedala via Wikimedia Commons—Own work, CC0, https://commons.wikimedia.org/w/index.php?curid=21725464 with added-on fundamental equations for current flowing through the lipid bilayer (I_c), current through a given ion channel (I_i) and total current through the membrane (I) for a cell with potassium (K^+) and sodium (Na^+) channels. V_m is the membrane potential, V_i is the reversal potential of the i-th *ion channel,* V_K *and* V_{Na} *are the potassium and sodium reversal potentials, respectively,* g_K *and* g_{Na} *are the potassium and sodium voltage-gated (nonlinear) conductances per unit area, respectively and* g_l *and* V_l *are the leak (linear) conductance per unit area and leak reversal potential, respectively.*

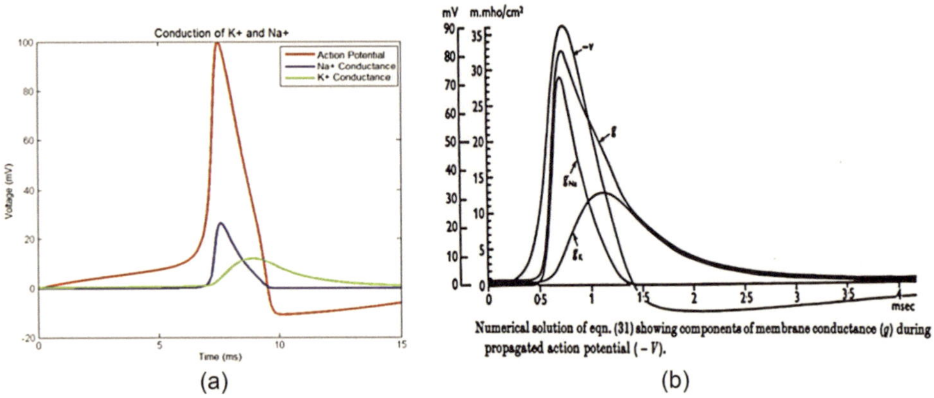

Figure 2.
(a) Main sodium and potassium conductance/current giving rise to the action membrane potential, reproduced under CC BY 4.0 from: Figure 6 of Johnson M & Chartier S (2017). Spike neural models (part I): The Hodgkin-Huxley model. The Quantitative Methods for Psychology [17]. (b) A graph of the sodium and potassium conductances (GNa and GK), their sum (Gm), and the membrane voltage (Vm) during a propagating nerve impulse, which is basically a numerical solution of the Eq. 4.32 published by Hodgkin AL & Huxley AF. A quantitative description of membrane current and its application to conduction and excitation in nerve. The Journal of Physiology [18].

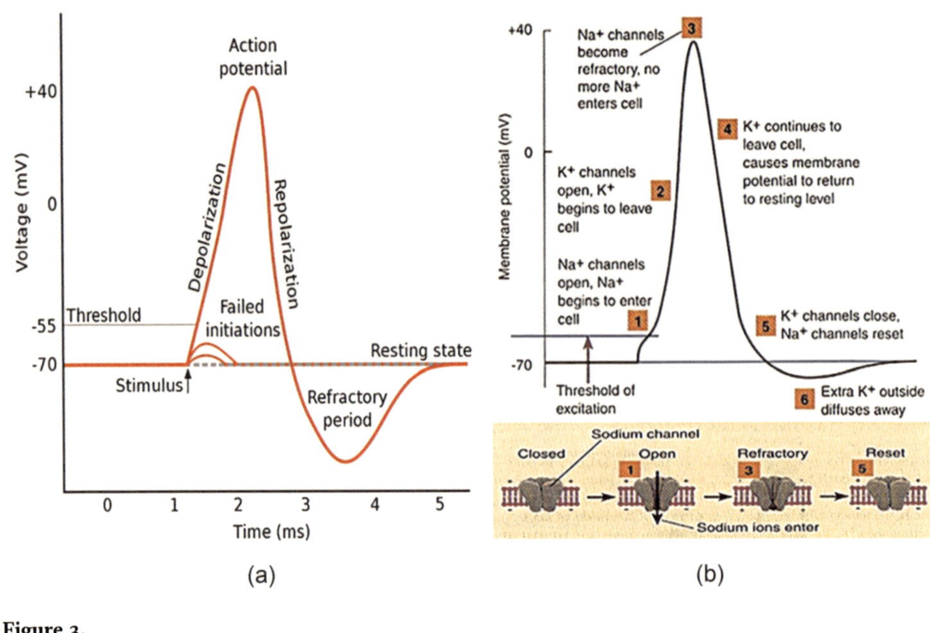

Figure 3.
(a) Approximate plot of a typical action potential shows its various phases as the action potential passes a point on a cell membrane. The membrane potential starts out at approximately −70 mV at time zero. A stimulus is applied at time = 1 ms, which raises the membrane potential above −55 mV (the threshold potential). After the stimulus is applied, the membrane potential rapidly rises to a peak potential of +40 mV at time = 2 ms. Just as quickly, the potential then drops and overshoots to −90 mV at time = 3 ms, and finally the resting potential of −70 mV is reestablished at time = 5 ms. By Original by en:User:Chris 73, updated by en:User:Diberri, converted to SVG by tiZom—Own work, CC BY-SA 3.0, https://commons.wikimedia.org/w/index.php?curid=2241513 (b) ion movements during an action potential, showing when channels open or close during the action potential. Bottom image shows the corresponding sodium channel states at specific points of the action potential. The diagram was created by: If Only and was retrieved online from: scioly.org/wiki/index.php/File:Image12.jpg (publicly available).

complex feature-extracting properties. Many pyramidal neurons with long/extensive arborization processes geometrically integrate in time all postsynaptic EPSP and IPSP on their somatodendrites into a postsynaptic amplitude-modulated potential (analogue signal) which is translated at the axon hillock via an *all-or-none* generated response to an action potential (digital signal). This is fundamentally a nonlinear process [20–22]. Neurons essentially communicate with each other via *all-or-nothing* action potentials (a binary code of 1 or 0). The code of this digital communication lies either in the instantaneous or average firing rates (frequency modulation) or in the critical timing or phase of the firing (time/phase modulation) [23–25]. The postsynaptic potential amplitude-modulation (AM) is basically converted into an instantaneous or average firing rate of cortical pyramidal neurons, a frequency-modulation (FM) code, often seen in primary sensory/visual or motor and other neocortical or allocortical areas. The majority of cortical neurons though seem to rely on a critical-timing and phase-modulation (PM) code for information processing. Furthermore, the firing frequency or firing rate patterns observed may reflect the level of synchrony or synchronization in the underlying spiking neurons [26–29].

2.2 Modelling postsynaptic potentials of neuronal networks

The highest contribution to Local Field Potentials (LFP) and the Electroencephalogram (EEG) signal comes from the postsynaptic membrane potentials of pyramidal

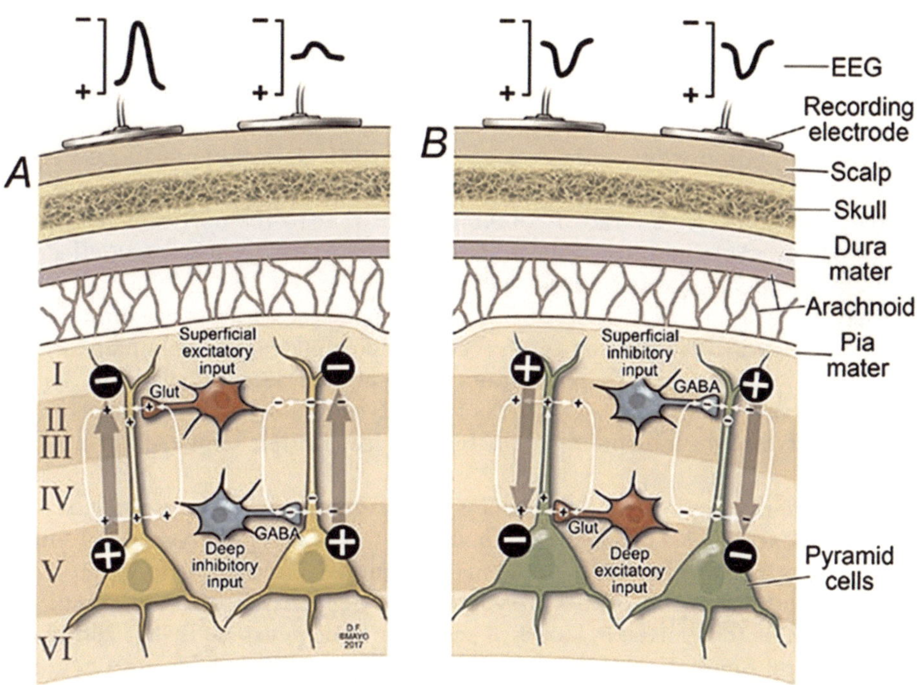

Figure 4.
Neurophysiological basis of the EEG potentials recorded on the scalp surface from the geometrical summation of minute current dipoles of millions of pyramidal cortical neurons with variable magnitude, polarity and orientation depending on the local dominant circuitries implicated each time. Reproduced with permission from Tatum et al. [30].

cells of cortical layer V creating a local current sink or rise, equivalent to a minute electrical dipole pointing at variable directions, depending on the local geometry (infoldings with gyri and sulci) and local circuitry of the cortical region, cortical neurons implicated and their spatial relationship with the scalp sensor (**Figure 4**) [29–32]. In addition, there is volume conducted electrical activity from every other region of the brain picked up at all scalp sensors, which is difficult to disentangle from the particular region in question, especially when other sources dominate the background (known as the reverse problem of EEG source localisation). To make things worse, the scalp and all the layers of tissue that intervene between the cerebral cortex and the scalp surface sensors act as low-pass filters, filtering out most of the high-frequency (mostly gamma range) activities of the EEG and severely contaminating the recorded EEG with muscle activity, eye or movement and other artefacts. The extracranial EEG may have excellent temporal resolution (down to the millisecond scale), but has fairly poor spatial resolution [29, 30, 32].

In principle, for an electrical signal deflection to be detected on the scalp sensor as an electrical potential difference between the underlying cortical region and a less active reference electrode, activities of millions of cortical/pyramidal neurons need to be spatiotemporally summated over an area of about 4–9 (2^2–3^2) cm^2 of cerebral cortex [30, 32]. The larger the numbers of cortical/pyramidal neurons and the more synchronous their depolarization (corresponding to firing rates or bursts of action potentials), the greater the deflection on the scalp EEG (if at right angles to the cortical surface and current lines are traveling in the same direction). Although there are asymmetries between the depolarizing and repolarizing regions of the pyramidal neurons as reflected in current lines and waveform morphologies, current lines are mainly created by the depolarizing wave front and fan out away from the current sink, reflecting the volume conductive properties of that brain region. In essence every depolarization wave-front creates a minute electrical dipole and depending on the direction of that (negative at the apical dendrites vs. positive at the somata or vice versa), the radial components of that to the cortical surface will constructively contribute towards a positive or respectively negative deflection on the scalp EEG signal (**Figure 4**) [29, 30, 32].

None of the shortfalls of scalp EEG is a limiting factor with the use of in-depth electrodes robotically implanted (with accurate coordinates) in the human brain for epilepsy surgery purposes, which can reach even the deepest regions of the brain and record intracranial stereo-EEG from about 1 mm^3 voxels of brain tissue (the issue is which specific points in the brain these multiple depth contacts should target) [29].

The hallmark of epilepsy on the EEG is the interictal epileptiform spike or sharp wave, excessive or hypersynchronous discharges and paroxysmal rhythmical activities or attenuation changes in the background EEG activity before, during and after the seizure. The paroxysmal or synchronous depolarisation of millions of cortical neurons (converted in bursts of spikes or tonic neuronal firing) allows for spatiotemporal summation of the electrical activity of millions of cortical neurons and underlies the excessive or hypersynchronous discharges and other epileptiform phenomena we detect on scalp EEG. The cellular neurophysiological correlate of the interictal and ictal epileptiform discharges is the paroxysmal depolarization shift (PDS) of individual cortical neurons. The PDS is a prolonged (calcium-dependent) neuronal depolarization that results in multiple sodium-channel (inward) current-mediated action potentials during depolarization, followed by a prominent after-hyperpolarization phase beyond the baseline resting membrane potential, mediated by calcium-dependent potassium-channel (outward) currents and/or gamma-aminobutyric acid (GABA)-activated chloride (Cl^-) currents [33–36].

2.3 Modelling action potentials of neuronal networks

Using biophysically accurate Hodgkin-Huxley-type models for action potentials of neuronal networks would be computationally so demanding [37] that we could simulate only few neurons in real time. Using an integrate-and-fire model is computationally very effective, but the model is unrealistically simple and incapable of producing the rich spiking and bursting dynamics exhibited by cortical neurons. On balance, we adopted the simple spiking model proposed by Izhikevich [38], which is as biologically plausible as the *Hodgkin-Huxley* model, yet as computationally efficient as the *integrate-and-fire* model.

The spiking model developed by Izhikevich [38] essentially represents the complex integration of depolarising forces (voltage-gated and leaky sodium channels/inward currents) and repolarising forces (voltage-gated and leaky potassium channels/outward currents) by determining few key parameters: (a) the rate of decay of the action potential, (b) the sensitivity threshold for triggering the action potential, (c) the reset level of the depolarisation potential of the membrane and (d) the reset level of the repolarisation potential of the membrane. Depending on these four parameters, the model reproduces spiking and bursting behavior of different known types of cortical neurons as shown in **Figure 5**, and allows for dynamic functional connectivity and emergence of waves and rhythms at different scales of neural organization [39].

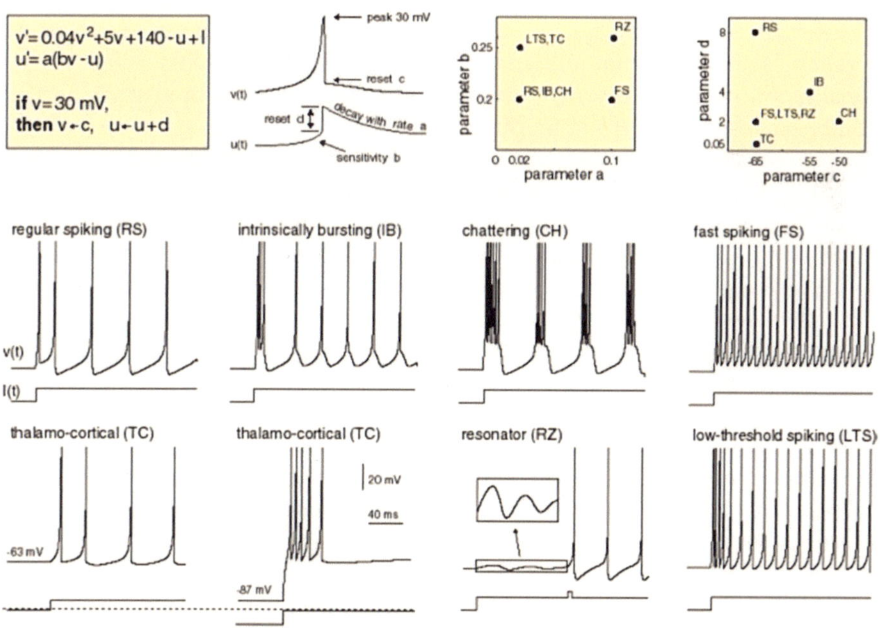

Figure 5.
Known types of cortical neurons correspond to different values of the parameters a, b, c, d in the model described by the equations in the left upper box. RS, IB, and CH are cortical excitatory neurons. FS and LTS are cortical inhibitory interneurons. Each inset shows a voltage response of the model neuron to a step of dc-current I = 10 (bottom). Time resolution is 0.1 ms. This figure has been reproduced with permission from Izhikevich [37–42].

2.4 Cortical neuronal spiking model

Neocortical neurons in the mammalian brain are classified into several types according to the pattern of spiking and bursting seen in intracellular recordings [38]. **Excitatory** cortical cells are divided into 4 different classes, with *RS (regular spiking)* neurons being the most typical neurons of the cortex. Upon a prolonged stimulus (injected DC-current, equivalent in a particular neuron to spatial convergence and temporal summation of more EPSP vs. IPSP) the neurons fire a few spikes with short interspike period and then the period increases (*spike frequency adaptation*). Increasing the strength of the injected DC-current increases the *interspike frequency*, though this can never become too fast because of large spike-after hyperpolarizations. **Inhibitory** cortical cells are usually divided into two classes, the most common being the *FS* (*fast spiking*) neurons that can fire periodic trains of action potentials with extremely high frequency, practically without any adaptation (slowing down). The other type of cortical inhibitory interneurons is the *LTS* (*low threshold spiking*) neurons that start spiking when a minimum low-threshold has been reached [37–42].

The best way to simulate the different dynamics of different neurons, is to assign for each excitatory cell $(a_i, b_i) = (0.02, 0.2)$ and $(c_i, d_i) = (-65, 8) + (15, -6)\, r_i^2$, where r_i is a random variable uniformly distributed on the interval [0, 1], and i is the neuron index. Thus, $r_i = 0$ corresponds to regular spiking (RS) cell, while $r_i = 1$ corresponds to the chattering (CH) cell. We use r_i^2 to bias the distribution toward RS cells. Similarly, each inhibitory cell has $(a_i, b_i) = (0.02, 0.25) + (0.08, -0.05)\, r_i$ and $(c_i, d_i) = (-65, 2)$ [38].

In addition, the Izhikevich model [39] can reproduce the behavior of **thalamocortical neurons**, which provide the major input to the cortex. *TC (thalamo-cortical)* neurons have two firing regimes: When at rest (v is around −60 mV) and then get depolarized, they exhibit tonic firing. However, if a negative current step is delivered and the membrane potential gets hyperpolarized (v is around −90 mV), the neurons fire a rebound burst of action potentials (bursting firing). The dynamics of other neuronal types, including those in **hippocampus**, basal ganglia, brainstem, and olfactory bulb, can also be simulated by this model [41].

The proposed Izhikevich simulation model [42] belongs to the class of pulse-coupled neural networks (PCNN). The synaptic connection weights among the neurons are described by a matrix $S = (s_{ij})$. Firing of the jth neuron instantaneously changes the variable v_i by s_{ij} [38]. Although the network is connected randomly and there is no synaptic plasticity included, the neurons tend to self-organize into assemblies that exhibit collective rhythmical behavior in a frequency range corresponding to that of the awake mammalian cortex. Changing the relative strength (weights) of synaptic connections (excitatory and inhibitory) and the strength of the thalamic drive can produce other types of collective behavior, including spindle waves and larger slow sleep oscillations [39]. Therefore, our spiking model is fairly adaptable and can reproduce the dynamics of many different known types of neocortical and allocortical neurons (biological plausibility), while its high computational efficiency allows us to observe and study collective cortical states at a global neuronal network level [37–42].

3. Epileptogenic brain networks

3.1 Cortico-thalamocortical neuronal networks

The reticular nucleus of the thalamus is part of the ascending reticular activating system (RAS) that modulates thalamocortical pathways and helps to synchronize and

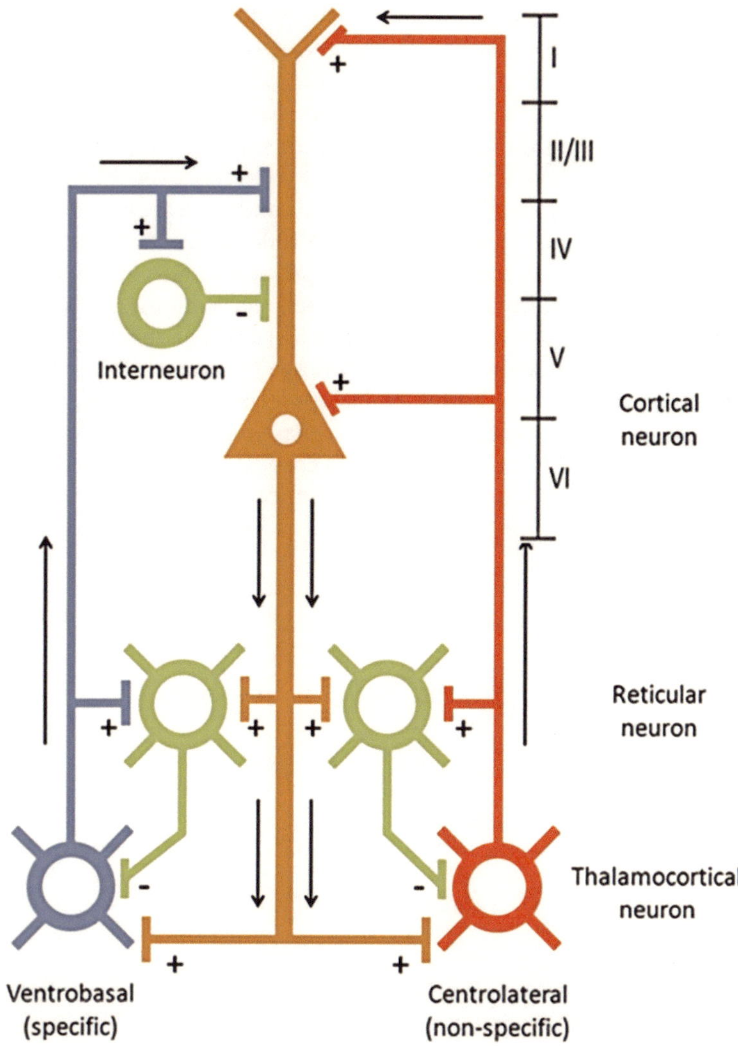

Figure 6.
The thalamocortical network. Please note the projection of specific and association thalamic nuclei project to layers III and IV of the neocortex and the projection of nonspecific thalamic nuclei to more superficial layers as well, layer I and II, allowing a temporal binding and integrative processing of information via cortical coincidence detection of specific and nonspecific thalamocortical inputs. By Zacharybarry (talk)—self-made, Public Domain, https://en.wikipedia.org/w/index.php?curid=33934402.

desynchronize the EEG. Typically, the EEG is synchronized when thalamic relay neurons are in burst mode, and desynchronized when they are in tonic mode. Stimulation of the RAS suppresses slower activity (e.g. delta or theta) and stimulates faster activity (e.g. beta and gamma) and vice versa. This modulation leads to remarkable changes in cerebral electrical activity during wakefulness and sleep [43].

Most regions of cortex receive converging inputs from both specific (or association) nuclei and nonspecific thalamic nuclei (**Figure 6**). Specific and association thalamic nuclei project most commonly to layers III and IV of the neocortex, whereas nonspecific thalamic nuclei may reach more superficial layers, including layer I and II. This arrangement allows for regions such as the reticular formation to alter cortical excitability levels by modulating the responsiveness of cortical neurons to inputs from specific or association thalamic nuclei. Furthermore, specific/association and nonspecific thalamic nuclei share reciprocal connections with the cerebral cortex. These provide a feedback mechanism from the cortex to the thalamus, which serves to control the amount of input that reaches a specific region of cortex at any moment (**Figures 6** and 7) [44–46].

Intracranial recordings in experimental animals and humans for brain surgery (including epilepsy surgery) suggest that gamma (30–70 Hz) rhythms are mostly generated in the superficial somatosensory cortical layers II/III, while high beta rhythms (20–30 Hz) are mostly generated in deep layer V mainly in association with activity in the fast-spiking interneurons (**Figure 6**) [47]. The majority of cortical pyramidal cells are firing at a lower frequency range (usually <25 Hz) and their information coding may be based more on instantaneous firing rates or the critical timing and phase of their activation (timing or phase modulation code). Certain neurons in primary sensory (e.g. in early visual cortical areas) [48] and other neocortical or allocortical areas (e.g. in the hippocampus) can fire at higher frequencies and

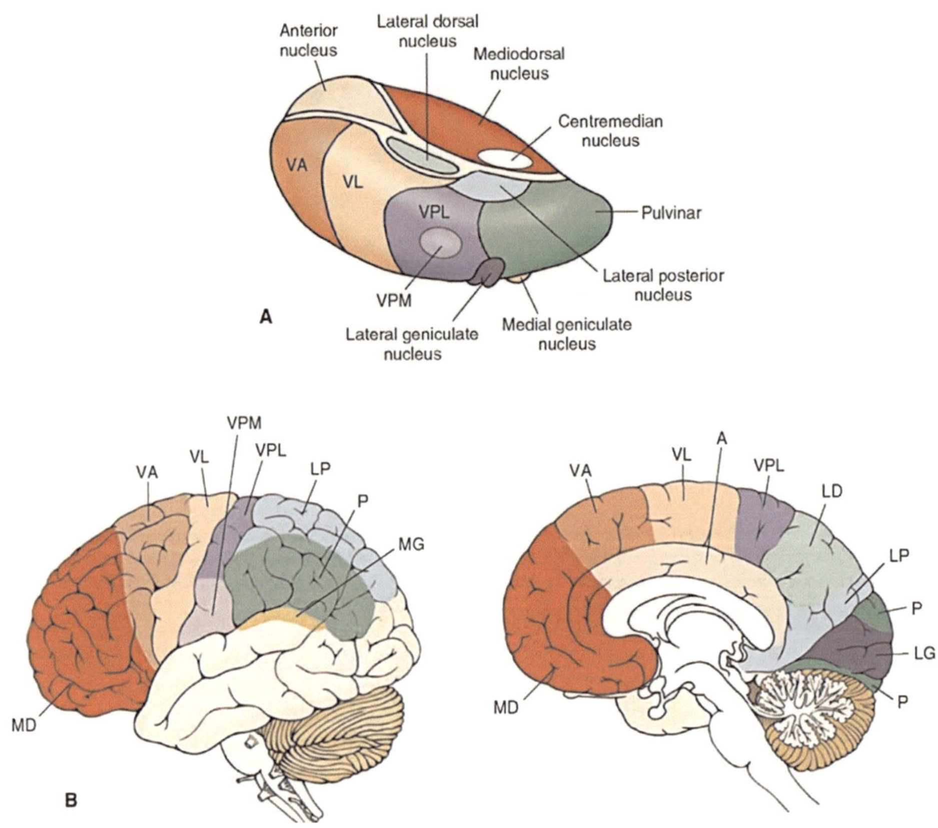

Figure 7.
Thalamocortical relationships. (A) The relative positions of thalamic nuclei. (B) Lateral (left) and medial (right) views of the cerebral cortex that demonstrate the projection targets of thalamic nuclei. Color-coding is to facilitate visualization of the reciprocal connections between thalamic nuclei and the cerebral cortex. VPM = ventral posteromedial nucleus; VPL = ventral posterolateral nucleus; VA = ventral anterior nucleus; VL = ventrolateral nucleus. Reproduced from http://what-when-how.com/neuroscience/the-thalamus-and-cerebral-cortex-integrative-systems-part-2/.

their information coding may be also based on the average firing rates (frequency modulation code) [46–49].

Gamma oscillations probably reflect a general rapid mechanism that 'synchronizes' neuronal networks in spatially disparate cortical areas to enable fast processing or coherent binding of visual, sensory or perceptual information across cortical and subcortical networks [50–52]. The integration of EPSP and IPSP into firing an action potential in a single neuron is essentially a nonlinear amplitude-modulation postsynaptic process and the majority of pyramidal neurons rely on an instantaneous or burst firing, critical-timing and phase-modulation code for information processing. Looking at neuronal networks from a macroscopic perspective (where intracranial LFP/EEG represents a collective mean-field measure), increased amplitude in higher-frequency oscillations may be considered as an index of neuronal spiking synchrony. That is, on a large scale there may be a quasi-linear relationship between the mean frequencies of oscillatory activities from postsynaptic integration of EPSPs and IPSPs on somatodendritic processes and the coordinated or synchronized firing rates of pyramidal cells, even if on the neuronal-cell microscale a non-linear relationship is seen through random or regular, bursting or tonic firing patterns of pyramidal cells [23, 25, 28, 29, 32, 37].

Although gamma oscillatory activities are also present during sleep and under anesthesia, fast oscillations are mainly associated with increased levels of alertness/wakefulness, in keeping with activity in cholinergic neurons of the brainstem and basal forebrain [50, 53]. Increase of beta activity has been demonstrated after finger or foot movement when the muscles relax, while increase of gamma activity (greater than 30 Hz) immediately preceding the finger movement is thought to be associated with activation of cortical motor neurons. Sometimes on the scalp EEG we record equivalent sensorimotor cortical mu-rhythms which undergo suppression (desynchronization) upon performing a motor action or motor imagery, even observing another person performing a motor action or abstract motion [54].

Light sleep transition is characterized by alpha drop-out and appearance of V-sharp waves, spindles (alpha and beta range) and K-complexes (represent a depolarizing–hyperpolarizing sequence within an oscillatory cycle) before further transition to the less organised large slow wave oscillations of sleep (down to 1 Hz). Sleep spindles must be generated within the thalamus as they persist in the thalamus after decortication and high brainstem transection. Sleep spindles must be driven by the reticular nucleus of the thalamus (GABAergic neurons), as they are abolished in the dorsal thalamus after disconnection from the reticular nucleus but are preserved in the rostral part of the reticular nucleus severed from the dorsal thalamus [55].

At the onset of sleep decreased activity of brainstem cholinergic neurons contributes to the overall hyperpolarization of thalamocortical cells, thus bringing their membrane potential in the range where bursting discharges can occur. Such clusters of high-frequency action potentials excite the dendrites of neurons in the reticular nucleus, and trigger a dendrodendritic avalanche leading to synchronization of the entire reticular nucleus. Bursting of reticular neurons causes powerful GABAergic inhibitory postsynaptic potentials in thalamocortical neurons which promote cortical deafferentation (slow wave oscillations). The end of this inhibitory cycle a rebound low threshold spike (LTS) is triggered, crowned by a high-frequency burst of action potentials, which in turn excites the target reticular cells (**Figure 6**) [56–58]. Although previously thought that the main functional correlate of sleep spindles was to block incoming sensory stimuli from the thalamus to the cortex, today we believe that sleep spindles also serve a process of memory consolidation during sleep [59].

3.2 Cortico-limbocortical neuronal networks

Theta activity in hippocampal networks (**Figure 8**) seems to represent a dynamic state emerging from engaging in the task of spatial navigation and memory retrieval processes. Larger theta activity has been seen in the left anterior hippocampus and parahippocampal cortex during goal-directed navigation compared to purposeless movements. Theta oscillations are also frequent during memory processing and more so during recall than learning tasks [60].

The major input to the Hippocampus is from the Entorhinal Cortex (EC) which is strongly and reciprocally connected with many cortical and subcortical structures. Different thalamic nuclei (anterior and midline groups), the medial septum (medial septal nucleus and diagonal band of Broca), the supramammillary nucleus of the hypothalamus, the raphe nuclei and locus coeruleus of the brainstem, are connected with the entorhinal cortex which serves as the interface between neocortical (e.g. parahippocampal gyrus and perirhinal cortex) and subcortical structures, and the hippocampus (**Figure 9**) [61].

The direct perforant pathway (axons from EC layer III) forms synapses on the very distal apical dendrites of CA1 neurons (monosynaptic circuit). The indirect perforant pathway (axons from EC layer II) reaches CA1 via the trisynaptic circuit: Granule cells in the Dentate Gyrus (first synapse), via the mossy fibers to pyramidal neurons in CA3 (second synapse), and via the Schaffer collaterals to pyramidal neurons in CA1 (third synapse). The major output of the Hippocampus is axons from CA1 projecting back directly and via the Subiculum to the Entorhinal Cortex, completing the trisynaptic circuit (**Figure 8**) [62]. The trisynaptic circuit of the hippocampus is organized transversely along the hippocampus. Association and commissural fibers are organized along the anteroposterior axis [63].

Hilar mossy cells and CA3 pyramidal cells give rise to ipsilateral hippocampal association fibers (to the dendrites of granule cells and GABAergic interneurons of the inner molecular layer of the dentate gyrus) and contralateral commissural fibers

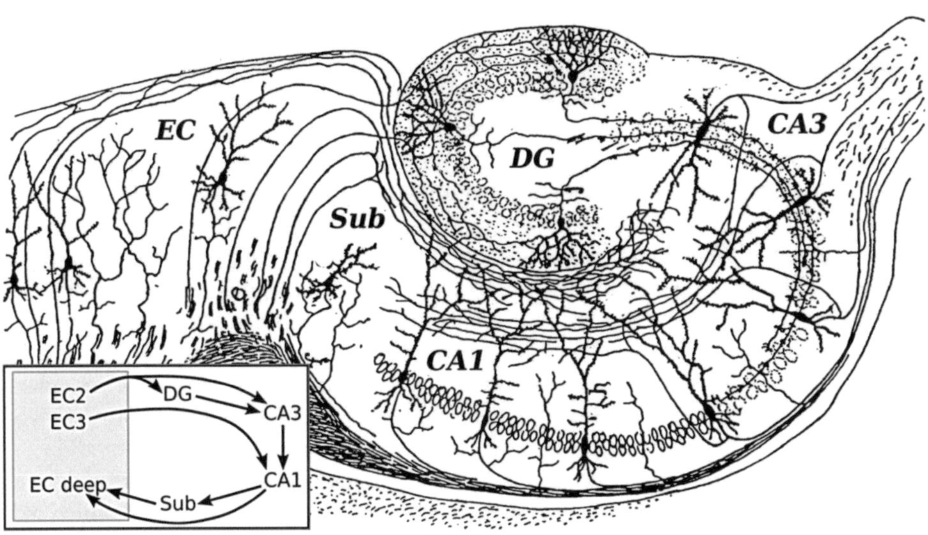

Figure 8.
By original: Santiago Ramón y Cajal (1852–1934) derivative = Looie496 - File:CajalHippocampus.jpeg from: Santiago Ramón y Cajal (1911) [1909] Histologie du Système nerveux de l'Homme et des Vertébrés, Paris: A. Maloine, Public Domain, https://commons.wikimedia.org/w/index.php?curid=3908039.

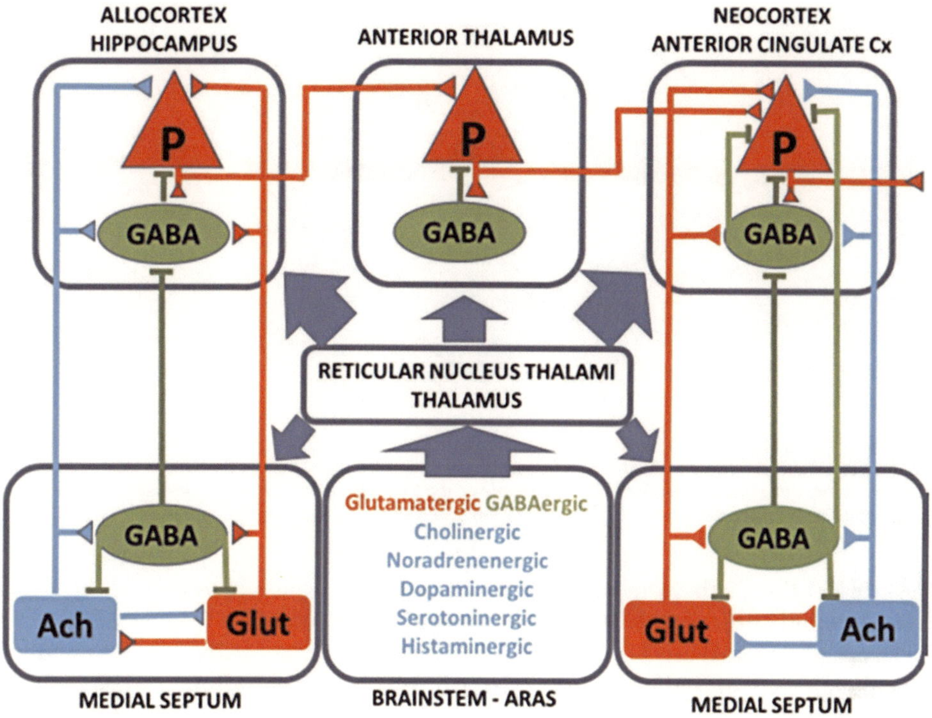

Figure 9.
The septal-hippocampal-thalamo-cortical axis. Overview of the complex interactive circuitry of excitatory, inhibitory and modulatory components, underlying some of the structural and functional connectivity of the limbo-cortical networks with clinical implications for the pharmacological and neurosurgical/neurostimulation treatment strategies of epilepsy.

(terminating on principal cells and interneurons of CA3, CA2 and CA1 regions), passing through the dorsal and ventral hippocampal commissures to reach the contralateral hippocampus and dentate gyrus. Hippocampal commissural connections are mainly excitatory and seem to be less abundant and evolutionary declined in monkey and humans [64].

Schaffer collaterals are axon collaterals from the CA3 pyramidal cells of the ipsilateral and contralateral hippocampi and project information to the pyramidal neurons of the CA1 hippocampal area. The Schaffer collateral synapses represent excitatory (positive feedback) glutamatergic synapses that assist activity-dependent plasticity and development of memory processing and formation [65]. Basket cells in CA3 receive excitatory input from the pyramidal cells and provide inhibitory feedback to the pyramidal cells. This recurrent inhibition is a powerful feedback circuit that can dampen excitatory responses and shape up the oscillatory activity of the hippocampus [66]. The hippocampal trisynaptic loop with so extensive feedforward and feedback projections constitutes a fundamental mechanism of recurrent controlled excitation underpinning memory, learning and emotion in the limbocortical networks (Papez circuit) (**Figure 10**).

There are many brain structures that transmit information to and from the trisynaptic circuit and their activity can be directly or indirectly modulated by the activity of the trisynaptic loop. A major output goes via the fornix to the lateral septal area and to the mammillary body of the hypothalamus and additional output pathways go to other cortical areas including the anterior cingulate and prefrontal cortex.

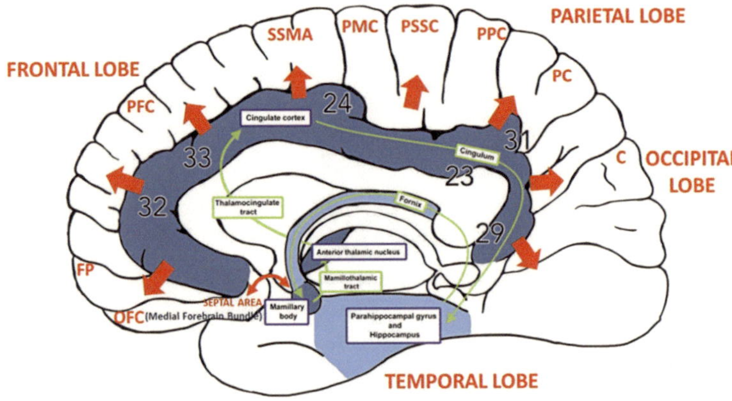

Figure 10.
Limbic system or circuit of Papez. Modified with permission from Weininger et al. [67]. OFC: orbitofrontal cortex, FP: frontopolar, PFC: prefrontal cortex, SSMA: supplementary sensorimotor area, PMC: primary motor cortex, PSSC: primary somatosensory cortex, PPC: posterior parietal cortex, PC: precuneus, C: cuneus.

The crura of the fornix form connections through the corpus callosum and the hippocampal commissures with bilateral hippocampal formations. The CA3 is connected to the lateral and medial septum via the alveus and the fimbria fornix. The CA1 and the subiculum are connected to parahippocampal regions, the entorhinal cortex and nucleus accumbens, and project to septal nuclei, preoptic nuclei, ventral striatum, orbital cortex and anterior cingulate cortex (via the precommissural fornix) and to the anterior and lateral dorsal nuclei of the thalamus, the mammillary bodies, and ventromedial hypothalamus (via the postcommissural fornix) [68].

The mammillary bodies receive information from the hippocampal formation via the fornix and relay information to the anterior nuclei of the thalamus and the anterior cingulate cortex (Brodmann 24, 32, 33) via the mammillothalamic tract. The subiculum relays information to the posterior cingulate cortex (Brodmann 23). Via the cingulate gyrus and the thalamic nuclei primary sensory and association cortical areas can be reached which integrate information at a higher level attending to complex stimuli in the external and internal environments (**Figure 10**). The temporal association cortex identifies the nature of stimuli (perception), while the frontal association cortex plans responses to the stimuli (behaviour). The association cortex projects to other association cortical areas and to subcortical structures including the hippocampus, thalamus, basal ganglia, cerebellum and brainstem [33]. The hippocampus receives modulatory input from the serotoninergic, noradrenergic and dopaminergic systems, the amygdala, the nucleus reuniens of thalamus and the medial septum. The supramammillary nucleus of the hypothalamus is strongly connected via the medial forebrain bundle to the medial septum, diencephalon and brainstem. The medial septum (medial septal nucleus and diagonal band of Broca) sends cholinergic (65%) and GABAergic fibers to CA1 and glutamatergic fibers to all parts of the hippocampus (**Figure 9**) [34].

Theta (4–12 Hz) high amplitude oscillations of the hippocampal CA1 pyramidal cells emerge during active exploration, voluntary movements, rapid eye movement (REM) sleep and certain brain states related to arousal [69]. Type 1 (fast theta oscillations) associated with spatial navigation and movement are driven by atropine-resistant inputs on the distal dendrites, whereas Type 2 (slow theta oscillations) associated with arousal and anxiety on sensory salience are driven by atropine-sensitive inputs on the somata, and chloride (Cl^{-})-mediated inhibitory postsynaptic potentials on pyramidal cells. Although in vitro experiments suggest that theta oscillations can be intrinsically generated in the hippocampal excitatory—inhibitory networks, in vivo studies suggest that the supramammillary-septal-hippocampal loop is crucial for the generation and modulation of theta local-field potential oscillations in the hippocampus [70]. The hippocampus is a highly epileptogenic structure and part of an extensive network (limbic system) that involves all 4 neocortical lobes, the diencephalon/thalamus and the brainstem (**Figure 10**). The medial septum and the anterior nucleus of the thalamus could critically influence and/or allow propagation of seizures from and into the neocortex (**Figure 9**) and therefore have both been studied as a site of neurostimulation in human [71] and in animals with optogenetics [72].

3.3 Neuronal network physiological dynamics

We start by simulating a rudimental recurrent neuronal network of randomly connected 1000 neurons in real time (**Figure 11**). In the raster plots of network neurons, every excitatory neuron is represented by a black dot and every inhibitory neuron is represented by a red dot. The neuronal activity is manifested between 200 ms and 700 ms (within a 0.5 s interval). The graphs below the raster plots depict the cumulative spiking activity of all neurons (mV vs. ms) during the same time interval; the spikes of the excitatory neurons are depicted with blue and the spikes of the inhibitory neurons are depicted with red.

In our initial recurrent neuronal network (the first raster plot and cumulative spikogram below on the left of **Figure 12**), we omitted all inhibitory neuronal activity; this resulted in a very unstable network which rapidly went out of control, with all

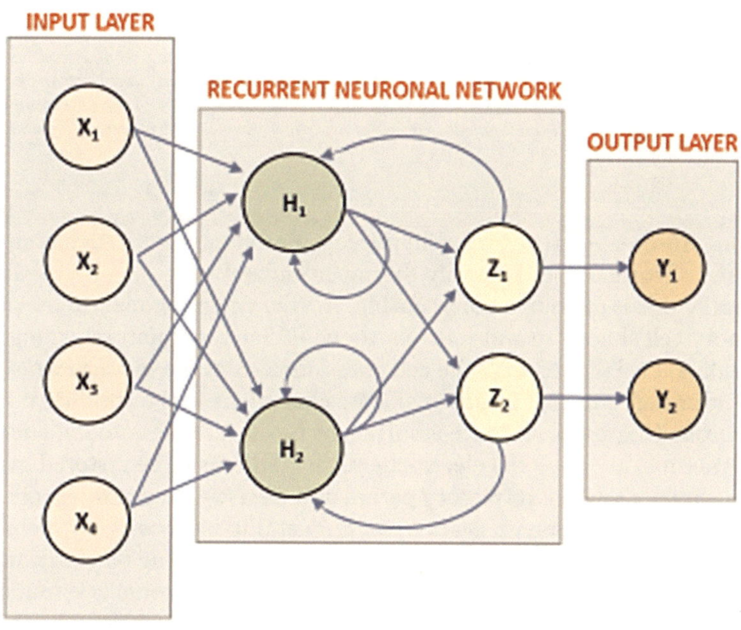

Figure 11.
Abstract state-space representation of the components of a recurrent neuronal network across different layers. The letters X, H, Z and Y represent sets of input [X], output [Y] and state variables [H, Z] in the form of state vectors related by differential or difference equations, whose values evolve over time depending on the values they have at any given time and the externally imposed values of input variables. The values of output variables depend on the state variables.

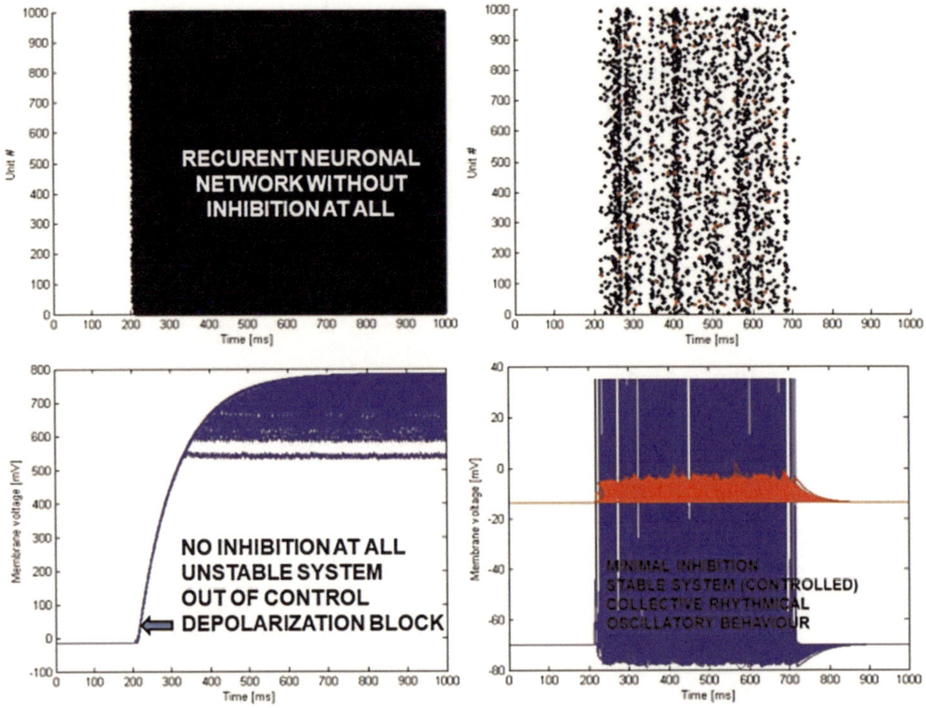

Figure 12.
Recurrent neuronal network without inhibition (on the left) and with inhibition (on the right). As soon as a tiny amount (1%) of recurrent inhibitory neurons were introduced into our random network, not only the spiking neuronal network gets effectively stabilized under such small inhibitory control, but starts also displaying quite complex rhythmical patterns and collective oscillatory behavior of self-organized neuronal assemblies. This is what we would describe as order emerging out of chaos. *At the meso/macroscopic level, quasi-linear phenomena (oscillations of delta, theta, alpha, beta and gamma range and the harmonics thereof) can emerge on the large scale of neuronal networks despite the complex non-linear and stochastic dynamics that govern the behavior of single neurons at the microscopic level.*

excitatory neurons depolarising up to hundreds of millivolts: with such high voltages a small volume of cortex would literally thermocoagulate!

Biologically this is of course not plausible. In vivo when the membrane potential of any excitatory cell shoots up and remains above −25 mV without returning to its resting baseline level of −65 mV, the cell goes into a so-called *depolarization block* where Na^+ channels enter an inactivated state equivalent to the absolute refractory state of neurons (**Figure 3**). If there is sufficient Na-K-ATPase activity, energy supply and adequate time, allowing the electrochemical gradients to be restored over time, the neuron enters a relative refractory period and the Na+ channels resume an active state in which they can open up again upon critical fluctuations in the membrane potential that reach/surpass threshold level. Therefore, ongoing *repolarization* of excitatory pyramidal cells (facilitated by inhibitory interneurons) is essential in maintaining the critical membrane potential fluctuations that drive both physiological and epileptogenic network oscillations.

Motivated by the anatomy of the mammalian cortex [28], we choose the ratio of excitatory to inhibitory neurons to be 4 to 1, and we make the inhibitory synaptic connections stronger about 4 times. Besides the synaptic input, each neuron receives a noisy external (thalamic) input. In principle, one can use *Regular Spiking* (RS) cells to model all excitatory neurons and *Fast Spiking* (FS) cells to model all inhibitory neurons [37]. Based on the *Izhikevich* model, our neuronal spiking network exhibits rich cortical-like asynchronous spiking dynamics (**Figure 5**); neurons fire Poisson spike-trains with random firing rates varying from 0 to 25 Hz (similar to the firing-rate range of most cortical pyramidal neurons). Darker vertical linear areas emerge in our raster plots (on the right of **Figure 12**) that indicate that there are episodes of synchronized firing in the alpha and gamma frequency range (around 10 and 40 Hz, respectively). Although the network is connected randomly without any synaptic plasticity, the neurons, whether excitatory or inhibitory, when firing tend to self-organize into assemblies and exhibit rather collective rhythmical behavior across a wide range of frequencies corresponding to physiological rhythmical EEG activities (delta, theta, alpha, beta and gamma range) in the awake and sleep state of the mammalian cortex.

As the number of external input neurons varies from 1 to 500 (external drive increases), so does the number of oscillatory cycles (**Figures 13** and **14**). For example, for <10 neuron input no collective rhythmical activity is generated, for 10–25 input neurons a 2 Hz (slow delta) oscillatory activity is generated, for 25–50 input neurons a 2–4 Hz (delta) oscillatory activity is generated, for 50–100 input neurons a 4–6 Hz (slow theta) oscillatory activity is generated, for 100–150 input neurons a 6–8 Hz (fast theta) oscillatory activity is generated, for 150–250 input neurons a 8–10 Hz (lower alpha) oscillatory activity is generated, for 250–400 input neurons a 10-12 Hz (higher alpha) oscillatory activity is generated, for 350–500 input neurons a 12–18 Hz (fast/beta) oscillatory activity is generated.

As we keep varying a few critical conditions in the continuous parameter space of our neuronal network model, a vast amount of variation can be introduced under scale transformation in the complex oscillatory behavior of the neuronal network. Our neuronal spiking network model can also simulate an allocortical septal-hippocampal network. The external driving input in a limbocortical neuronal network would be coming from the medial septum (integral part of the supramammillary-septal-hippocampal loop). Such a simulation requires a different ratio of excitatory to inhibitory neurons with perhaps stronger synaptic strengths, reflecting the strong perisomatic inhibition by parvalbumin-positive basket cells (generating Sharp-Wave Ripples and Gamma Oscillations in hippocampal models) [34, 49, 66, 68, 70].

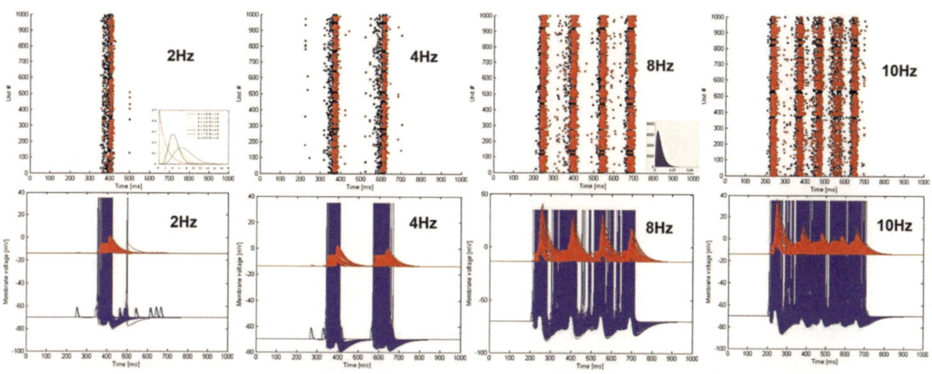

Figure 13.
Complex rhythmical patterns and collective oscillatory behaviour of self-organised neuronal assemblies (oscillations of delta, theta, and alpha range).

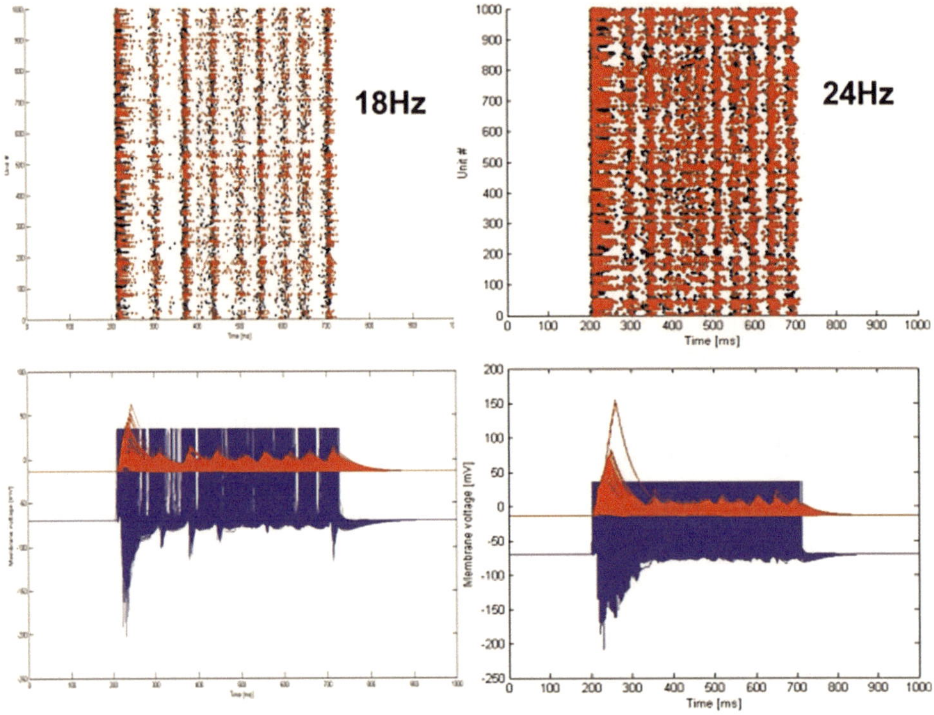

Figure 14.
Complex rhythmical patterns and collective oscillatory behavior of self-organized neuronal assemblies (oscillations in the beta and low gamma frequency range).

3.4 Neuronal network epileptogenic dynamics

While varying the aforementioned network parameters in our *local neocortical or allocortical* network simulation, the most interesting and unexpected behaviour of our network model was the emergence of erratic stochastic bursting of transient (spiking) neuronal activity outside the anticipated interval of 200-700 ms. This was initially an isolated event, quickly brought under control by a concomitant surge of local inhibitory activity (please see raster plot on the left and cumulative spikogram under in **Figure 15**). This stochastic type of bursting activity was reminiscent of an isolated and fairly limited paroxysmal depolarization [73] that died out as a result of sufficient or effective *local inhibitory control* [74]. On a macroscopic scale a *paroxysmal depolarization shift* could appear like an *interictal epileptiform discharge* on intracranial (iEEG) [74] and extracranial (scalp EEG) recordings [73].

At that stage the thalamic or septal input into our network was random noise or a Poisson train-like noise input. Slightly tuning the network further in the critical parameter space as follows: 200 *thalamo-cortical* or *septo-hippocampal* neuron input, overall synaptic weight/strength constant at 0.14, gamma probability distribution for the inhibition of scale-factor $\theta = 2.0$ and shape-parameter $k = 0.015$, and intrinsic membrane excitability with resting membrane potential at −75 mV, sensitivity threshold at −65 mV and resting membrane recovery threshold at −15 mV, were sufficient to trigger the most remarkable and unpredictable behaviour observed in our neuronal network.

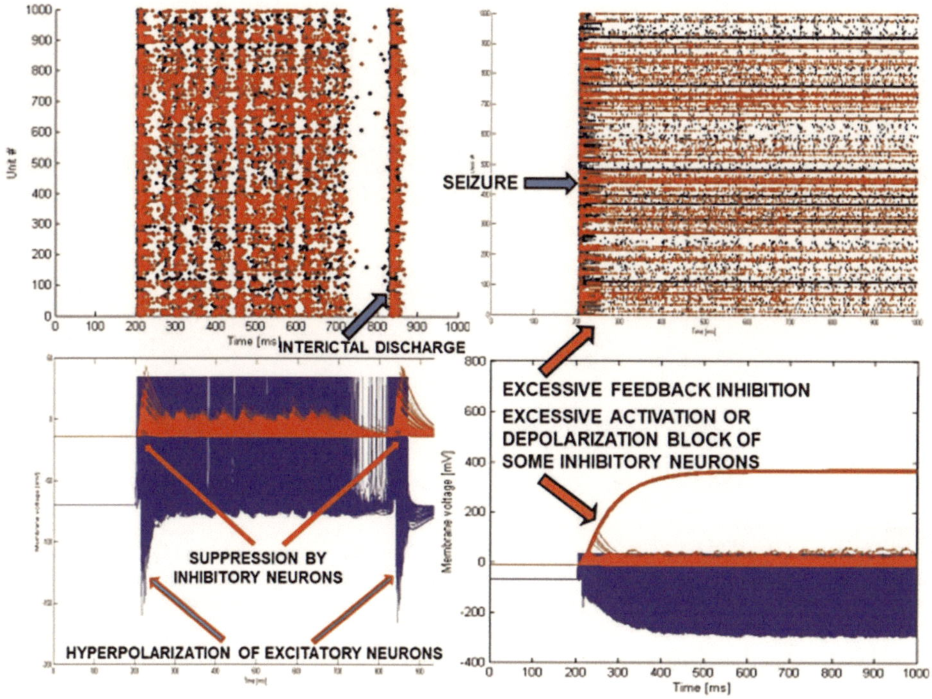

Figure 15.
Interictal and ictal paroxysmal depolarization shifts for critical parameters of our neuronal network (200 neuron input, overall synaptic weight/strength constant at 0.14, gamma probability distribution for the inhibition with scale-factor θ = 2.0 and shape-parameter k = 0.015, intrinsic membrane excitability with resting membrane potential at −75 mV, sensitivity threshold at −65 mV and resting membrane recovery threshold at −15 mV).

This was the generation of sustained *pathological high frequency oscillatory activity* that emerged from stochastic or random noise like other previously described physiological high frequency (beta/gamma range) oscillatory activities [75]. This was amplified in amplitude and frequency beyond any limit. It could not be checked upon by the local inhibitory network and eventually went completely out of control in what could be the equivalent of a *seizure*. The model allows for incredible predictions of the underlying processes taking place during epileptogenesis/ictogenesis (the process of seizure generation and development).

As can be appreciated from the figure above (raster plot and cumulative spikograms at the bottom on the right of **Figure 15**) ictogenesis starts with a *hyperexcitation wave-front* through an initial process of *stochastic resonance* [75] similar to previously described high-frequency oscillatory generation processes [76]. This is immediately followed by a massive feedback burst of inhibitory activity (*phasic inhibition*) [77] that desynchronizes or brings the underlying unit oscillators to a transient halt, resulting on a network scale in transient attenuation/suppression of background rhythms (**Figure 16**). Depending then on the level of sustained or effective background inhibition (*tonic inhibition*) [78] a process of *synchronization* [35] starts and recurrent cycles of alternating excitation and inhibition go on to entrain larger and larger ensembles of unit oscillators into a common resonating high frequency [36]. This maximizes the amplitude of the oscillation and recruits a massive number of unit oscillators into excessive and hypersynchronous oscillatory activity.

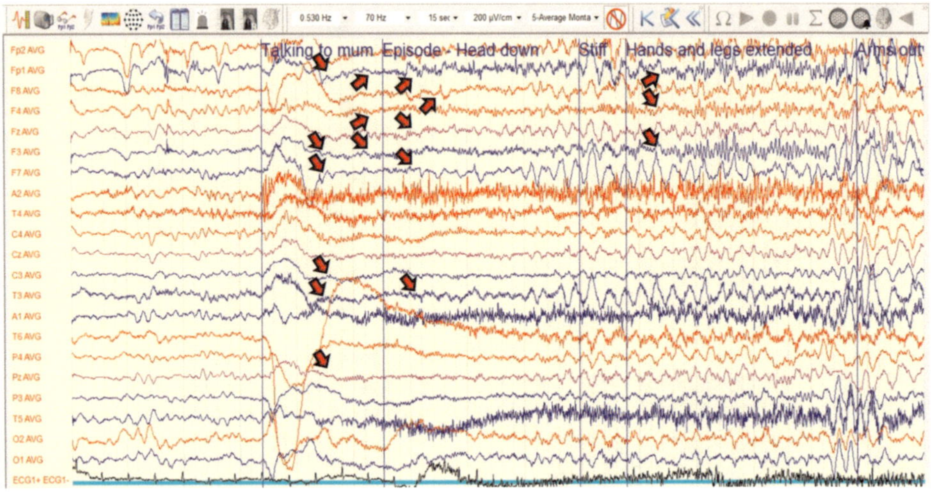

Figure 16.
Initial background rhythm attenuation (intermixed with muscle artefact on some channels) followed by rapid build-up of fast oscillatory activities. This is likely the result of widespread, initial phasic and then tonic inhibition through an extensive cortico-ganglio/thalamo-cortical (left > right) network. This focal-onset left mesial frontal/ sylvian tonic seizure manifested suddenly and evolved rapidly (secondary generalized).

In both physiological and pathological high-frequency activities modelled, there was always a massive inhibitory response to the initial overexcitation wave front, something that is of note as it suggests that for any organized large-scale oscillatory activity to develop a critical level of *synchronization* of unit oscillators is required through recurrent cycles of excitation and inhibition [79]. This is achieved with the early excessive feedback wave of inhibition that follows the initial huge wave of excitation, resulting in an excessive hyperpolarization of the excitatory neurons. This is pivotal in uniformly suppressing or phase-resetting all unit oscillators and synchronously restarting them or coupling/forcing them into a *synchronization process* [78] that gradually entrains larger and larger neuronal ensembles. Excessive *afterhyperpolarization potentials* [80–82] are also crucial in shifting a huge number of inactivated sodium channels from the inactivated to the closed state, rendering them available again for repeated waves of overexcitation during a sustained *paroxysmal depolarization shift* [83, 84].

Resonance is an oscillation of maximal amplitude with all unit oscillators oscillating in a synchronous or synchronized manner, observed when the frequency of a periodically applied *depolarization force* (or a Fourier component of it) is equal or close to a natural frequency of the system. When a small oscillating or periodic *depolarization force* is applied near a resonant frequency of a dynamic system, the system will oscillate at a higher amplitude than when the same force is applied at other, non-resonant frequencies [35]. Obviously, the resonant frequencies of local neocortex or allocortex are defined by the critical combination of local structural and functional connectivity, periodic thalamic or septal input, synaptic weights, interaction of excitatory (recurrent excitatory synapses) and inhibitory (loss of inhibition or disinhibition) components and intrinsic neuronal excitability (endogenous bursting) [78, 79].

The inhibitory oscillatory components of our neuronal network manifest another interesting phenomenon. Driven by the initial overdepolarization of pyramidal cells, some of the inhibitory interneurons excessively 'depolarized' to hundreds of millivolts, a rather unique prediction of our model (right bottom of **Figure 15**). Translating this early abnormality in a critical subpopulation of inhibitory cells in biological terms, raises two possibilities. These would allow an initial wave of over-excitation due to stochastic noise and critical time coincidences in a potentially hyperexcitable system of unit oscillators to rapidly or progressively attain an oscillation of maximal amplitude (resonance) in depolarizing and hyperpolarizing directions while going through consecutive synchronization cycles of excitation and inhibition [85–89]. One possibility is that some of the inhibitory interneurons can go into an actual *depolarization block*, leaving effectively unchecked the excitatory oscillators (unopposed EPSPs on pyramidal cells) to rapidly or progressively evolve into a paroxysmal depolarization (this is likely to occur towards the end of a seizure due to metabolic depletion). The other possibility is that excessive initial phasic inhibitory GABAergic activity, sometimes in combination with GABA transporter (GAT-1) malfunction in astrocytes, releases too much extrasynaptic GABA which facilitates concurrent extrasynaptic $GABA_BR$ activation. This results in enhanced/sustained tonic $GABA_A$ currents that persistently hyperpolarize and increase membrane Cl^- conductance causing bursts of IPSPs to override the depolarizing currents in thalamocortical [85, 86] or pyramidal neurons [90]. Such rhythmical bursts of IPSPs can entrain thalamocortical/gangliocortical (**Figure 16**) and limbocortical (**Figures 17-19**) networks to paroxysmal and/or hypersynchronous activity.

The neuronal network attains a paroxysmal depolarization state when the different excitatory and inhibitory oscillatory components reach maximal

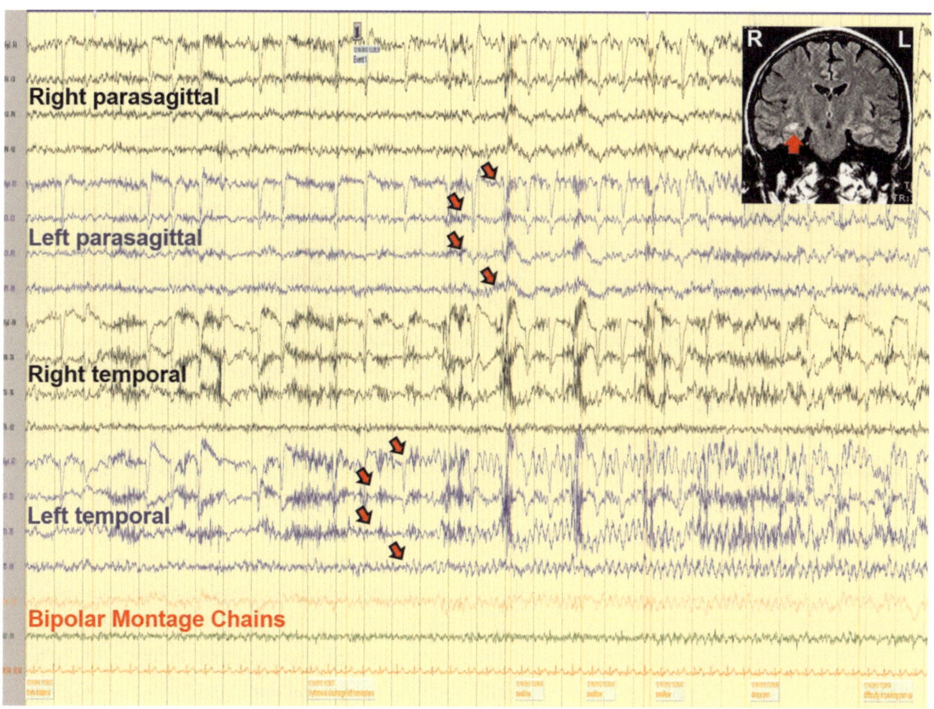

Figure 17.
A left-side mesial temporal onset seizure on the EEG with a rhythmical theta activity build-up over the left temporal region. The MRI scan at the right top corner of the figure indicates severe right mesial temporal lobe sclerosis. Because of lack of concordance between scalp Electroencephalography and MR imaging, we had to undertake intracranial iEEG recordings in this case to determine the exact ictal-onset (potential epileptogenic zone).

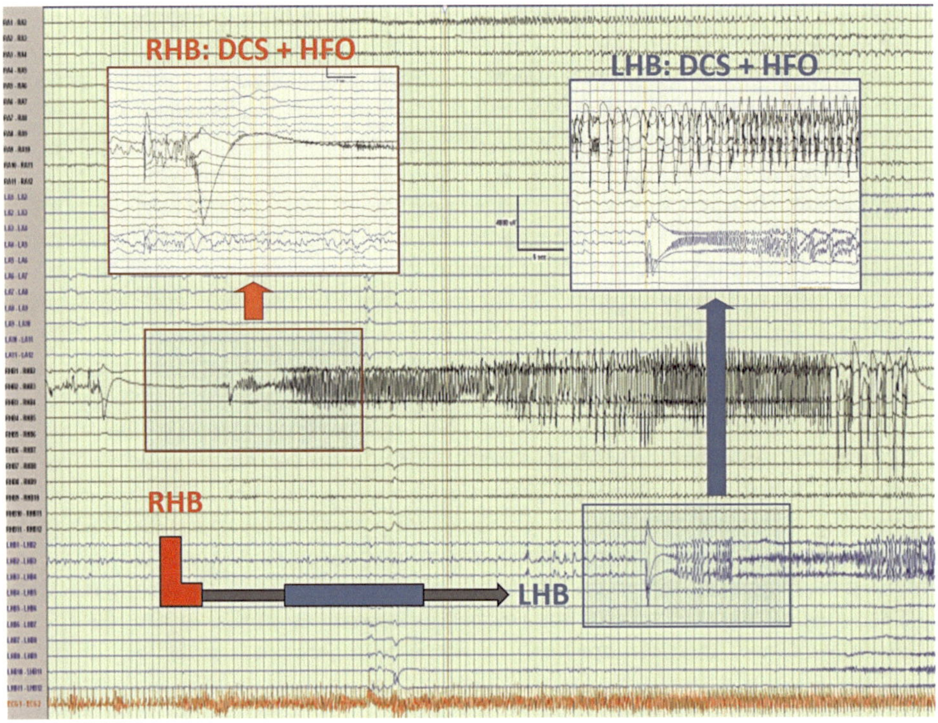

Figure 18.
Human intracranial EEG recordings of bilateral hippocampi to demonstrate baseline (DC) shifts and high-frequency oscillations (HFOs) as surrogate markers of the ictal onset zone. In this patient they always started from the most atrophic and sclerotic (burnt out) right hippocampus and subsequently spread to the left hippocampus. Resection of the right hippocampus in this case conferred seizure freedom and confirmed the epileptogenic zone.

depolarization and hyperpolarization potentials or when excitatory and inhibitory oscillatory units synchronize via depolarizing and hyperpolarizing currents into an oscillation of maximal amplitude. Epileptogenic depolarizing currents can emerge from abnormal or excessive ionic Na^+ [91] and Ca^{++} [92] channel-conductances or excessive excitatory (Glutamate) and neuromodulatory (Acetylcholine, Noradrenaline, Dopamine, Serotonin, etc) neurotransmitter release and receptor function [87–89]. Epileptogenic afterhyperpolarizing currents can emerge from abnormal, insufficient or excessive ionic Cl^- [93] and K^+ [94–96] channel-conductances or abnormal, insufficient or excessive inhibitory (GABA) neurotransmitter release and receptor function [85, 86]. Multiple combinations of the above epileptogenic mechanisms are plausible.

Critical synchronization, resonant oscillation and massive depolarization of excitatory and inhibitory neurons seem to account for a massive release of potassium ions (K^+) from the principal (pyramidal) and supportive glial cells [97]. These shift the resting electrochemical/equilibrium gradients of the cell membrane from −65 mV probably closer to −50 mV, where voltage-gated Na^+ channels are still active and much more likely to open (they are also less likely to be inactivated because of increased Cl^- conductance and sustained tonic GABAergic currents). The membrane conductances essentially change to levels that allow for a massive sustained depolarization shift of the principal/pyramidal cells (massive influx of Na^+ and slower Ca^{2+} inward currents) to take place [83, 84]. Paroxysmal depolarization shifts manifest

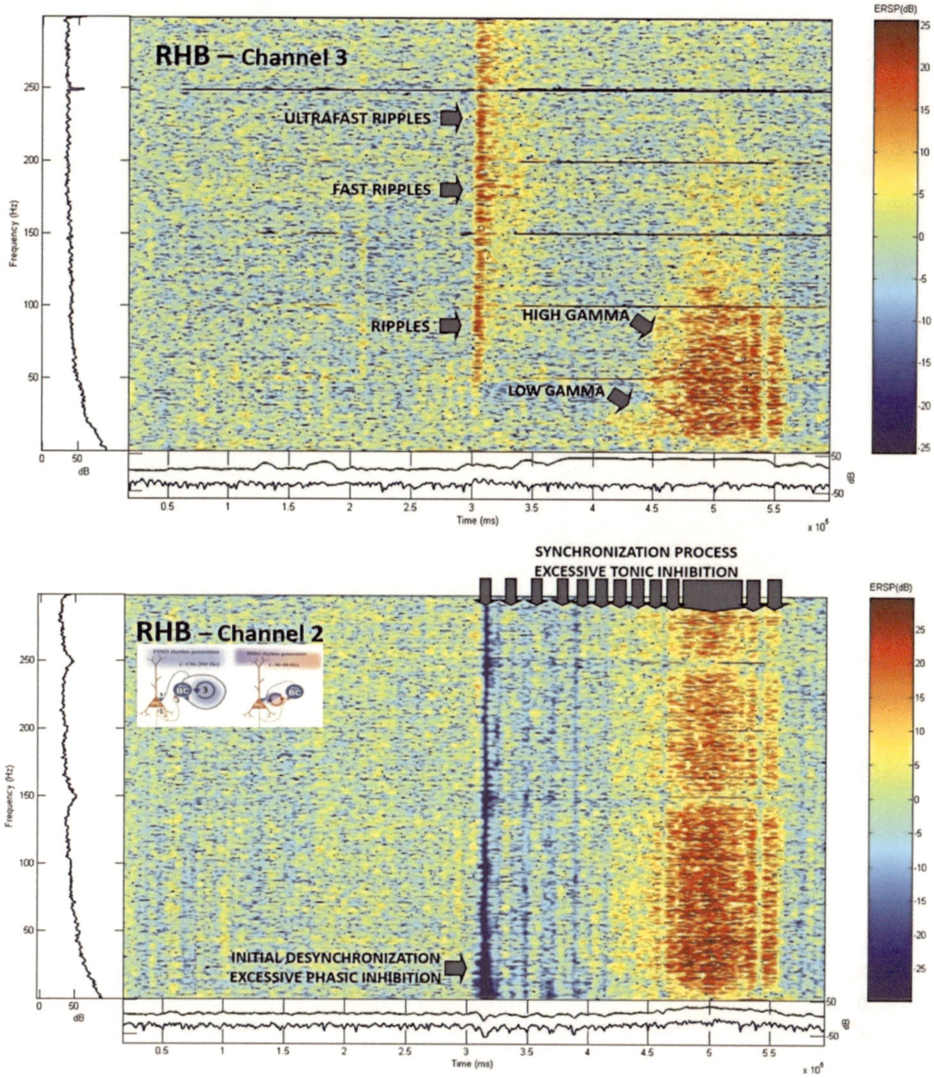

Figure 19.
Within the right posterior hippocampus (Channel 3 Spectrogram) an initial wave of over-excitation (high-frequency ripple synchronization) with ripples/fast ripples followed by excessive feedback inhibition and brief period of attenuation (broadband suppression/desynchronization or phase-resetting of unit oscillators) of the right middle hippocampus (Channel 2 Spectrogram) and recurrent excitation-inhibition oscillatory cycles (synchronization process) before a sustained paroxysmal depolarization shift develops with pathological low/high gamma, ripple and fast ripple oscillations. The inset figure shows the interaction of hippocampal pyramidal cells with parvalbumin-positive basket-cells (interneurons) generating high-frequency gamma-oscillations (PING mechanism) and ripple-frequency phase-modulations (FINO mechanism), reproduced from [66].

electrophysiologically with huge baseline (DC) shifts and very high amplitude pathological high-frequency oscillations (pHFOs) known as pathological beta/gamma oscillations (15–80 Hz), (fast) ripples (80–150 Hz) and (ultra)fast ripples (150–500 Hz) (**Figures 18–20**) [66, 98–100].

Obviously, as this excessive and hypersynchronous overdrive of excitatory and inhibitory neurons goes on, the membrane depolarization shifts towards a more 'toxic'

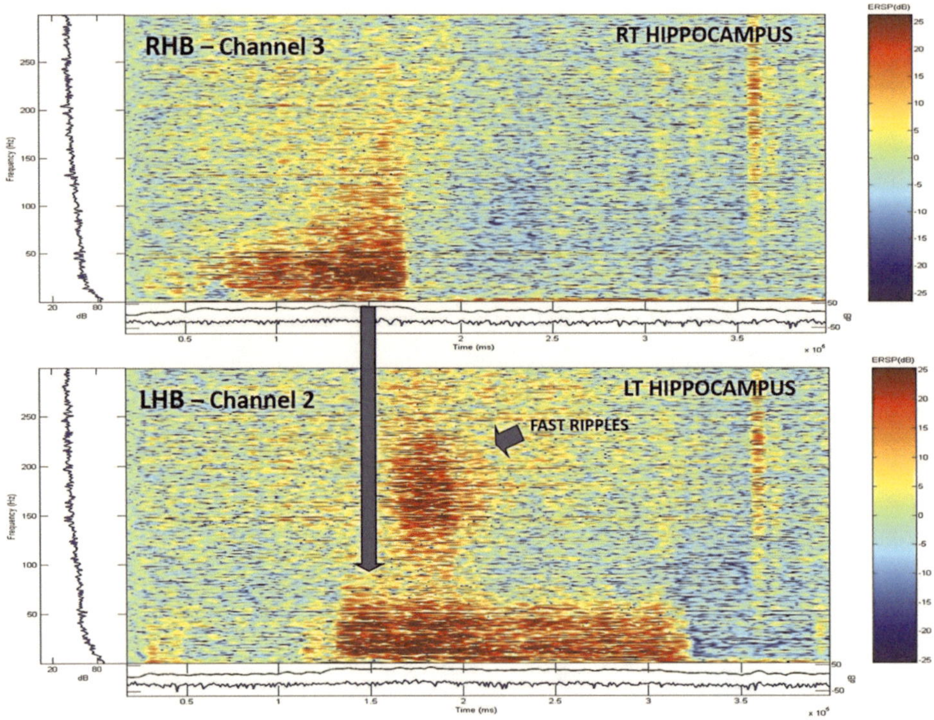

Figure 20.
Spreading of pathological/epileptogenic gamma oscillatory activity in man from the right hippocampus (top spectrogram) to the left hippocampus (bottom spectrogram) with subsequent fast ripples in the left hippocampus. Hilar mossy cells and CA3 pyramidal cells give rise not only to ipsilateral associational hippocampal fibers but also some contralateral commissural fibers (terminating on principal cells and interneurons to CA3, CA2 and CA1 regions), passing through the posterior and anterior hippocampal commissures to reach the contralateral hippocampus and dentate gyrus. More extensive polysynaptic pathways via the entorhinal and perirhinal cortex integrate the hippocampus with the ipsilateral and contralateral hemispheres and homologous hippocampal network. This particular focal motor unaware seizure died out in the right hippocampus/hemisphere but propagated and continued in the left hippocampus/hemisphere, declaring itself on the scalp EEG as seen in ***Figure 15****.*

range of less negative potentials (around −35 mV or above) which will eventually render all voltage-gated Na-channels inactivated. At that point, as the Na-K-ATPase pumps and ion-transporters require sufficient energy and time to restore the electrochemical membrane gradients, a combination of inactivated Na-channels and metabolic depletion will bring the activity of the excitatory and inhibitory cells to sub physiological levels or to a halt (depolarization block and metabolic depletion) during the postictal phase (**Figure 21**) [73, 74].

4. Epileptogenic networks: Aetiology, pathogenesis, pathophysiology and clinical implications

4.1 Epileptogenic cortico-thalamo/ganglio-cortical networks (pathophysiology of typical absences, myoclonic jerks, primary generalized tonic-clonic seizures and photoparoxysmal responses)

Idiopathic (genetic) generalized epilepsy is the prototypical phenotype of primary generalized epilepsies manifesting with a variable combination of absence, myoclonic

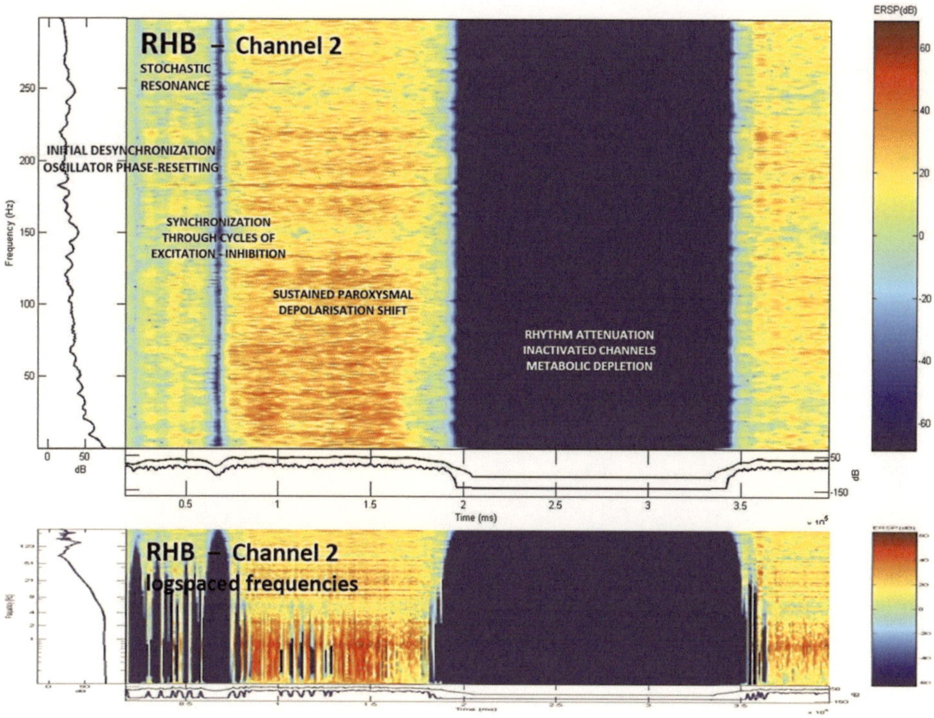

Figure 21.
This particular seizure started in the right hippocampus with a brief period of broadband attenuation (desynchronization or phase-resetting of unit oscillators). After an initial synchronization phase (through recurrent cycles of depolarization and repolarization) it transitioned to a sustained broadband paroxysmal depolarization shift and lasted altogether for about 2 min and 15 s. It ended with a prolonged postictal period of broadband suppression of background rhythms for another 2.5 min. The spectrogram at the top shows the power per 1 Hz individual frequency component from 1 to 300 Hz (nonorthogonal continuous 3-cycle wavelet decomposition) versus time in x10^5 ms, while the spectrogram at the bottom shows log-spaced frequencies (pseudo-orthogonal decomposition) to help appreciate the concentration of power which is mostly <4 Hz but more sustained over time between 12 and 64 Hz (beta and low/high gamma range).

and generalized tonic-clonic seizures, and photoparoxysmal responses (PPR) in the young population (<25 years). Absence epilepsy is one of the most common generalized epilepsies of childhood. It has a unique endo-phenotype based on perturbed cortico-thalamocortical circuitry with genetic and developmental features. It can appear in childhood (CAE) or juvenile ≥10 years (JAE) and can modify or improve its phenotype in adolescence. Typical absences can be triggered in more than >80% of patients with hyperventilation. CAE can even regress with or without treatment, while JAE that is more likely to present with generalized tonic-clonic seizures (up to 80% GTCS) compared to CAE (up to 20%) may also be more resistant to treatment [101].

Another relative endophenotype is that of juvenile myoclonic epilepsy (JAE) and of epilepsy with GTCS seizures alone or on awakening (EGTCS-a, Janz) with perhaps more permanent or pharmaco-resistant features, characterised more by motor myoclonic seizures (MJ) and generalized tonic-clonic seizures (GTCS) with or without absences and perhaps different engagement or imbalance of fronto-central (thalamo-basoganglio-cortical) networks. Photosensitivity (PS) may coexist in roughly 1/5 of CAE, 1/4 of JAE or 1/3 of JME epilepsies [101].

The EEG hallmark of this wide spectrum of genetic/developmental epilepsies varying from less motor manifestations (absences) to more motor manifestations

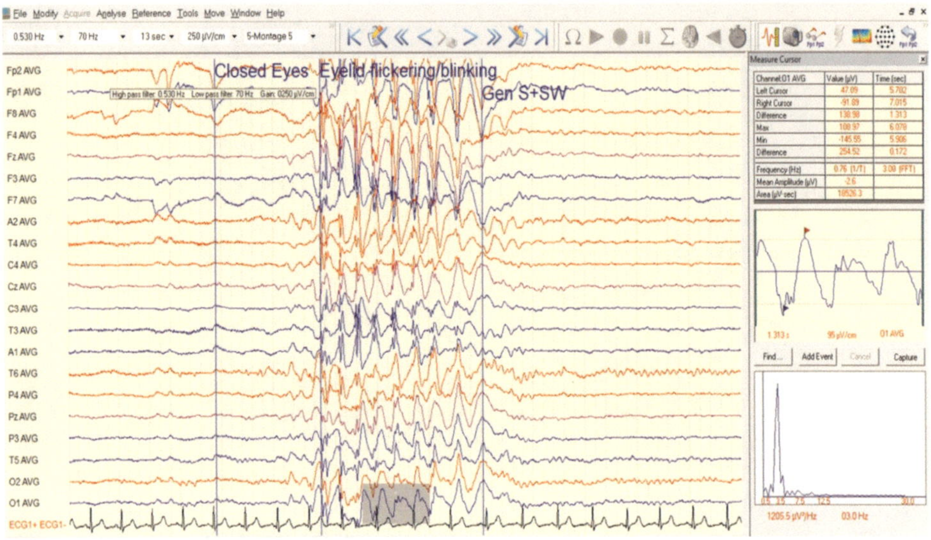

Figure 22.
Spike/polyspike and slow wave discharges in a patient with Juvenile Myoclonic Epilepsy. The spike is a recruiting wave associated with excitatory postsynaptic potentials (EPSPs) and the slow wave is an inhibitory wave associated with hyperpolarizing postsynaptic potentials (IPSPs) in cortical cells, essentially preventing the development of a generalized convulsive seizure.

(myoclonic and generalized tonic-clonic seizures) is the epileptiform generalized spike-and-wave discharges (GSWD) (**Figure 22**). These can also vary in morphology across the different subtypes from a typical regular 3 Hz spike-and-slow-wave (often seen in CAE albeit some variations) to more irregular 3–6 Hz spikes/polyspikes-and--slow-wave patterns (often seen in JAE, JME and EGTCS-a) against a fairly normal background EEG (perhaps with some exceptions occasionally of more focal spikes, sharp waves, slow waves, OIRDA, etc.). The spike is a recruiting wave associated with excitatory postsynaptic potentials (EPSPs) from thalamic relay neurons on cortical cells and consequent phasic inhibition of the thalamic relay neurons. The slow wave is associated with tonic inhibition and hyperpolarizing postsynaptic potentials (IPSPs) on thalamic relay neurons resulting in secondary tonic inhibition of cortical cells (deafferentation), overriding cortical excitation and entraining pyramidal cortical neurons and thalamo-cortical networks in hypersynchronous paroxysmal activity.

The thalamocortical interaction implicated in primary or idiopathic (genetic) generalized epilepsies is one of the most studied epileptogenic networks. The basic thalamocortical circuitry is composed of pyramidal neocortical neurons, thalamic relay neurons, and neurons from the reticular nucleus of the thalamus (NRT) (**Figure 6**). The thalamic relay neurons receive ascending inputs from the Reticular Activation System (ARAS) and project to neocortical pyramidal neurons. Cholinergic pathways from the forebrain and ascending serotonergic, noradrenergic, and cholinergic brainstem pathways regulate the excitability of the thalamic relay neurons and the thalamocortical circuitry (**Figure 9**) [102]. As a result, thalamic relay neurons manifest oscillations in their resting membrane potential, which increase the probability of synchronous activation of neocortical pyramidal neurons (EPSPs) during depolarization states and lower the probability of neocortical activation (IPSPs) during hyperpolarization states. This generates thalamocortical oscillatory rhythms and induces slow coherent oscillations in the cortex (resonance phenomena), characterised by periods of relatively increased excitation (up-states) and periods of relatively increased inhibition (down-states), such as the fast oscillations in sleep spindles and larger slower oscillations observed in sleep [103, 104].

Within a potential cortical area (sensorimotor cortex) medium-amplitude 5–9 Hz oscillations secondary to decreased phasic ($GABA_AR$) inhibition [105–107] may entrain other cortical areas and the thalamus leading to a strong and synchronous cortical output that excites the GABAergic neurons of the Nucleus Reticularis of the Thalamus (NRT). The thalamic relay neurons have GABA-B receptors and receive GABAergic tonic inhibition from the neurons of the Nucleus Reticularis of the Thalamus (NRT) [108]. Also increased ambient (extrasynaptic) GABA levels around thalamic relay neurons due to reduced GABA uptake by GAT-1 (malfunction of thalamic astrocytes GABA transporter), may further enhance extrasynaptic $GABA_AR$ tonic inhibition [78]. Enhanced tonic inhibition persistently hyperpolarizes thalamic relay neurons and increases their membrane Cl^- conductance.

The hyperpolarization of thalamic relay neurons due to excessive or sustained GABAergic tonic inhibition of thalamic relay neurons shifts the T-calcium channels from the inactivated to the closed state and permits the synchronous opening of a large population of the T-calcium channels (about every 100 milliseconds). The rhythmic IPSP bursts on thalamic relay neurons, driven by a transient low-threshold calcium channel (transient T-calcium current) with intrinsic bursting behaviour, induce a widespread burst of excitation on neocortical pyramidal cells giving rise to the spike and secondary tonic cortical inhibition (following excessive excitatory bursting) which causes widespread cortical deafferentation, a phenomenon we macroscopically observe on scalp EEG as generalized slow wave complexes following the spike(s) (**Figure 22**) and as *absences* in patients' behaviour [85, 86, 90, 97]. A functional mutation in the CACNA1H gene encoding the Cav3.2 low-voltage activated Ca^{+2} channel has been found in the Genetic Absence Epilepsy Rats from Strasbourg (GAERS animal model of absence epilepsy) [109]. Alterations or mutations in the chloride channel subunits or molecules that regulate their function can increase membrane conductance of Cl^-. Increased Cl^--mediated hyperpolarizing currents (IPSPs) increase the number of T-calcium channels available for activation, resulting in imbalanced networks of excitatory and inhibitory components with increased synchronization in the thalamocortical circuit and decreased seizure-threshold [94, 101–117]. Animal models of absence seizures have demonstrated that GABA-B receptor antagonists can suppress absence seizures, whereas GABA-B agonists can worsen them [111].

This explains why antiepileptic medications such as Ethosuximide, Valproic acid, Lamotrigine, Levetiracetam and Zonisamide, by blocking or suppressing the T-calcium channel currents, are more effective in preventing absence seizures. On the other hand, antiepileptic medications that indiscriminately increase GABA levels (e.g. Tiagabine, Vigabatrin) or Phenobarbital (prolongs Cl^--channel opening duration) are associated with worsening of absence seizures. The effect of benzodiazepines (Diazepam, Lorazepam, Clonazepam and partial agonist: Clobazam) may be slightly more selective (increased frequency of Cl^--channel opening), also manifested on the EEG often with increased fast cortical oscillatory activity (increased intracortical inhibition and synchronization) with more variable effects on the degree of synchronization (usually desynchronization) of the thalamocortical circuit [101–111].

4.2 Epileptogenic focal neocortical and allocortical/limbocortical networks (pathogenesis and pathophysiology of focal-onset, aware or unaware, seizures with secondary propagation, bilateral spreading or generalization

Focal seizures manifesting with a variable combination of auras, sensory, motor, limbic or autonomic, aware or unaware seizures with secondary unilateral or bilateral propagation and generalization represent the collective phenotype of focal-onset, propagated or secondarily generalized epilepsy. They can occur at any age, but would be more common among the adult-onset epilepsies or in patients with an apparently normal brain development, who have never had previously any seizures in young life. Although there are no obvious neurological deficits or abnormal brain development, thorough investigations may reveal a range of subtle focal brain abnormalities or insults (structural, ischaemic/vascular, inflammatory, infectious, metabolic, autoimmune, neoplastic, degenerative, epigenetic, etc) which could be part of localised or more widespread epileptogenic networks [1–6]. On the other hand, focal/multi-focal or generalized symptomatic epilepsies are usually associated with some kind of focal or generalized brain dysfunction, injury or developmental abnormality. People with symptomatic epilepsies have neurological or cognitive deficits and a higher chance of intellectual disability, cerebral palsy, Lennox-Gastaut syndrome and other neurodevelopmental conditions/problems. Nowadays, as a result of widespread applications of epilepsy surgery with direct intracranial EEG recordings, focal neocortical and allocortical/limbocortical epileptogenic networks have been more thoroughly studied and better understood [1–6].

Please see the example below of a focal-onset musicogenic seizure with progressive ipsilateral propagation, bilateral spreading and secondary generalization in a patient who turned out to have an autoimmune (GAD65 + ve antibody-mediated) limbic encephalitis (**Figures 23–25**).

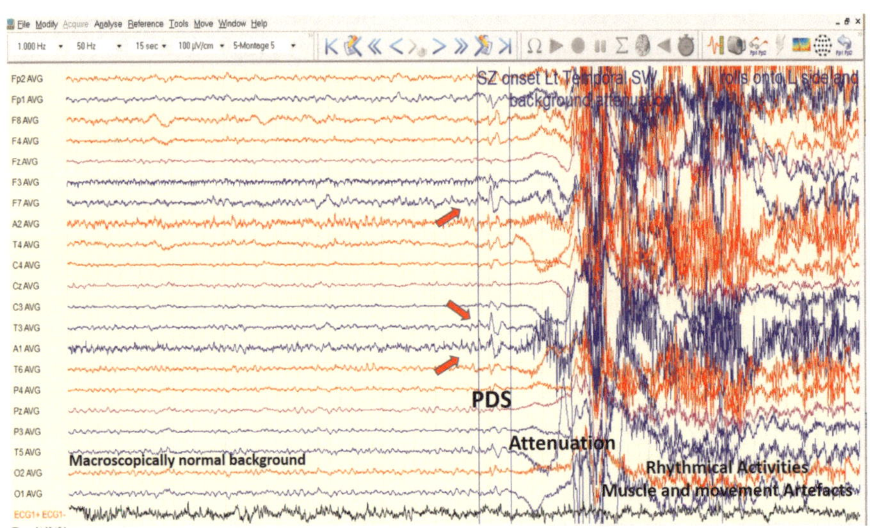

Figure 23.
Against a normal background upon the patient listening to one of her favourite songs from her childhood (previously she had been exposed to all sort of different music styles, including the most dysharmonic/atonality scales of Schoenberg's dodecaphony) a single high-amplitude sharp wave appeared over the left frontotemporal (maximum at anterior temporal F7 electrode) region, followed by a widespread desynchronization/attenuation of the EEG for 1–2 s.

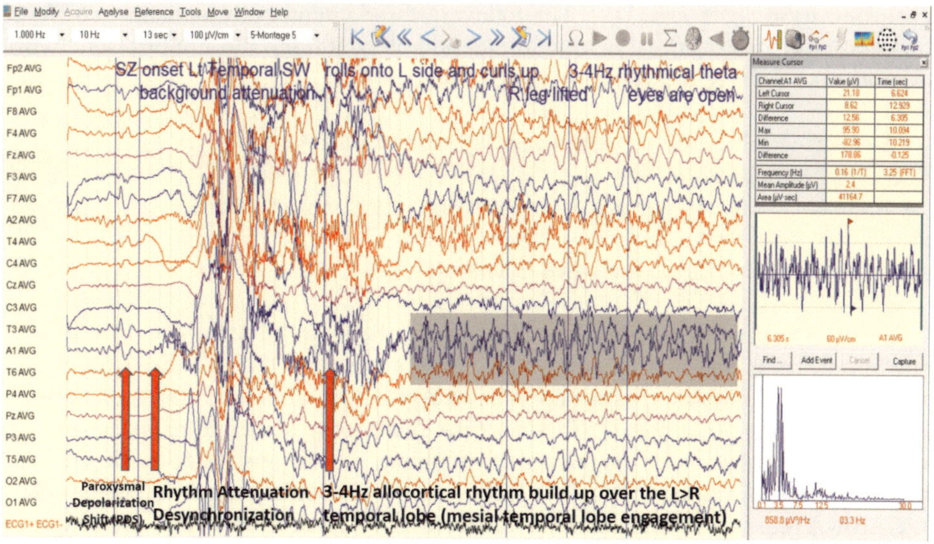

Figure 24.
Following the background EEG attenuation lasting for 1–2 s, a widespread frontotemporal (maximum at mid-temporal A1 electrode) rhythmical 3–4 Hz theta-range activity (with opening of her eyes), started building up indicating engagement of left limbo-cortical networks (left hippocampus and limbic system).

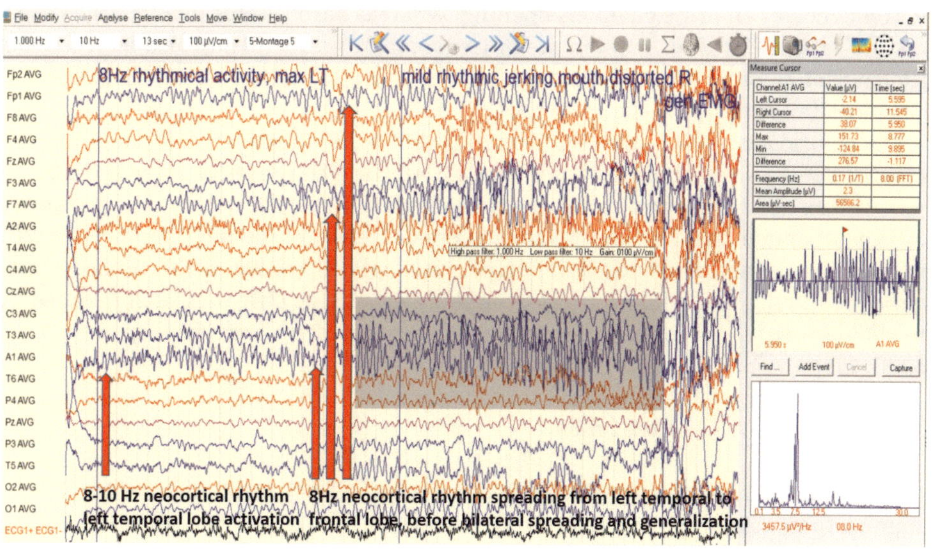

Figure 25.
Following the engagement of left limbo-cortical networks, there is an acceleration of the background activity over the left hemisphere into a rhythmical 8–9 Hz alpha-range (neocortical) activity predominantly over the left frontotemporal (maximum at A1 temporal electrode) region, before spreading to bilateral frontal regions and clinically manifesting with a generalized tonic-clonic seizure (the EEG gets obscured by muscle and movement artefact).

In the focal type of epilepsies, there is either decreased inhibition and/or increased excitation and/or structural and functional connectivity changes (acquired or cumulated through life) in the local network that can alter the dynamic interaction of excitatory and inhibitory network components and/or result in increased network synchronization processes [118]. The following fundamental mechanisms in different combinations can alter the dynamic interaction of excitatory and inhibitory network components resulting in focal-onset seizures:

- Decreased inhibition or defective activation of GABAergic neurons
- Increased excitation or increased network synchronization processes

4.2.1 Mechanisms leading to decreased inhibition

The release of γ-amino-butyric acid (GABA), the main inhibitory brain neurotransmitter, from presynaptic neuron terminals binds to GABA-A and GABA-B receptors and inhibits the postsynaptic neuron either by direct induction of an inhibitory postsynaptic potential (IPSP) via GABA-A receptor-mediated chloride currents, or by indirect inhibition of the release of excitatory neurotransmitter in the presynaptic afferent projection, with a GABA-B receptor-mediated potassium current [119].

GABA-A receptors (made of 2 alpha, 2 beta and 1 gamma subunits) are coupled to chloride channels which are modulated by several mechanisms, such as changes in the 3-dimensional conformation of subunits/proteins or phosphorylation at different sites of the channel. For example, chloride channels are modulated by benzodiazepines (e.g. diazepam, lorazepam, clonazepam, clobazam), barbiturates (e.g. phenobarbital, pentobarbital), Topiramate or Cenobamate or Stiripentol. Benzodiazepines increase the frequency of chloride channels opening, whereas barbiturates increase the duration of channels opening. Topiramate or Cenobamate or Stiripentol also increase the frequency of chloride channel opening, but they bind to a different site from benzodiazepines (allosteric modulators) [120].

The chloride equilibrium potential is roughly the same as the resting membrane potential equilibrium, about −70 mV. Therefore, the electromotive force for net chloride flux during the resting potential is minimal. As the summation of excitatory postsynaptic potentials (EPSPs) results in depolarization of the membrane potential, the influence of inhibitory postsynaptic potential (IPSP) mediated chloride currents on the membrane potential becomes even more crucial, as only these can increase the threshold for firing an action potential and decrease neuronal excitability [80–82, 85, 86, 90, 121].

4.2.1.1 Defective GABA-A inhibition

Mutations or lack of expression of appropriate GABA-A receptor complex subunits, their assembly molecules or the molecules that modulate their electrical properties can cause decreased inhibition [120, 122]. For example, hippocampal pyramidal neurons may not be able to assemble alpha-5, beta-3, gamma-3 receptors because of deletion of chromosome 15 (i.e. Angelman syndrome) [123]. Animal models of focal-onset epilepsy based on pilocarpine models, electrical or chemical kindling, have shown changes in the distribution of subunits of the GABA-A receptor complex [124].

4.2.1.2 Defective GABA-B inhibition

The GABA-B receptor complex, often located in the presynaptic excitatory nerve terminals, consists of 2 subunits (with 7 transmembrane domains each), coupled to potassium channels modulated via G proteins. Upon activation it drives a potassium current with longer latency and duration of action compared to the chloride current generated by activation of the GABA-A receptor. Thus, alterations in the GABA-B receptor complex may be crucial for ictal transformation [108, 125].

4.2.1.3 Defective network function of GABAergic interneurons

As we have demonstrated, in complex neuronal networks with recurrent feedforward and feedbackward projections from excitatory to inhibitory neurons, feedforward and feedbackward inhibition emerge from the critical time activation of GABAergic inhibitory neurons relative to the output of the Glutamatergic excitatory neurons of the network [35, 76–78, 126, 127]. The hippocampal model has been extensively investigated as the prototype neuronal network of focal epileptogenesis. Schaffer collateral axons from the CA3 pyramidal neurons (main afferent input) activate the CA1 principal neurons (hippocampal pyramidal cells). At the same time collateral feedforward projections to GABAergic inhibitory interneurons activate their somata, before or during activation of the apical dendrites of the CA1 pyramidal neurons [66, 79–82, 100, 128].

As a result of this crucial structural and functional connectivity, during passive transmission of the excitatory postsynaptic potential (EPSP) from the apical dendrites to the axon hillock of the CA1 pyramidal neurons, a concurrent GABAergic inhibitory postsynaptic potential (IPSP) inhibits the soma or axon hillock of the CA1 pyramidal neurons. This feedforward inhibitory projection simultaneously hinders pyramidal cell depolarization and firing of an action potential [129]. Recurrent axon collaterals from the CA1 pyramidal neurons activate GABAergic interneurons after the pyramidal neurons have fired an action potential. This creates a feedbackward inhibitory system (multiple inhibitory circuits with built-in time-lags) that allows GABAergic cells to control repetitive firing in principal neurons (CA1 pyramidal cells) and also inhibit the surrounding hippocampal pyramidal cells. The critical timing of these excitatory-inhibitory cycles (push-and-pull mechanism) accounts for the generation of normal gamma-oscillatory and hippocampal sharp-wave ripples, or abnormal fast ripples and ultra-fast ripples [35, 36, 66, 76–79, 98–100, 126, 127, 130, 131].

The mossy cells of the hilar polymorphic region of the dentate gyrus of the hippocampus (which receive feedforward input from the Entorhinal Cortex and feedback activation from CA3) appear to activate GABAergic neurons and gate-control the inhibitory tone of the network. The mossy cells may be susceptible to seizure-related neuronal death [132, 133]. The loss of mossy cells results not only in impairment of GABAergic interneuron activation (deafferentation), but also in synaptic reorganization and changes in network plasticity, with formation of newly sprouted circuits of excitatory and inhibitory cells in an attempt to restore inhibition. However, with epilepsy progression the sprouted synaptic contacts also create recurrent excitatory circuitries that permanently alter the balance between excitatory and inhibitory tone in the hippocampal network [133, 134].

4.2.1.4 Defective intracellular buffering of calcium

In rodent hippocampal experiments, recurrent seizures can result in progressive loss of hyperpolarized resting membrane potentials in the hilar polymorphic region of the dentate gyrus and eventually loss of interneurons that lack the calcium-binding proteins parvalbumin and calbindin [135, 136]. Further experiments showed the critical role of adequate concentrations of calcium-binding proteins for neuronal survival in settings with sustained increases in intracellular calcium under neuronal cellular stress [137], such as in status epilepticus, febrile convulsions, brain hypoxia and other metabolic, toxic, ischaemic and inflammatory brain insults. Interindividual differences in these calcium-binding proteins may explain the variable susceptibility of different patients and with advancing age to epileptogenesis via the premature loss of critical interneurons, a process that alters inhibitory controls of local neuronal networks in favour of excitation [138–140].

4.2.2 Mechanisms leading to increased excitation

Similar concepts of structural and functional organisation to the chloride channels, with crucial electrophysiological implications, also hold for the voltage-gated sodium, potassium and calcium channels. Alterations or mutations in the chloride, potassium or sodium and calcium channel subunits or in the molecules that regulate their function may increase or decrease the membrane permeability and conductance of chloride, potassium or sodium and calcium ions forming the chloride-mediated and potassium-mediated hyperpolarizing currents (IPSPs) that counterbalance the sodium-mediated and calcium-mediated depolarizing currents created by the summation of EPSPs [94, 112–114]. The overall network balances and imbalances attained in excitatory and inhibitory components critically modulate the seizure-threshold or the tendency to seizures [33, 35, 36, 66, 69, 76–79, 85–90, 93, 97, 115].

A lower seizure-threshold and thus increased epileptogenesis may result from inappropriate activation of fast or long-acting NMDA channels or reduced intracellular calcium-buffering proteins (parvalbumin and calbindin), increasing the vulnerability of neurons to cellular stress-injury and death [134, 137, 140]. The release of the excitatory amino acid Glutamate from presynaptic neuron terminals mediates excitatory potentials (EPSPs) in the postsynaptic neuron membrane via: N-methyl-D-aspartic acid (NMDA), α-amino-3-hydroxy-5-methyl-4-isoxazole propionic acid (AMPA)/Kainate, and Metabotropic Glutamate receptors. These receptors are coupled by means of different postsynaptic membrane mechanisms to several depolarizing channels [33, 66, 69, 87–89, 116].

4.2.2.1 Increased activation of NMDA receptors

Fast glutamatergic neurotransmission is based on activation of AMPA/Kainate and NMDA receptors. The AMPA/Kainate receptors are coupled to channels that create currents of monovalent cations (sodium and potassium), whereas the NMDA receptors open channels that allow also divalent cations to pass through (calcium). Slow glutamatergic neurotransmission is also possible via metabotropic receptors, which alter postsynaptic membrane excitability with late-onset and more prolonged postsynaptic changes in phosphorylation and gene expression by means of a second-messenger system which uses calcium as a catalyst for various intracellular reactions. Normal NMDA receptor function is thought to be associated with learning and memory, whereas increased activation has been observed in several animal models of focal epilepsy, such as kindling, kainic acid, pilocarpine, etc. [116, 117, 141, 142].

4.2.2.2 Increased synchronization induced by ephaptic phenomena

Widespread electrical fields emerging from synchronous activation of many pyramidal neurons in laminar hippocampal structures or local changes in extracellular ionic concentrations of potassium and calcium [97] or increased neuronal coupling due to more permanent changes in functional gap junctions [143] may further increase the excitability of neuronal assemblies by nonsynaptic (ephaptic) interactions, predisposing to focal-onset seizures or status.

4.2.2.3 Increased synchronization and/or activation from recurrent excitatory collaterals

Intractable focal-onset epilepsies are frequently accompanied by abnormalities in the limbic system, particularly in the hippocampal formation. Hippocampal atrophy and sclerosis are common lesions associated with neuronal loss and gliosis, particularly affecting the hilar polymorphic region and CA1 pyramidal regions, with relative sparing of the CA2 pyramidal region and only intermediate severity lesions in the CA3 pyramidal region and dentate granule neurons. About 2/3 of patients with intractable temporal-lobe epilepsy have marked hippocampal sclerosis, while animal models with >100 brief convulsions or epileptic status showed similar changes, suggesting that epileptogenesis or hippocampal/limbic system kindling can be a self-perpetuating process [132, 138, 139].

Perhaps subtler and more common than overt hippocampal sclerosis may be Mossy-fiber sprouting. The mossy fibers are the axons of the dentate granule neurons that mostly project to the hilar polymorphic region and CA3 pyramidal neurons. Progressive loss of neurons in the hilar polymorphic region and degeneration of their synaptic projections on dentate granule neurons, induce sprouting of the neighbouring mossy-fiber axons and formation of recurrent excitatory collaterals, with an overall increase in the excitatory drive of dentate granule neurons [133, 134, 140].

4.3 Epileptogenicity: critical conditions and clinical implications

As we have demonstrated in previous sections, epileptogenicity seems to be intricately related to the mechanisms that vary the level of consciousness and transition through sleep (cortico-thalamocortical circuits) and the processes of learning, memory, emotion and complex behaviour (cortico-limbocortical circuits). In our simulations of neocortical, thalamocortical and allocortical neuronal networks the following parameters or processes have emerged as most critical for epileptogenesis:

4.3.1 Excitability of individual neurons and entire networks

The intrinsic/inherent cell membrane and synaptic membrane excitability properties, influence the electrochemical ionic gradients/equilibriums and ionic transmembrane conductances (presynaptic, synaptic, extrasynaptic and postsynaptic receptors, ionic channels and ligands/neurotransmitters, transporters, ion pumps and exchangers, channelopathies and antiepileptic drug effects). These ultimately determine the intrinsic excitability and oscillatory dynamics of the individual neuron and its interactions with other structurally/functionally interconnected neurons.

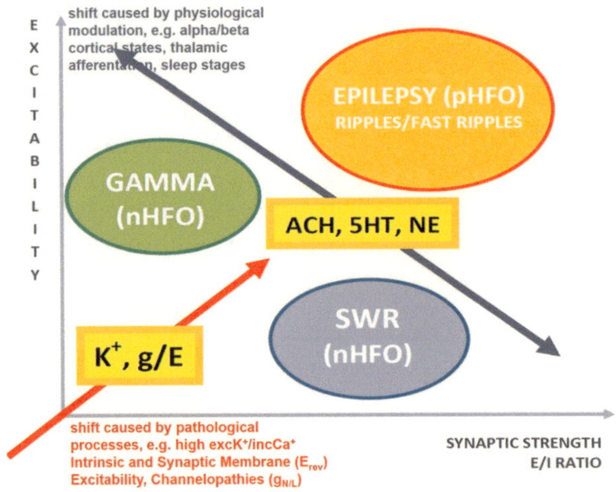

Figure 26.
A summary of the interaction of multiple neurophysiological epileptogenic mechanisms.

The membrane excitability characteristics and shortening of time integration constant (via a "push-and-pull" mechanism) of synchronized, coincidental or critically interacting excitatory (EPSP) and inhibitory (IPSP) postsynaptic potentials can increase or decrease the excitability of the entire network (**Figure 26**) [79, 85–97].

The intrinsic excitability of the neuron directly increases excitation or the number of fast or transient oscillations by reducing the relative refractory period of the firing neurons and/or increase the probability of spatiotemporal summation or integration of synchronous/coincidental EPSPs and/or IPSPs. These mechanisms can drive the firing rates of neurons broadband, the amplitude of the postsynaptic oscillations high and the time/phase dynamics of their firing complex. They can thus generate a range of physiological transient and rhythmical cortical neuronal activities (in delta, theta, alpha, beta and gamma frequencies as reflected in extracellular recordings, local field potentials, intracranial and scalp surface EEG) and can drive local or widespread networks in paroxysmal/hypersynchronous activity [85–97, 126, 127].

4.3.2 Structural connectivity of neuronal networks

The structural connectivity (spatial network geometry) of neurons is critical for the dynamic interaction of excitatory and inhibitory network components. The spatial distribution of excitatory and inhibitory neurons reflects how sharpened or spread out the inhibition may be around excitatory neurons. Changing the relative distribution of the inhibition, that is, how locally vs. widespread the inhibition acts, shapes up and critically determines individual neuronal and collective oscillatory network behavior, generating a range of physiological transient and rhythmical cortical neuronal activities and can drive local or widespread networks in paroxysmal or hypersynchronous activity (**Figures 13–15**) [76, 79, 126, 127, 144].

Early developmental and life-long brain changes induce progressive small-to-large scale structural changes in cortical networks (and thus in epilepsies with age) via migration and branching-off patterns of neurons, plasticity changes via sprouting or pruning of neuronal processes, reinforcement or attenuation of synaptic contacts.

Cortical network development and life-long plasticity changes determine long-term episodic memory formation or retrieval and operational learning memory, and may be responsible for the development of thalamocortical (e.g. idiopathic/genetic generalized epilepsies) or build-up of neocortical and limbocortical epileptogenic networks (e.g. abnormal excitatory recurrent collaterals) [1–6, 101, 126, 127, 144].

4.3.3 Functional connectivity of neuronal networks

The functional connectivity (synaptic strengths/weights) of the network is critical for the dynamic interaction of excitatory and inhibitory network components. Changing the relative strength of synaptic connections, critically determines the oscillatory behavior of the network, and can produce a range of physiological transient and rhythmical cortical neuronal activities and collective behaviors, from local spatiotemporal to systemic synchronization phenomena. These can sustain the excitatory up-states of cortical neurons, shape and enhance plasticity, memory, structural and functional connectivity of thalamocortical, limbocortical and neocortical neuronal networks (**Figures 13–15**) [43–53, 126, 127, 144].

Memory and learning (new memory formation and retrieval, short-term working memory and operational learning), diurnal variations in brain function (spindle waves and slow sleep oscillations) and longer lasting plasticity changes emerge from the short to long-term potentiation and/or depression of synaptic connections. Developmental brain changes are also associated with progressive plasticity changes, affecting brain rhythms and epilepsies with age including the build-up (e.g. through abnormal excitatory recurrent collaterals) or shape up of epileptogenic networks (e.g. through recurrent seizures, high firing-rate and/or hypersynchronous synaptic activations) [1–6, 126, 127, 144].

4.3.4 Critical global and local transient brain states (microstates)

Neocortical activation is driven by thalamic/reticular input and the level of consciousness rises or drops with a varying reticular, thalamic, septal or neocortical drive. As higher numbers of thalamic input neurons engage larger numbers of cortical neurons, this physiologically brings about thalamocortical arousal, thalamosensory afferentation and amplifies cortical cognitive processing upon multi-modal sensory stimulation of the cortex. Cortical rhythms speed up and modulate from slow oscillatory (delta and theta range) activities during sleep and low consciousness states to faster oscillatory activities (alpha with closed eyes and beta/gamma with open eyes) in the awake and alert brain (**Figures 13–15**) [43–53].

In a similar fashion, limbocortical activation is externally driven by varying septal, reticular, thalamic, or neocortical input. While internally limbocortical activation is driven by CA3/modulatory Mossy-fiber input into CA1/Subiculum and further into entorhinal cortex that interfaces with the neocortical parahippocampal gyrus and perirhinal cortex, anterior nucleus of the thalamus, the posterior and anterior cingulate cortex, temporal association cortex (for stimulus perception) and frontal association cortex (for planning behavioural responses) and other subcortical structures. This gives rise to physiological high-frequency gamma oscillations (nHFO) and sharp-wave ripples (SWR) for memory storage/retrieval and arousal respectively (**Figure 26**) [33, 34, 49, 60–66, 68, 69].

When critical local and global conditions are met, transitions across different global and local brain states (microstates) become a powerful modulator of small-to-large scale neocortical, ganglio-thalamocortical and/or limbocortical networks that elicit or unmask epileptogenic network activity in the form of pathological gamma-oscillations (pHFO), ripples and fast ripples, paroxysmal depolarizations and DC shifts at the microscopic/mesoscopic level (extracellular and local-field potentials) or spike-and-waves discharges, sharp and rhythmical fast and slow waves or attenuation/electrodecremental responses at the macroscopic level (EEG) (**Figure 26**) [33, 34, 60–66, 68, 69, 98–100, 126, 127]

4.3.5 Epileptogenicity and clinical implications

The above have crucial clinical implications for our current practice and future approach to epileptic disorders. We need to identify focal lesions or more widespread abnormalities of structural and/or functional connectivity (focal cortical dysplasias, developmental dysplasias/malformations, previous or perinatal brain injuries, hypoxic or metabolic and toxic insults, ischaemic or vascular lesions and malformations, tumours, space-occupying lesions, infiltrative, (para)neoplastic, inflammatory, infective/postinfective, autoimmune, vasculitic, (epi)genetic, neurodegenerative, etc) and modify the local and global, structural and/or functional connectivity and network excitability [1–6, 98–100, 127].

A common approach to modifying the local and global functional connectivity and network excitability is by means of antiepileptic medications (**Figure 27**). For this

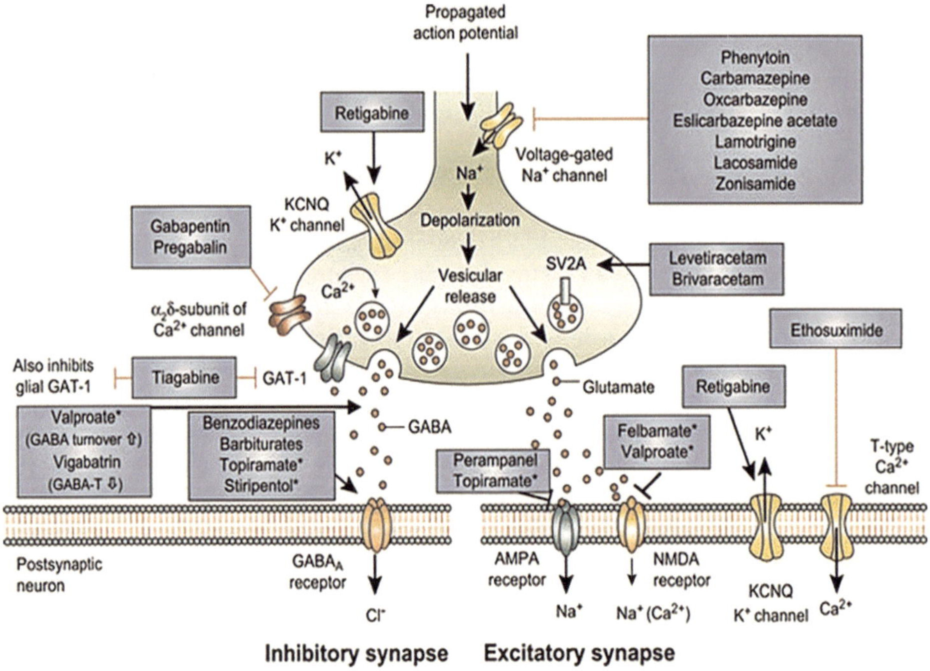

Figure 27.
Mechanism of action of clinically approved anti-seizure drugs. Published in Löscher et al. [145] under CC BY-NC 4.0 license. The updated and modified figure has been reproduced with permission from Löscher and Schmidt [146]. The initial figure was modified with permission from Macmillan Publishers Ltd © Bialer, M. & White, H. S. Nat. Rev. Drug Discov. 9, 68–82 (2010). Drugs marked with asterisks indicate that these compounds act by multiple mechanisms (not all mechanisms shown here). GABA-T: GABA aminotransferase, GAT: GABA transporter, SV2A: synaptic vesicle protein 2A, GABA: gamma-aminobutyric acid, NMDA: N-methyl-D-aspartate, AMPA: α-amino-3-hydroxy-5-methyl-4-isoxazolepropionic acid, KCNQ: a family of voltage-gated potassium channels (also known as the Kv7 family).

purpose, we employ different antiseizure medications as monotherapy or in various optimal combinations (polytherapy):

1. fast voltage-gated Na^{+}-channel blockers (e.g. Carbamazepine, Eslicarbazepine, Oxcarbazepine, Phenytoin, Lamotrigine acting also on L-type Ca^{+2}-channels, Zonisamide acting also on T-type Ca^{+2}-channels, Cenobamate acting also on GABA-A receptors, Valproate acting also on Ca^{+2} and K^{+} channels and having anti-NMDA and indirect GABAergic effects, etc.) and/or slow voltage-gated Na^{+}-channel blockers (e.g. Lacosamide)

2. SVPA2 receptor inhibitors (e.g. Brivaracetam, Levetiracetam acting also on N-Type Ca^{+2}-channels)

3. AMPA glutamate receptor antagonists (e.g. Perampanel, Topiramate acting also on GABA-A receptors), with all the above used to control/suppress the intrinsic local and global network excitability

4. GABA-A receptor agonists (e.g. Clobazam, Clonazepam, Midazolam, Lorazepam, Diazepam, Phenobarbitone, Primidone, Stiripentol, etc) used to enhance local and global network inhibition (**Table 3**) [1–11].

An alternative approach to modifying the local and global, structural and functional network connectivity and excitability is by means of neurosurgery (with resection or thermocoagulation of highly epileptogenic lesions/zones and disruption or disconnection of epileptogenic networks, etc) [98] and neurostimulation (vagus nerve, electrical/magnetic or optogenetic cortical stimulation or deep brain stimulation of the thalamus [anterior, centromedian, subthalamic nuclei]/basal ganglia, hippocampus, etc.) [71].

Current antiseizure medications (**Figure 27**) are mostly effective at preventing initiation, propagation, spreading or generalization of epileptic seizures. Modelling epileptogenesis across all scales of neuronal organization will further our understanding of the mechanisms of epileptogenesis, leading to better pharmacological and neurosurgical or neurostimulation treatment strategies and the development of new antiepileptic and epileptogenesis-modifying medications [1–11, 71, 98].

Special thanks

I would like to thank Dr. T. A. T. Hughes, Prof. K. Hamandi, Prof. W. Gray, Prof. P. Smith, Dr. G. Payne, from University Hospital of Wales (Cardiff) and Cardiff University Brain Research Imaging Center (CUBRIC), Prof. M. Richardson, Dr. L. Nashef, Dr. F. Brunnhuber, Dr. R. Elwes, Dr. N. Mullatti, Dr. A. Valentin, Mr. R. Selway from King's College Hospital (London), Dr. S. Boyd from Great Ormond Street Hospital (London), Prof. C. P. Panayiotopoulos from St. Thomas' Hospital (London), Prof. A. Lang from Krembil Neuroscience Center (Toronto), Prof. I. Bodis-Wollner from SUNY Downstate Medical Center (New York), Dr. J. P. D. Bland and Dr. N. Moran from Kent University Hospitals NHS Trust (Canterbury), Prof. P. Halász from the National Institute of Neuroscience (Budapest), Prof. L. Vécsei, Prof. Z. Janka and Prof. G. Benedek from University Hospital and Medical School of Szeged, Prof. G. Kostopoulos and Prof. A. Bezerianos from Medical

University of Patras, who have influenced my development in Neurology, Epileptology and Clinical Neurophysiology through discussions and collaborations we have had over the years. Most importantly, I would like to thank all the patients I care for, working every day on improving their health and quality of life. Last but not least, I would like to thank my parents and my family for their endless support all these years. I hope this novel approach to epilepsy will open the way to a better understanding of the mechanisms of epileptogenesis, improved pharmacological, neurosurgical and neurostimulation treatment strategies, and the development of new antiepileptic and epileptogenesis-modifying medications.

Author details

Nick Tsarouchas[1,2]

1 East Kent University Hospitals Foundation NHS Trust, UK

2 King's College Hospital London, UK

*Address all correspondence to: nick_tsarouchas@yahoo.com

References

[1] Available from: https://www.epilepsydiagnosis.org/

[2] Luders HO, Noachtar S, editors. Epileptic Seizures: Pathophysiology and Clinical Semiology. New York: Churchill Livingstone; 2000

[3] Engel J Jr. A proposed diagnostic scheme for people with epileptic seizures and with epilepsy: Report of the ILAE Task Force on Classification and Terminology. Epilepsia. 2001;**42**(6): 796-803

[4] Loddenkemper T, Kellinghaus C, Wyllie E, Najm IM, Gupta A, Rosenow F, et al. A proposal for a five-dimensional patient-oriented epilepsy classification. Epileptic Disorders. 2005;7 (4):308-316. PMID: 16338673

[5] Panayiotopoulos CP. Chapter 4, Principles of therapy in epilepsies. In: The Epilepsies: Seizures, Syndromes and Management. Oxfordshire (UK): Bladon Medical Publishing; 2005. Available from: https://www.ncbi.nlm.nih.gov/books/NBK2607/

[6] Berg AT, Berkovic SF, Brodie MJ, et al. Revised terminology and concepts for organization of seizures and epilepsies: Report of the ILAE Commission on Classification and Terminology, 2005-2009. Epilepsia. 2010;**51**(4):676-685

[7] Bressler SL, Menon V. Large-scale brain networks in cognition: Emerging methods and principles. Trends in Cognitive Sciences. 2010;**14**(6): 277-290

[8] Panayiotopoulos CP. Principles of therapy in the epilepsies. In: A Clinical Guide to Epileptic Syndromes and their Treatment. London: Springer; 2010

[9] Rudzinski LA, Shih JJ. The classification of seizures and epilepsy syndromes. In: Foyaca-Sibat H, editor. Novel Aspects on Epilepsy. London, UK, London, UK: InTech; 2011

[10] Marson A, Burnside G, Appleton R, Smith D, Leach JP, Sills G, et al. The SANAD II study of the effectiveness and cost-effectiveness of valproate versus levetiracetam for newly diagnosed generalized and unclassifiable epilepsy: An open-label, non-inferiority, multicenter, phase 4, randomised controlled trial. Lancet. 2021;**397** (10282):1375-1386

[11] Marson A, Burnside G, Appleton R, Smith D, Leach JP, Sills G, et al. The SANAD II study of the effectiveness and cost-effectiveness of levetiracetam, zonisamide, or lamotrigine for newly diagnosed focal epilepsy: An open-label, non-inferiority, multicenter, phase 4, randomised controlled trial. Lancet. 2021;**397**(10282):1363-1374

[12] Koch C, Segev I, editors. Methods in Neuronal Modeling: From Synapses to Networks. Cambridge, Massachusetts: The MIT Press; 1989

[13] Catterall WA, Raman IM, Robinson HP, Sejnowski TJ, Paulsen O. The Hodgkin-Huxley heritage: From channels to circuits. The Journal of Neuroscience. 2012;**32**(41):14064-14073

[14] Stiles PJ, Gray CG. Improved Hodgkin-Huxley type model for neural action potentials. European Biophysics Journal. 2021;**50**(6):819-828

[15] Gray DJ, Wu SM. Foundations of Cellular Neurophysiology. 3rd ed. Cambridge, Massachusetts [u.a.]: MIT Press; 1997

[16] Jaslove SW. The integrative properties of spiny distal dendrites. Neuroscience. 1992;**47**(3):495-519

[17] Johnson MG, Chartier S. Spike neural models (part I): The Hodgkin-Huxley model. The Quantitative Methods for Psychology. 2017;**13**(2):105-119. DOI: 10.20982/tqmp.13.2.p105

[18] Hodgkin AL, Huxley AF. A quantitative description of membrane current and its application to conduction and excitation in nerve. The Journal of Physiology. 1952;**117**(4):500-544. DOI: 10.1113/jphysiol.1952.sp004764

[19] Raghavan M, Fee D, Barkhaus PE. Generation and propagation of the action potential. Handbook of Clinical Neurology. 2019;**160**:3-22

[20] Parajuli LK, Urakubo H, Takahashi-Nakazato A, Ogelman R, Iwasaki H, Koike M, et al. Geometry and the organizational principle of spine synapses along a dendrite. eNeuro. 2020; **7**(6):ENEURO.0248-20.2020

[21] Bachatene L, Bharmauria V. The dendritic tree: A mathematical integrator. Frontiers in Bioscience. 2017; **9**:55-61

[22] Gulledge AT, Carnevale NT, Stuart GJ. Electrical advantages of dendritic spines. PLoS One. 2012;**7**(4):e36007

[23] Freeman WJ, Rogers LJ. Fine temporal resolution of analytic phase reveals episodic synchronization by state transitions in gamma EEGs. Journal of Neurophysiology. 2002;**87**(2):937-945

[24] Freeman WJ, Burke BC, Holmes MD. Aperiodic phase-resetting in scalp EEG of beta-gamma oscillations by state transitions at alpha-theta rates. Human Brain Mapping. 2003;**19**(4):248P272

[25] Freeman WJ. Mass Action in the Nervous System. New York: Academic Press; 1975

[26] Softky WR, Koch C. The highly irregular firing of cortical cells is inconsistent with temporal integration of random EPSPs. The Journal of Neuroscience. 1993;**13**(1):334-350

[27] König P, Engel AK, Singer W. Integrator or coincidence detector? The role of the cortical neuron revisited. Trends in Neurosciences. 1996;**19**(4): 130-137

[28] Kisley MA, Gerstein GL. The continuum of operating modes for a passive model neuron. Neural Computation. 1999;**11**(5):1139-1154

[29] Buzsáki G, Anastassiou CA, Koch C. The origin of extracellular fields and currents—EEG, ECoG, LFP and spikes. Nature Reviews Neuroscience. 2012;**13** (6):407-420

[30] Tatum WO et al. Clinical utility of EEG in diagnosing and monitoring epilepsy in adults. Clinical Neurophysiology. 2018;**129**(5):1056-1082

[31] Sinha M, Narayanan R. Active dendrites and local field potentials: Biophysical mechanisms and computational explorations. Neuroscience. 2021;**S0306-4522**(21): 00450-00454

[32] Brienza M, Mecarelli O. Neurophysiological basis of EEG. In: Mecarelli O, editor. Clinical Electroencephalography. Cham: Springer; 2019

[33] Rolls ET. The cingulate cortex and limbic systems for action, emotion, and memory. Handbook of Clinical Neurology. 2019;**166**:23-37

[34] Hangya B, Borhegyi Z, Szilágyi N, Freund TF, Varga V. GABAergic neurons of the medial septum lead the hippocampal network during theta activity. The Journal of Neuroscience. 2009;**29**(25):8094-8102

[35] Jiruska P, de Curtis M, Jefferys JG, Schevon CA, Schiff SJ, Schindler K. Synchronization and desynchronization in epilepsy: Controversies and hypotheses. The Journal of Physiology. 2013;**591**(4):787-797

[36] Moss F, Ward LM, Sannita WG. Stochastic resonance and sensory information processing: A tutorial and review of application. Clinical Neurophysiology. 2004;**115**(2):267-281

[37] Izhikevich EM. Dynamical Systems in Neuroscience: The Geometry of Excitability and Bursting. Cambridge, Massachusetts: The MIT Press; 2007

[38] Izhikevich EM. Which model to use for cortical spiking neurons? IEEE Transactions on Neural Networks. 2004; **15**(5):1063-1070

[39] Izhikevich EM, Edelman GM. Large-scale model of mammalian thalamocortical systems. Proceedings of the National Academy of Sciences of the United States of America. 2008;**105**(9): 3593-3598

[40] Izhikevich EM. Simple model of spiking neurons. IEEE Transactions on Neural Networks. 2003;**14**(6):1569-1572

[41] Izhikevich EM. Resonate-and-fire neurons. Neural Networks. 2001;**14**(6–7):883-894

[42] Izhikevich EM. Hybrid spiking models. Philosophical Transactions of the Royal Society A: Mathematical, Physical and Engineering Sciences. 1930; **2010**(368):5061-5070

[43] Murata Y, Colonnese MT. Thalamic inhibitory circuits and network activity development. Brain Research. 2019; **1706**:13-23

[44] Llinas RR, Leznik E, Urbano FJ. Temporal binding via cortical coincidence detection of specific and nonspecific thalamocortical inputs: A voltage-dependent dye-imaging study in mouse brain slices. Proceedings of the National Academy of Sciences of the United States of America. 2002;**99**(1): 449-454

[45] Traub RD, Contreras D, Cunningham MO, Murray H, LeBeau FE, Roopun A, et al. Single-column thalamocortical network model exhibiting gamma oscillations, sleep spindles, and epileptogenic bursts. Journal of Neurophysiology. 2005;**93**(4): 2194-2232

[46] Hughes SW, Lörincz M, Cope DW, Blethyn KL, Kékesi KA, Parri HR, et al. Synchronized oscillations at alpha and theta frequencies in the lateral geniculate nucleus. Neuron. 2004;**42**(2): 253-268

[47] Hughes SW, Errington A, Lorincz ML, Kékesi KA, Juhász G, Orbán G, et al. Novel modes of rhythmic burst firing at cognitively-relevant frequencies in thalamocortical neurons. Brain Research. 2008;**1235**:12-20

[48] Rolls ET, Tovee MJ. Processing speed in the cerebral cortex and the neurophysiology of visual masking. Proceedings of the Royal Society of London—Series B: Biological Sciences. 1994;**257**:9-15

[49] Buzsáki G. Hippocampal sharp wave-ripple: A cognitive biomarker for episodic memory and planning. Hippocampus. 2015;**25**(10):1073-1188

[50] Steriade M. Synchronized activities of coupled oscillators in the cerebral cortex and thalamus at different levels of vigilance. Cerebral Cortex. 1997;7(6): 583-604

[51] Rodriguez R, Kallenbach U, Singer W, Munk MH. Short- and long-term effects of cholinergic modulation on gamma oscillations and response synchronization in the visual cortex. The Journal of Neuroscience. 2004;**24**(46): 10369-10378

[52] Tallon-Baudry C, Bertrand O, Peronnet F, Pernier J. Induced gamma-band activity during the delay of a visual short-term memory task in humans. The Journal of Neuroscience. 1998;**18**(11): 4244-4254

[53] Bollimunta A, Mo J, Schroeder CE, Ding M. Neuronal mechanisms and attentional modulation of corticothalamic α oscillations. The Journal of Neuroscience. 2011;**31**(13): 4935-4943

[54] Shibuya S, Unenaka S, Shimada S, Ohki Y. Distinct modulation of mu and beta rhythm desynchronization during observation of embodied fake hand rotation. Neuropsychologia. 2021;**159**: 107952

[55] Timofeev I, Steriade M. Low-frequency rhythms in the thalamus of intact-cortex and decorticated cats. Journal of Neurophysiology. 1996;**76**(6): 4152-4168

[56] Contreras D, Steriade M. Cellular basis of EEG slow rhythms: A study of dynamic corticothalamic relationships. The Journal of Neuroscience. 1995;**15**(1 Pt 2):604-622

[57] Steriade M, Contreras D, Curró Dossi R, Nuñez A. The slow (<1 Hz) oscillation in reticular thalamic and thalamocortical neurons: Scenario of sleep rhythm generation in interacting thalamic and neocortical networks. The Journal of Neuroscience. 1993;**13**(8): 3284-3299

[58] Cona F, Lacanna M, Ursino M. A thalamo-cortical neural mass model for the simulation of brain rhythms during sleep. Journal of Computational Neuroscience. 2014;**37**(1):125-148

[59] Krishnan GP, Rosen BQ, Chen JY, Muller L, Sejnowski TJ, Cash SS, et al. Thalamocortical and intracortical laminar connectivity determines sleep spindle properties. PLoS Computational Biology. 2018;**14**(6):e1006171

[60] Nuñez A, Buño W. The Theta Rhythm of the Hippocampus: From Neuronal and Circuit Mechanisms to Behavior. Frontiers in Cellular Neuroscience. 2021;**15**:649262

[61] Catani M, Dell'acqua F, Thiebaut de Schotten M. A revised limbic system model for memory, emotion and behaviour. Neuroscience and Biobehavioral Reviews. 2013;**37**(8):1724-1737

[62] Anderson P, Morris R, Amaral BT, O'Keefe J, editors. The Hippocampus Book. 1st ed. New York: Oxford University Press; 2006. ISBN 978-0-19-510027-3

[63] Schlichting ML, Gumus M, Zhu T, Mack ML. The structure of hippocampal circuitry relates to rapid category learning in humans. Hippocampus. 2021; **31**(11):1179-1190

[64] Scharfman HE, Myers CE. Hilar mossy cells of the dentate gyrus: A historical perspective. Frontiers in Neural Circuits. 2013;**9**(6):106

[65] Schultz SR, Rolls ET. Analysis of information transmission in the Schaffer

collaterals. Hippocampus. 1999;**9**(5): 582-598

[66] Schlingloff D, Káli S, Freund TF, Hájos N, Gulyás AI. Mechanisms of sharp wave initiation and ripple generation. The Journal of Neuroscience. 2014;**34**(34):11385-11398

[67] Weininger J, Roman E, Tierney P, Barry D, Gallagher H, Murphy P, et al. Papez's forgotten tract: 80 years of unreconciled findings concerning the thalamocingulate tract. Frontiers in Neuroanatomy. 2019;**13**:14. DOI: 10.3389/fnana.2019.00014

[68] Müller C, Remy S. Septo-hippocampal interaction. Cell and Tissue Research. 2018;**373**(3):565-575

[69] Buzsáki G. The hippocampo-neocortical dialogue. Cerebral Cortex. 1996;**6**(2):81-92

[70] Buzsáki G. Theta oscillations in the hippocampus. Neuron. 2002;**33**(3): 325-340

[71] Fisher R, Salanova V, Witt T, Worth R, Henry T, Gross R, et al. Electrical stimulation of the anterior nucleus of thalamus for treatment of refractory epilepsy. Epilepsia. 2010;**51**(5):899-908

[72] Takeuchi Y, Harangozó M, Pedraza L, Földi T, Kozák G, Li Q, et al. Closed-loop stimulation of the medial septum terminates epileptic seizures. Brain. 2021;**144**(3):885-908

[73] Arnal-Real C, Mahmoudzadeh M, Manoochehri M, Nourhashemi M, Wallois F. What triggers the interictal epileptic spike? A multimodal multiscale analysis of the dynamic of synaptic and non-synaptic neuronal and vascular compartments using electrical and optical measurements. Frontiers in Neurology. 2021;**596926**

[74] Tryba AK, Merricks EM, Lee S, Pham T, Cho S, Nordli DR Jr, et al. Role of paroxysmal depolarization in focal seizure activity. Journal of Neurophysiology. 2019;**122**(5):1861-1873

[75] Bulsara A, Jacobs EW, Zhou T, Moss F, Kiss L. Stochastic resonance in a single neuron model: Theory and analog simulation. Journal of Theoretical Biology. 1991;**152**(4):531-555

[76] Whittington MA, Traub RD, Faulkner HJ, Stanford IM, Jefferys JG. Recurrent excitatory postsynaptic potentials induced by synchronized fast cortical oscillations. Proceedings of the National Academy of Sciences of the United States of America. 1997;**94**(22): 12198-12203

[77] Farrant M, Nusser Z. Variations on an inhibitory theme: Phasic and tonic activation of GABA(A) receptors. Nature Reviews. Neuroscience. 2005;**6**(3):215-229

[78] Crunelli V, Cope DW, Terry JR. Transition to absence seizures and the role of GABA(A) receptors. Epilepsy Research. 2011;**97**(3):283-289

[79] Sanchez-Aguilera A, Navas-Olive A, Valero M. Feedback and Feedforward Inhibition May Resonate Distinctly in the Ripple Symphony. Journal of Neuroscience. 25 Jul 2018;**38**(30): 6612-6614

[80] Bryson A, Hatch RJ, Zandt BJ, Rossert C, Berkovic SF, Reid CA, et al. GABA-mediated tonic inhibition differentially modulates gain in functional subtypes of cortical interneurons. Proceedings of the National Academy of Sciences of the United States of America. 2020;**117**(6): 3192-3202

[81] Williamson A, Alger BE. Characterization of an early

afterhyperpolarization after a brief train of action potentials in rat hippocampal neurons in vitro. Journal of Neurophysiology. 1990;**63**(1):72-81

[82] Song I, Savtchenko L, Semyanov A. Tonic excitation or inhibition is set by GABA(A) conductance in hippocampal interneurons. Nature Communications. 2011;**2**:376

[83] Hotka M, Kubista H. The paroxysmal depolarization shift in epilepsy research. The International Journal of Biochemistry & Cell Biology. 2019;**107**:77-81

[84] Kubista H, Boehm S, Hotka M. The paroxysmal depolarization shift: Reconsidering its role in epilepsy, epileptogenesis and beyond. International Journal of Molecular Sciences. 2019;**20**(3):577

[85] Cherubini E. Phasic $GABA_A$-mediated inhibition. In: Noebels JL, Avoli M, Rogawski MA, Olsen RW, Delgado-Escueta AV, editors. Jasper's Basic Mechanisms of the Epilepsies [Internet]. 4th ed. Bethesda (MD): National Center for Biotechnology Information (US); 2012

[86] Walker MC, Kullmann DM. Tonic $GABA_A$ receptor-mediated signaling in epilepsy. In: Noebels JL, Avoli M, Rogawski MA, Olsen RW, Delgado-Escueta AV, editors. Jasper's Basic Mechanisms of the Epilepsies [Internet]. 4th ed. Bethesda (MD): National Center for Biotechnology Information (US); 2012

[87] Dingledine R. Glutamatergic mechanisms related to epilepsy: Ionotropic receptors. In: Noebels JL, Avoli M, Rogawski MA, Olsen RW, Delgado Escueta AV, editors. Jasper's Basic Mechanisms of the Epilepsies [Internet]. 4th ed. Bethesda (MD): National Center for Biotechnology Information (US); 2012

[88] Hanada T. Ionotropic glutamate receptors in epilepsy: A review focusing on AMPA and NMDA receptors. Biomolecules. 2020;**10**(3):464

[89] Bianchi R, Wong RKS, Merlin LR. Glutamate receptors in epilepsy: Group I mGluR-mediated epileptogenesis. In: Noebels JL, Avoli M, Rogawski MA, Olsen RW, Delgado-Escueta AV, editors. Jasper's Basic Mechanisms of the Epilepsies [Internet]. 4th ed. Bethesda (MD): National Center for Biotechnology Information (US); 2012

[90] Crunelli V, Leresche N, Cope DW. GABA-A receptor function in typical absence seizures. In: Noebels JL, Avoli M, Rogawski MA, Olsen RW, Delgado-Escueta AV, editors. Jasper's Basic Mechanisms of the Epilepsies [Internet]. 4th ed. Bethesda (MD): National Center for Biotechnology Information (US); 2012

[91] Catterall WA. Sodium channel mutations and epilepsy. In: Noebels JL, Avoli M, Rogawski MA, Olsen RW, Delgado-Escueta AV, editors. Jasper's Basic Mechanisms of the Epilepsies [Internet]. 4th ed. Bethesda (MD): National Center for Biotechnology Information (US); 2012

[92] Cain SM, Snutch TP. Voltage-gated calcium channels in epilepsy. In: Noebels JL, Avoli M, Rogawski MA, Olsen RW, Delgado-Escueta AV, editors. Jasper's Basic Mechanisms of the Epilepsies [Internet]. 4th ed. Bethesda (MD): National Center for Biotechnology Information (US); 2012

[93] Shen KF, Yang XL, Liu GL, Zhu G, Wang ZK, Shi XJ, et al. The role of voltage-gated chloride channels in the

epileptogenesis of temporal lobe epilepsy. eBioMedicine. 2021;**70**:103537

[94] Nikitin ES, Vinogradova LV. Potassium channels as prominent targets and tools for the treatment of epilepsy. Expert Opinion on Therapeutic Targets. 2021;**25**(3):223-235

[95] Cooper EC. Potassium channels (including KCNQ) and epilepsy. In: Noebels JL, Avoli M, Rogawski MA, Olsen RW, Delgado-Escueta AV, editors. Jasper's Basic Mechanisms of the Epilepsies [Internet]. 4th ed. Bethesda (MD): National Center for Biotechnology Information (US); 2012

[96] Poolos NP. Hyperpolarization-activated cyclic nucleotide-gated (HCN) ion channelopathy in epilepsy. In: Noebels JL, Avoli M, Rogawski MA, Olsen RW, Delgado-Escueta AV, editors. Jasper's Basic Mechanisms of the Epilepsies [Internet]. 4th ed. Bethesda (MD): National Center for Biotechnology Information (US); 2012

[97] Dudek FE, Snow RW, Taylor CP. Role of electrical interactions in synchronization of epileptiform bursts. Advances in Neurology. 1986;**44**:593-617

[98] Wu S, Veedu HPK, Lhatoo SD, Koubeissi MZ, Miller JP, Lüders HO. Role of ictal baseline shifts and ictal high-frequency oscillations in stereo-electroencephalography analysis of mesial temporal lobe seizures. Epilepsia. 2014;**55**(5):690-698

[99] Zijlmans M, Jiruska P, Zelmann R, Leijten FSS, Jefferys JGR, Gotman J. High-frequency oscillations as a new biomarker in epilepsy. Annals of Neurology. 2012;**71**(2):169-178

[100] Sullivan D, Csicsvari J, Mizuseki K, Montgomery S, Diba K, Buzsáki G. Relationships between hippocampal sharp waves, ripples and fast gamma oscillation: Influence of dentate and entorhinal cortical activity. The Journal of Neuroscience. 2011;**31**(23):8605-8616

[101] Elmali AD, Auvin S, Bast T, Rubboli G, Koutroumanidis M. How to diagnose and classify idiopathic (genetic) generalized epilepsies. Epileptic Disorders. 2020;**22**(4):399-420

[102] McCormick DA. Cellular mechanisms underlying cholinergic and noradrenergic modulation of neuronal firing mode in the cat and guinea pig dorsal lateral geniculate nucleus. The Journal of Neuroscience. 1992;**12**(1): 278-289

[103] Connelly WM, Crunelli V, Errington AC. Passive synaptic normalization and input synchrony-dependent amplification of cortical feedback in thalamocortical neuron dendrites. The Journal of Neuroscience. 2016;**36**(13):3735-3754

[104] Sinha N, Wang Y, Dauwels J, Kaiser M, Thesen T, Forsyth R, et al. Computer modelling of connectivity change suggests epileptogenesis mechanisms in idiopathic generalized epilepsy. NeuroImage: Clinical. 2019;**21**:101655

[105] Hirose S. Mutant GABA(A) receptor subunits in genetic (idiopathic) epilepsy. Progress in Brain Research. 2014;**213**:55-85

[106] Riaz M, Abbasi MH, Sheikh N, Saleem T, Virk AO. GABRA1 and GABRA6 gene mutations in idiopathic generalized epilepsy patients. Seizure. 2021;**93**:88-94

[107] Benedek K, Berényi A, Gombkötő P, Piilgaard H, Lauritzen M. Neocortical gamma oscillations in idiopathic generalized epilepsy. Epilepsia. 2016; **57**(5):796-804

[108] Sanchez-Vives MV, Barbero-Castillo A, Perez-Zabalza M, Reig R. GABA-B receptors: Modulation of thalamocortical dynamics and synaptic plasticity. Neuroscience. 2021;**456**: 131-142

[109] Powell KL, Cain SM, Ng C, Sirdesai S, David LS, et al. A Cav3.2 T-type calcium channel point mutation has splice-variant-specific effects on function and segregates with seizure expression in a polygenic rat model of absence epilepsy. Journal of Neuroscience. 2009;**29**:371-380

[110] Hosford DA, Clark S, Cao Z, et al. The role of GABA-B receptor activation in absence seizures of lethargic (lh/lh) mice. Science. 1992;**257** (5068):398-401

[111] Crunelli V, Lőrincz ML, McCafferty C, Lambert RC, Leresche N, Di Giovanni G, et al. Clinical and experimental insight into pathophysiology, comorbidity and therapy of absence seizures. Brain. 2020;**143**(8):2341-2368

[112] Nolan D, Fink J. Genetics of epilepsy. Handbook of Clinical Neurology. 2018;**148**:467-491

[113] Brunklaus A, Du J, Steckler F, Ghanty II, Johannesen KM, Fenger CD, et al. Biological concepts in human sodium channel epilepsies and their relevance in clinical practice. Epilepsia. 2020;**61**(3):387-399

[114] Niday Z, Tzingounis AV. Potassium channel gain of function in epilepsy: An unresolved paradox. The Neuroscientist. 2018;**24**(4):368-380

[115] Kramer MA, Cash SS. Epilepsy as a disorder of cortical network organization. The Neuroscientist. 2012; **18**(4):360-372

[116] Traynelis SF, Wollmuth LP, McBain CJ, Menniti FS, Vance KM, Ogden KK, et al. Glutamate receptor ion channels: Structure, regulation, and function. Pharmacological Reviews. 2010;**62**(3): 405-496

[117] Hansen KB, Wollmuth LP, Bowie D, Furukawa H, Menniti FS, Sobolevsky AI, et al. Structure, function, and pharmacology of glutamate receptor ion channels. Pharmacological Reviews. 2021;**73**(4):298-487

[118] Jasper's Basic Mechanisms of the Epilepsies [Internet]. 4th ed. Bethesda (MD): National Center for Biotechnology Information (US); 2012

[119] Jembrek MJ, Vlainic J. GABA receptors: Pharmacological potential and pitfalls. Current Pharmaceutical Design. 2015;**21**(34):4943-4959

[120] Galanopoulou AS. GABA(A) receptors in normal development and seizures: Friends or foes? Current Neuropharmacology. 2008;**6**(1):1-20

[121] Mulley JC, Scheffer IE, Petrou S, Berkovic SF. Channelopathies as a genetic cause of epilepsy. Current Opinion in Neurology. 2003;**16**(2): 171-176

[122] Ghit A, Assal D, Al-Shami AS, Hussein DEE. GABA-A receptors: Structure, function, pharmacology, and related disorders. Journal, Genetic Engineering & Biotechnology. 2021; **19**(1):123

[123] Galanopoulou AS. Mutations affecting GABAergic signaling in seizures and epilepsy. Pflügers Archiv. 2010;**460**(2):505-523

[124] Jiang Y, Han CL, Liu HG, Wang X, Zhang X, Meng FG, et al. Abnormal hippocampal functional network and

related memory impairment in pilocarpine-treated rats. Epilepsia. 2018; **59**(9):1785-1795

[125] Chalifoux JR, Carter AG. GABAB receptor modulation of synaptic function. Current Opinion in Neurobiology. 2011;**21**(2):339-344

[126] Timofeev I, Bazhenov M, Seigneur J, Sejnowski T. Neuronal synchronization and thalamocortical rhythms in sleep, wake and epilepsy. In: Noebels JL, Avoli M, Rogawski MA, Olsen RW, Delgado-Escueta AV, editors. Jasper's Basic Mechanisms of the Epilepsies [Internet]. 4th ed. Bethesda (MD): National Center for Biotechnology Information (US); 2012

[127] Jefferys JGR, Jiruska P, de Curtis M, Avoli M. Limbic network synchronization and temporal lobe epilepsy. In: Noebels JL, Avoli M, Rogawski MA, Olsen RW, Delgado-Escueta AV, editors. Jasper's Basic Mechanisms of the Epilepsies [Internet]. 4th ed. Bethesda (MD): National Center for Biotechnology Information (US); 2012

[128] Kajiwara R, Wouterlood FG, Sah A, Boekel AJ, Baks-te Bulte LT, Witter MP. Convergence of entorhinal and CA3 inputs onto pyramidal neurons and interneurons in hippocampal area CA1—An anatomical study in the rat. Hippocampus. 2008;**18**(3):266-280

[129] Martínez A, Lübke J, Del Río JA, Soriano E, Frotscher M. Regional variability and postsynaptic targets of chandelier cells in the hippocampal formation of the rat. The Journal of Comparative Neurology. 1996;**376**(1): 28-44

[130] Fernández-Ruiz A, Makarov VA, Benito N, Herreras O. Schaffer-specific local field potentials reflect discrete excitatory events at gamma frequency that may fire postsynaptic hippocampal CA1 units. The Journal of Neuroscience. 2012;**32**(15):5165-5176

[131] Taxidis J, Coombes S, Mason R, Owen MR. Modeling sharp wave-ripple complexes through a CA3-CA1 network model with chemical synapses. Hippocampus. 2012;**22**(5):995-1017

[132] Sloviter RS. Status epilepticus-induced neuronal injury and network reorganization. Epilepsia. 1999;**40**(suppl 1):S34-S39; discussion S40-1

[133] Botterill JJ, Lu YL, LaFrancois JJ, Bernstein HL, Alcantara-Gonzalez D, Jain S, et al. An excitatory and epileptogenic effect of dentate gyrus mossy cells in a mouse model of epilepsy. Cell Reports. 2019;**29**(9):2875-2889.e6

[134] Scharfman HE. The dentate gyrus and temporal lobe epilepsy: An "exciting" era. Epilepsy Currents. 2019; **19**(4):249-255

[135] Mody I, Reynolds JN, Salter MW, Carlen PL, MacDonald JF. Kindling-induced epilepsy alters calcium currents in granule cells of rat hippocampal slices. Brain Research. 1990;**531**(1–2):88-94

[136] Verkhratsky A, Shmigol A, Kirischuk S, Pronchuk N, Kostyuk P. Age-dependent changes in calcium currents and calcium homeostasis in mammalian neurons. Annals of the New York Academy of Sciences. 1994;**747**: 365-381

[137] Scharfman HE, Schwartzkroin PA. Protection of dentate hilar cells from prolonged stimulation by intracellular calcium chelation. Science. 1989;**246** (4927):257-260

[138] Cavazos JE, Das I, Sutula TP. Neuronal loss induced in limbic

pathways by kindling: Evidence for induction of hippocampal sclerosis by repeated brief seizures. The Journal of Neuroscience. 1994;**14**(5 pt 2): 3106-3121

[139] Coulter DA. Chronic epileptogenic cellular alterations in the limbic system after status epilepticus. Epilepsia. 1999; **40**(Suppl 1):S23-S33; discussion S40-1

[140] Sutula T, Cascino G, Cavazos J, et al. Mossy fiber synaptic reorganization in the epileptic human temporal lobe. Annals of Neurology. 1989;**26**(3) :321-330

[141] Curia G, Longo D, Biagini G, Jones RS, Avoli M. The pilocarpine model of temporal lobe epilepsy. Journal of Neuroscience Methods. 2008;**172**(2): 143-157

[142] Lévesque M, Biagini G, de Curtis M, Gnatkovsky V, Pitsch J, Wang S, et al. The pilocarpine model of mesial temporal lobe epilepsy: Over one decade later, with more rodent species and new investigative approaches. Neuroscience and Biobehavioral Reviews. 2021;**130**: 274-291

[143] Wu XL, Ma DM, Zhang W, Zhou JS, Huo YW, Lu M, et al. Cx36 in the mouse hippocampus during and after pilocarpine-induced status epilepticus. Epilepsy Research. 2018;**141**:64-72

[144] Nadler JV. Plasticity of glutamate synaptic mechanisms. In: Noebels JL, Avoli M, Rogawski MA, Olsen RW, Delgado-Escueta AV, editors. Jasper's Basic Mechanisms of the Epilepsies [Internet]. 4th ed. Bethesda (MD): National Center for Biotechnology Information (US); 2012

[145] Löscher W, Gillard M, Sands ZA, Kaminski RM, Klitgaard H. Synaptic vesicle glycoprotein 2A ligands in the treatment of epilepsy and beyond. CNS Drugs. 2016;**30**(11):1055-1077

[146] Löscher W, Schmidt D. Epilepsy: Perampanel—New promise for refractory epilepsy? Nature Reviews Neurology. 2012;**8**:661-662

Chapter 2

Neurophysiology Involved in Neuroplasticity: Mechanisms of Forgetting

Jose Rodrigo Carrillo-Marquez and Jose Damian Carrillo-Ruiz

"You appear to be astonished," he said, smiling at my expression of surprise. "Now that I do know it, I shall do my best to forget it."

"To forget it!"

"You see," he explained, "I consider that a man's brain originally is like a little empty attic, and you must stock it with such furniture as you choose. Now the skillful worker is incredibly careful indeed as to what he takes into his brain-attic. He will have nothing but the tools which may help him in doing his work, but of these, he has a large assortment and all in the most perfect order. It is a mistake to think that little room has elastic walls and can distend to any extent. Depending upon it there comes a time when for every addition of knowledge, you forget something that you knew before. It is of the highest importance, therefore, not to have useless facts elbowing out the useful ones."

The Science of Deduction in A Study in Scarlet.

Sir Arthur Conan Doyle, 1887.

Abstract

Neuroplasticity is the brain's ability to adapt to new stimuli, with the objective to overcome and learn how to deal with novel situations. In this chapter, it will be explained the new neurophysiological mechanism that entitles the processes of brains' plasticity. The intriguing phenomena that surround cognitive mechanisms will be described on a morphological and molecular scale, aiming to understand some of the brains' functions. The principal objective is to clarify and explain that neuroplasticity can take place in different complex tasks such as adaptative behaviors, memory, learning, and automatic conducts. Also, the evolutionary advantage of forgetting will be deeply discussed. The work will describe the functioning of the brain when adapting to new circumstances that affect the procedures of memory. It will be explained why applying biotechnology and neurobioethics is crucial for merging basic and clinical sciences.

Keywords: pruning, forgetting, forgetfulness, neuroplasticity, sprouting

IntechOpen

1. Introduction

Forgetting or forgetfulness is not only a passive activity of neurons but also an active process, in which, specific brain areas play role in the neural circuits of temporal and frontal lobes to eliminate information that does not need to be stored in memory. Its importance is sustained by the fact of avoiding saturation of the neural circuitry; knowledge that could be irrelevant or shallow for the diary living activities or in its case, accumulation of chronic unnecessary skills.

This chapter is constructed on the basic principles of neuroplasticity and memory, in order to understand how the information is obtained from the senses and then the perception that might be needed to be kept in the brain, depending on its modality. After this point, which is the molecular and cellular mechanism used by the neurons to filter the relevant juxtaposing the purposeless. It is being added to the clinical feature of neurological illness to distinguish amnesia from forgetting. It will be expressed the future of treating patients that need to ameliorate forgetfulness in their daily life.

Although many new areas are being researched, others are kept apart from the interest of professionals. That is why this book chapter is presented, aiming to inform the new physiological mechanisms that have been described in the forgetting process, so as its relationship with neuroplasticity. Also, there will be an in-depth explanation of some pathologies and affections that trigger this neurophysiological response.

2. Why science takes a fundamental role in neuroplasticity?

Understanding part of the physiology of the nervous system has been one of the most outstanding achievements of the 20th century. It is indisputable that these achievements could not have happened if there was no research involved, so as the interest of the scientific community to try to explain the mechanisms of the brain. The decade from 1990 to 2000 was established by the American Academy of Neurology (AAN) as the "decade of the brain" [1], and it was a transition between elemental bases of neuroscience to a more specific so-called era. It is important to mention that in the 21st century, the growing curiosity to investigate and explain neurophysiology has reached unprecedented milestones due to emerging technologies that have been crucial for research and clinical practice. Topics such as memory, neuroplasticity, neurobiochemistry, neuronal tracts, sprouting, and neuroimmunology are reaching new horizons because of the molecular approach [2]. This is the reason why it should be investigated the neuroplasticity and forgetting. Not just as a new trend, but rather as fields of neuroscience that must be discussed by the community in order to achieve progress. Without the new biotechnology, and in this case neurobiotechnology; discovering and knowing the molecular mechanisms of plasticity, memory, and forgetting would be impossible.

3. Neuroplasticity

3.1 An approach from neuroembryology

To fully understand neuroplasticity, it is first necessary to classify and define what this concept stands for. Several authors have tried to state a formal definition, yet many of them have not taken into consideration this phenomenon over time. Neuroplasticity is a complex process that starts in the embryo neurogenesis, beginning with neurulation. It is important to state that the closure of the neural tube happens between the 21st and 25th day [3]. From the formation of the neural tube until the moment in which the fetus has a complete morphological nervous system many molecular signals initiate and finish the growth, connectivity, and cellular differentiation within the brain and the rest of the nervous system. Between the 8th week of pregnancy (time by which every human system is formed) to the moment of labor, the nervous system shifts drastically [4].

3.2 The notch pathway and neuroplasticity

One of the most important mechanisms of these changes is the Notch differentiation which is a gene-mediated regulatory cascade also known as the "notch signaling pathway". Its importance comes from its direct participation in cell proliferation and fate, so as differentiation and apoptotic mechanisms. This process assures the development and distinction of the nervous system from small animals (*i.e., Drosophila melanogaster*) to humans [5, 6].

Although this process elucidates how neuroblasts tend to differentiate into neurons and both glial cells, it should be said that the whole process is more complex than the intervention between the "Hes family (1, 5 & 6)", RBPJ and the ligands with Jagged and DII [7]. What can be said is that an incorrect step in the process might change future physiological functions within the nervous system, opening the possibility of mental illnesses and congenital neurological diseases which are not going to be described in this manuscript. Information regarding the topic has been published, but the reality is that studies just explain the correlation between multiple genes, transcripts, and proteins that take part in the complex process of neurological plasticity [8].

Also, epigenetic factors can alter the expression of this mechanism that takes place in embryogenesis and after-birth processes. Lasky states that memory and learning are two of the most important and multifactorial processes that involve a relationship with Notch, so as diverse pathologies that may arise from this process in postnatal development [9]. This evidence has also been restated a few years ago by Engler, adding the participation of NOTCH1 and NOTCH2 in this phenomenon, as well as adding a description of how all these factors take place in adults [10].

3.3 Fresh perspectives regarding neuroplasticity

The reality is that younger individuals have a better capacity than older ones to form neuronal connections, learn and perform new tasks, due to the capacity of younger neurons to arborize and form new circuits [11]. Adult neurogenesis, and thus neuroplasticity, is possible to occur in the hippocampus. This theory has been deeply discussed, but the reality is that more research and more substantial medical evidence should be published in order to get to a consensus [12]. Despite age, individuals that make exercise tend to have an increase in neurotrophic factors (*i.e.,* Brain-Derived Neurotrophic Factor, Nerve Growth Factor, and Glial Cell line-derived Factor) that brings a more efficient functionality in comparison with those who do not exercise [13]. The relationship between oxytocin and the increase of pro-social behaviors, neuronal excitability, and plasticity has been described by Froemke [14]. It is important to acknowledge that the research made by this group determined that mammals use oxytocinergic modulation in the process of maternal care and social bonding. It is a prolific contribution to the field because of the hormonal and neurological modulation involved in physiological and pathophysiological processes.

3.4 A new proposal in defining neuroplasticity

As said before, neuroplasticity is one of the most intriguing, and yet not fully known mechanisms that occur within the nervous system. It has been said that neuroplasticity consists of changes in a morphological, biochemical, and pharmacological adaptation in neuronal networks. The main objective of this physiological process is the adaptative response of the brain to diverse stimuli. Moreover, neuroplasticity and its physiology are fundamental concepts when approaching the pathophysiology of several neurologic illnesses, based on the ideas proposed by de Olveira in 2020 [15]. Understanding neuroplasticity as a process that relates directly to behavioral patterns and pharmacology, so as the interaction of these fields with pathology.

As a contribution to this field, it is being proposed to define neuroplasticity as a temporal and active mechanism related to neurogenesis and nervous cell apoptosis, starting in the neurulation, and finished until the subject's death, having characteristics of neurochemical and morphological changes that can be altered by other substances and environmental factors. This definition is based upon neurobioethics in order to understand this phenomenon not just at a basic and scientific level but comprehending these neural occurrences in a medical and anthropological way.

It will be described how neuroplasticity is related to memory formation, forgetfulness, and the differentiation between a physiological and pathological operation (**Figure 1**).

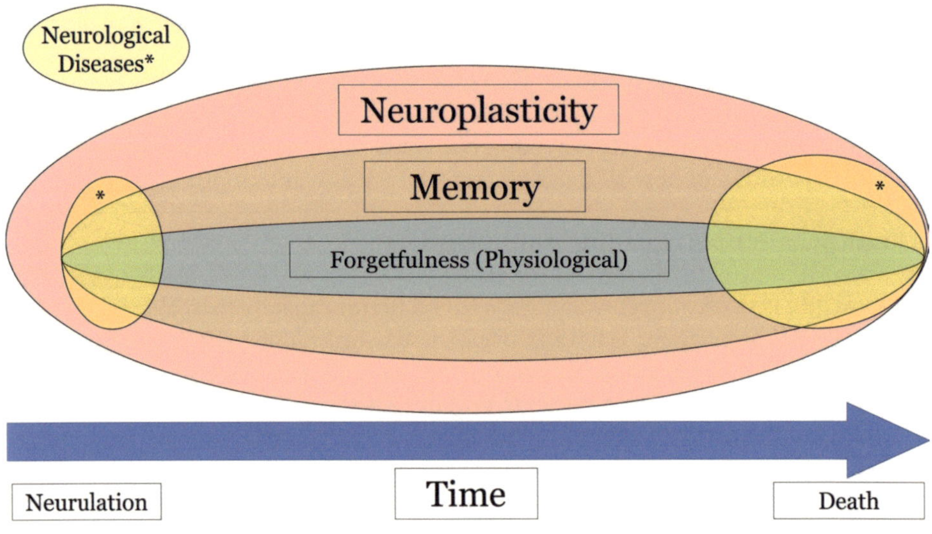

Figure 1.
This graphic represents the relationship between memory and forgetfulness, involved in all the process, the neuroplasticity. This is evident from the formation of the nervous system in neurulation and continues through the time to the moment of the subject's death. Also, the presence of neurological illness modifies both functions.
** = sporadic diseases during lifetime.*

4. Into the memory

Memory as well known is a neurophysiological mechanism that is complex and subdivided into many different subclassifications. The aim of this document is not to discuss nor describe exhaustively each category [16], but rather to mention them and comprehend how memories are formed. Also, this will help to establish in the following pages the criteria to understand what the concept of forgetfulness really means (**Figure 2**).

4.1 Mechanisms of memory formation

According to human physiology, memory is a phenomenon that takes place in the following structures: the hippocampus and the cortex (specifically the neocortex) [17]. Moreover, memory cannot be described as an isolated phenomenon since the cortex and limbic system takes place in memory consolidation. The amygdala plays a fundamental role in emotional response linked with memory being fear, one of the principal factors for plasticity and neurochemical activities in that anatomic region [18, 19]. It is important to state that the relationship between the amygdala and hippocampus balances the encoding processes of memory [20]. Also, it has been determined that the cingulate cortex is involved in action-outcome learning, and it also is an important place of brain connectivity [21].

It is important to mention that memory has also a neurophylogenetic component. Chittka described that many animals such as wasps, bees, and cockroaches have the ability of spatial learning for surviving [22]. Decapod crustaceans have the same ability as well, it has been published that crabs can adapt and be conditioned to solve mazes having an improvement in comparison with those who did not have time to explore the trial [23]. Krichbaum reported that dogs have the capability of episodic memory for accomplishing tasks, even though seven out of ten dogs tested positive in conduct

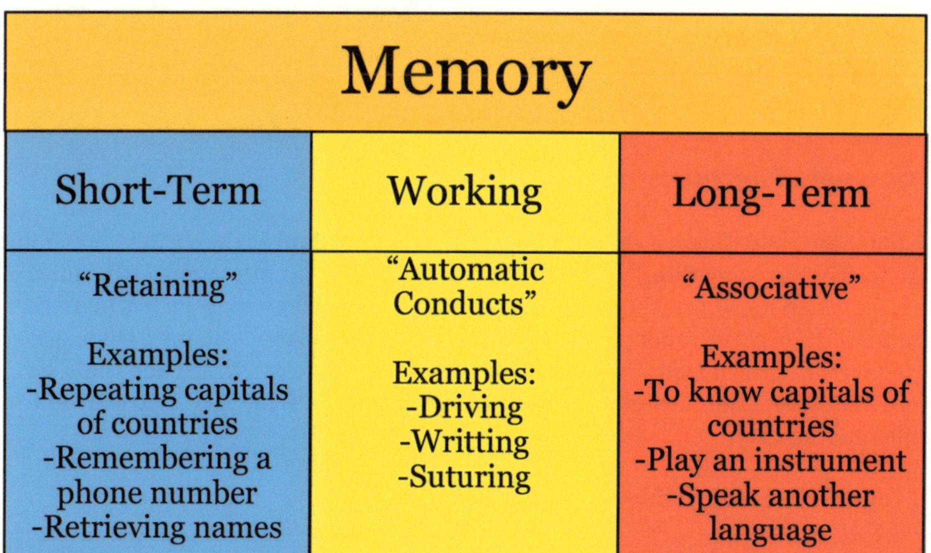

Figure 2.
It is demonstrated the memory function: short-term, working memory and long-term with its main characteristics and examples of each subtype.

experiments, more research should be done to clarify this proposal [24]. According to scientific literature, monkeys can retrieve information from short to medium periods of time, use working memory, and have a limited self-perception when answering cognitive tests [25]. Humans are at top of this evolutionary scale due to their complex brain structure and functional morphology, that no other species own.

It might sound coherent to the fact that the cortex is involved in decoding the diverse stimuli that pass through the senses (*i.e.,* sight, touch, smell, taste, hearing, cognitive estimation, and reasoning). Memory is determined by several factors such as chemicals, hormones, neurotrophic factors, neurotransmitters, external stimuli, and sleep.

Regarding sleep, Klinzing determined that Long-Term Memory is formed during slow-wave sleep in the hippocampus of rodents, so as in humans [26]. Speaking about hormones, it has been described by Acosta that the circadian rhythm plays an important role in starting and stopping the gene and cellular structures that are regulated by an extensive number of hormones (*i.e.,* Cortisol, Digestive Hormones) [27]. Also, the relationship between hormone regulation and the gut-brain axis (GBA) has a fundamental role with the hypothalamus in maintaining a healthy metabolism and a correct function of glutamatergic, cholinergic, calcium-calmodulin, mitogen-activated protein kinase (MAPK), extracellular signal-regulated kinases (ERK's), phosphoinositide 3-kinase (PI3K), ghrelin, leptin, insulin, and glucagon signaling [28].

Also, one of the most recent studies has determined that smoking cannabis containing the active molecule of tetrahydrocannabinol (THC) disrupts short memory, but unclear evidence of neurogenesis in the hippocampus also has been reported [29]. Alcohol abuse in big quantities can prevent the brain from forming new memories, and after the withdrawal, incapacity to remember correctly what happened when drunk. Also, it has other metabolic consequences [30].

Moreover, the hippocampus builds coherent long-lasting memories, merging exploratory visual exploration and memory formation [31]. Adding that also the other senses can take place in this process. The reality is that the theory of engrams "basic biological unit of memory" has been merged with neuroplasticity to understand the physiological mechanism of forming memories [32]. As it has been stated, reducing complex processes: in this case, a "memory to an engram" also can bring detrimental consequences for scientific research.

5. Concept of forgetting or forgetfulness

Forgetting is defined as the loss of memory or the failure of the brain to remember. It is the fact that something concrete or a special person, is not present in the mind of a subject. Although he or she might have known them well previously. It is interesting that forgetfulness is a synonym for forgetting in a colloquial or scientific language, based on the ideas of Della Sala and Cubelli [33]. Nevertheless, forgetting could be more punctual to the "fact of forget", and forgetfulness is the process and state of not remembering [34].

5.1 Objectives of forgetting

It is relevant for the brain and the mind to experience forgetfulness. In the same way a person could be memorizing events, skills, situations, and circumstances in a natural process; it is necessary for a similar path to forget all the things that are not essential in the activity's life. Indeed, one cannot exist if the other one is abolished. It erases the shallow and specific things dispensable to work. This allows the new experience or knowledge, even though opening the gate for creativity or innovation, inclusively, it could be considered a paradoxical effect because one person forgets insignificant information and could store an important one [35, 36].

Furthermore, forgetting helps to ameliorate traumatic or anxious situations that could be dangerous to mental health. This indicates a capability to grow and promote personality maturation coupled with emotional intelligence.

5.2 Microglia and macroglia in forgetting

The microglia cells are the caregivers of the neurons because they support, nourish, and attack the estrange and harmful cells. This last activity is based on the activation of the complement system: for the attack including C1q, C3, and CR3; and for the cell, preservation uses CD47 (cluster of differentiation 47), SIRPα (signal-regulatory protein), and CX3CR (chemokine receptor 1). Indeed, these components of complement resting the microglia to prune synapsis during neural activity and plasticity [37]. Neuroplasticity involves the stability of the long-term potential (LTP), a neural mechanism based on the dependent- Ca++ gates to produce the learning and memory phenomena [38]. It is in the same direction the memory needs the microglia to consolidate that the use of complement in the inverse sense to get forgetting, undoubtedly with the use of microglia. In a double mechanism with gradual and activity-dependent, the retraction of synapsis mediated by the glia contributes to forgetting [39] (**Figure 3**).

Although there is no substantial evidence of the participation of macroglia (*i.e.,* oligodendrocytes and astrocytes) in forgetting mechanisms [40, 41], further investigation must be required. Based on the proposition that astrocytes play an important role in neurotransmission uptake and neural guidance, so as their immunological properties as cells [42].

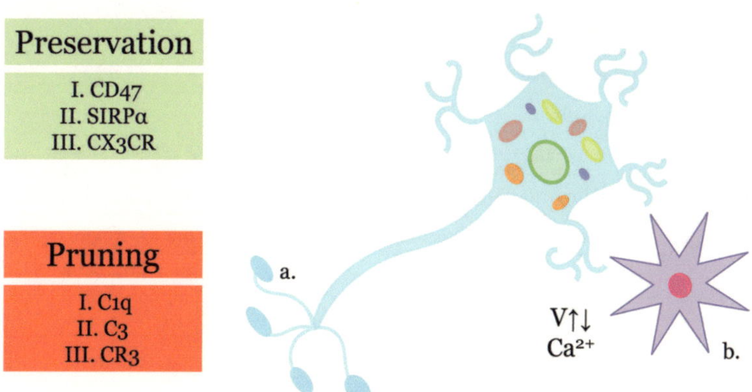

Figure 3.
It is shown the cellular mechanism of forgetting. Preservation and pruning in the neuron (a) and the micro/macroglia (b). V↑↓Ca^{2+}= Regulation in voltage-gated calcium channels; CD47 = Cluster of Differentiation 47, SIRPα = Signal-Regulatory Protein Alpha, CX3CR = Chemokine Receptor 1; C1q = Complement component 1q, C3 = Complement component 3, CR3 = Complement receptor 3 [37].

6. Neurological diseases

6.1 Amnesia or forgetting?

The first point to consider is determining if exists amnesia or forgetting: It is not the same. Although in both circumstances it shows the absence of retrieval of the elements lived by the person and how they get stored in the brain, the amnesia is produced by a real element damaging the neurons of the memory circuit. There are several conditions that could be presented in different illnesses that produced amnesia [43]. The next are the most common causes to produce it (**Table 1**):

There is much information about each one of the neurologic diseases, but this is not the discussion matter of this chapter. Those have been mentioned to show the difference between amnesia and forgetfulness.

On the other hand, forgetting is the absence of reuptake of the skills or memory elements without lesion to the brain neurons, and neuronal circuitry is intact. Sometimes normal forgetting is considered like a pathological memory problem, and this is not the real image of what exactly happens. An example could be a geriatric patient with failures in memory that could be attributed to incipient dementia. The process of forgetting is explained due to perception and/or attention alteration. Indeed, the simple past of the time could produce a diminution of the capability to retain the information entered into the brain [44, 45].

Some of the concrete neurologic problems are described below (**Table 2**).

6.2 Aging

It is well known that neuroplasticity is produced during the initial stages of the human being after-birth, when the brain is immature. It is needed for establishing

I.	Moderate to severe cranial-brain trauma
II.	Chronic and evolutive dementia (*e.g.*, Alzheimer or Parkinson Disease)
III.	Cardiac and brain stroke
IV.	Nutritional deficiency (*e.g.*, Korsakoff Disease)
V	Brain tumors
VI.	Seizures and epilepsy
VII.	Cerebral infections
VIII.	Thyroid hormones alterations (*i.e.*, hypothyroidism and hyperthyroidism).
IX.	Psychotropic drugs (*e.g.*, antidepressants, antiepileptic, antipsychotic…).
X.	Poisoning interfering central nervous system
	Carbon monoxide (CO)
	Animal or vegetal substances
	Alcohol
	Cannabinoids
	Hard drugs

Table 1.
Amnesia and its causes.

I.	Aging	
II.	Sleep deprivation	
III.	Hypermnesia	
IV.	Severe mental stress	
	Anxiety	Depression

Table 2.
Concrete neurologic problems affecting memory.

connections and making neural circuits that could regulate basic and important levels of diverse functions in the Central Nervous System (CNS) [46]. In this sense, the neuroplasticity of a newborn, and infant, and inclusive an adolescent is more relevant compared to the neuroplasticity of adult and geriatric subjects and patients. The person develops forgetting when the lifetime is passing through his eyes. It is the expected, so it is an important event, but not only the oldest people experiment it. As it has been stated, forgetting is necessary to leave irrelevant knowledge, learn and memorize new ones in each stage of his life [36, 47].

6.3 Sleep deprivation

Patient with moderate-to-severe problems of sleep with different etiologies could present more frequent number of forgetting memories compared to normal subjects. The presence of insomnia is a determinant factor for the lack of concentration because it affects the attention and a correct state of consciousness. It is important the state of anatomical and functional integrities of the Reticular Ascendent Activating System (RAAS), which takes part in several neurotransmitters involved in the arousal-sleep cycle. In this matter, the orexin-neuron groups are involved in this regulation, according to what Chambers stated in 2017 [48]. To consolidate the knowledge, the hippocampus needs to have the presence of a neurophysiological plateau with the intervention of calcium (Ca^{++}) to produce the hippocampal delta waves. Such manifestations demonstrate the consolidation of a set of skills. This is mediate by acetylcholine. It is possible that forgetting could use the same anatomical point, with the performance of inhibitory neurotransmitters like GABA or glycine [49].

6.4 Hypermnesia

One of the most interesting situations of medicine is the presence of photographic memory. It is rare to find people with this extraordinary capability. For the person who had this gift could be used in a social and economic success [50, 51]. The person with this "excess of memory" has an enormous advantage to learn and to extract information from their neuronal circuits. It demonstrated that it exists a hyper-connectivity and optimization of the function of neurotransmitters like noradrenaline, as described by Hurlemann [52]. Nevertheless, when a subject has this situation, it could be difficult to destroy superficial information, and every single moment could be remembered, everything from the big picture to the small details, depending on the individual's capability. This illness does not allow the use of forgetting, so in this person, this could give birth to a paradoxical and noxious psychological effect [53].

6.5 Severe mental stress

One of the most important defense mechanisms against brutal stress is the presence of forgetfulness when some life events, like rape or war survival, could have been lived. This produces a tremendous shock in subjects or in patients, opening the possibility of being diagnosed with "post-traumatic stress disorder" (PTSD). For lucky individuals who forget what happened, the brain uses these psychological mechanisms, described by psychologist and psychiatrist, to bury the unpleasant experience in the unconsciousness or to delete it completely from the memory [54]. It is certainly that powerful inhibitory system in the limbic and memory circuits that had been involved including dopamine, as reported by Sabanda in 2021 [55]. Patients with this trauma had mild to severe anxiety and depression [56]. These add a motive to avoid remembering painful events and could potentiate forgetting. Some studies with war veterans are performed with PET or functional MRI to know what happened in the brain lobes with stress.

7. Therapeutic aspects

Nowadays, there is not an effective treatment for exclusively the absence or the excess of forgetting. There are two possibilities for therapeutic goals: using drugs and not using them. In the first one, all the strategies to avoid forgetting. At this point, it can be used all the drugs involved in the treatment of dementia. Although how it was explained, the mechanism of action is not the same, because in amnesia there is clear damage to neurons, in forgetting not. All the cholinesterase inhibitors like donepezil [57], rivastigmine, or galantamine [58]; glutamate regulators like memantine [59]; or more recently, with the use of monoclonal antibodies treatment like aducanumab could be considered [60], all the studies published in 2022. It is desirable the use of cognitive-behavioral psychotherapy using strategies and exercises to promote memory.

It exists also the use of neuromodulation to promote memory in patients with Alzheimer's disease (AD) [61]. It is still an experimental tool for this proposal. The main target is Deep Brain Stimulation (DBS) to avoid amnesia. The use of DBS could improve the fall of memory centered in the fornix of patients with AD [62, 63]. Not forgetting (how it is considered in this chapter) has already been evaluated.

On the other hand, there are no punctual drugs to increment forgetting. In patients with depression or anxiety, the treatment is focused on its disease, like antidepressants or anxiolytics. Insomnia patients could use anxiolytics. Also, specific medications, like suvorexant, a drug that acts in the orexin neurotransmitter in the hypothalamus to regulate the sleep-arousal cycle, according to Rahman in 2022 [64].

There is no choice to improve forgetfulness in a precise manner, as DBS does with AD patients. It must be explored this area, thinking in a brain surgical target: a list of possible targets is each part of the verbal memory, like the hippocampus, fornix, or amygdala. It could be considered, and indeed, they are used with DBS in patients with epilepsy, aggressiveness, or AD.

Depressive or anxious patients could be treated with Transcranial Magnetic Stimulation (TMS) to diminish the symptoms [65, 66]. Also, it is possible for future protocols to evaluate forgetfulness.

8. Conclusions

Forgetting or forgetfulness is a neurologic active state in which the subject or the patient has an impossibility to obtain old information stored in their memory. Neuroplasticity is a fundamental concept when approaching forgetfulness. Macroscopically, the neuronal circuitry is not damaged, although its molecular scene changes the opportunity to retrieve the information. It is important to mention the existence of drugs that improve amnesia (which is not the same as forgetting), and it is possible to act over forgetfulness. There are no medications to help forget, in cases of sleepiness, mental trauma, or aging. The use of DBS or TMS could be a possibility to increase or decrease, in other words, neuromodulate forgetting.

Conflict of interest

The authors declare no conflict of interest.

Thanks

The authors would like to thank Anahuac University Mexico, so as Alpha Health Sciences Leadership Program, for all their support in the realization of this manuscript. So as the Stereotactic and Functional Neurosurgery Unit of Mexico General Hospital.

Author details

Jose Rodrigo Carrillo-Marquez[1,2*] and Jose Damian Carrillo-Ruiz[1,3]

1 Faculty of Health Sciences, Anahuac Mexico University, Mexico City, Mexico

2 Alpha Health Sciences Leadership Program, Mexico Anahuac University, Mexico City, Mexico

3 Stetereotactic and Functional Neurosurgery, Mexico General Hospital, Mexico

*Address all correspondence to: carrillomarquezjr@gmail.com

References

[1] Goldstein M. The decade of the brain. Neurology. 1990;**40**(2):321. DOI: 10.1212/WNL.40.2.321

[2] Südhof TC. Molecular neuroscience in the 21st century: A personal perspective. Neuron. 2017;**96**(3):536-541. DOI: 10.1016/j.neuron.2017.10.005

[3] Tang L. Modeling neurulation. Nature Methods. 2019;**16**:1205. DOI: 10.1038/s41592-019-0680-8

[4] Elshazzly M, Lopez MJ, Reddy V, Caban O. Embryology, central nervous system. In: StatPearls. Treasure Island (FL): StatPearls Publishing; 2021

[5] Jeibmann A, Paulus W. Drosophila melanogaster as a model organism of brain diseases. International Journal of Molecular Sciences. 2009;**10**(2):407-440. DOI: 10.3390/ijms10020407

[6] Perveen FK, editor. Fly, Drosophila melanogaster: Modeling of human diseases (part II). In: Drosophila melanogaster - Model for Recent Advances in Genetics and Therapeutics. London: IntechOpen; 2018. DOI: 10.5772/intechopen.73199

[7] Zhang R, Engler A, Taylor V. Notch: An interactive player in neurogenesis and disease. Cell and Tissue Research. 2018;**371**(1):73-89. DOI: 10.1007/s00441-017-2641-9

[8] Bray SJ. Notch signalling: A simple pathway becomes complex. Nature Reviews. Molecular Cell Biology. 2006;7(9):678-689. DOI: 10.1038/nrm2009

[9] Lasky JL, Wu H. Notch signaling, brain development, and human disease. Pediatric Research. 2005;**57**(7):104-109. DOI: 10.1203/01.PDR.0000159632.70510.3D

[10] Engler A, Zhang R, Taylor V. Notch and neurogenesis. Molecular Mechanisms of Notch Signaling. 2018;**1066**:223-234. DOI: 10.1007/978-3-319- 89512-3_11

[11] McLaughlin CN, Broihier HT. Keeping neurons young and foxy: FoxOs promote neuronal plasticity. Trends in Genetics. 2018;**34**(1):65-78. DOI: 10.1016/j.tig.2017.10.002

[12] Kempermann G, Gage FH, Aigner L, Song H, Curtis MA, Thuret S, et al. Human adult neurogenesis: evidence and remaining questions. Cell Stem Cell. 2018;**23**(1):25-30. DOI: 10.1016/j.stem.2018.04.004

[13] de Sousa Fernandes MS, Ordônio TF, Santos GC, Santos LE, Calazans CT, Gomes DA, et al. Effects of physical exercise on neuroplasticity and brain function: A systematic review in human and animal studies. Neural Plasticity. 2020;**2020**:8856621. DOI:10.1155/2020/8856621

[14] RC I, Young LJ. Oxytocin, neural plasticity, and social behavior. Annual Review of Neuroscience. 2021;**44**:359-381. DOI: 10.1146/annurev-neuro-102320-102847

[15] de Oliveira RM. Neuroplasticity. Journal of Chemical Neuroanatomy. 2020;**108**:101822. DOI: 10.1016/j.jchemneu.2020.101822

[16] Cowan N. What are the differences between long-term, short-term, and working memory? Progress in Brain Research. 2008;**169**:323-338. DOI: 10.1016/S0079-6123(07)00020-9

[17] Sekeres MJ, Winocur G, Moscovitch M. The hippocampus and related neocortical structures in memory transformation. Neuroscience Letters. 2018;**680**:39-53. DOI: 10.1016/j.neulet.2018.05.006

[18] Lövblad KO, Schaller K, Vargas MI. The fornix and limbic system. Seminars in Ultrasound, CT, and MR. 2014;**35**(5):459-473. DOI: 10.1053/j.sult.2014.06.005

[19] Ressler RL, Maren S. Synaptic encoding of fear memories in the amygdala. Current Opinion in Neurobiology. 2019;**54**:54-59. DOI: 10.1016/j.conb.2018.08.012

[20] Roesler R, Parent MB, LaLumiere RT, McIntyre CK. Amygdala-hippocampal interactions in synaptic plasticity and memory formation. Neurobiology of Learning and Memory. 2021;**184**:107490. DOI: 10.1016/j.nlm.2021.107490

[21] Rolls ET. The cingulate cortex and limbic systems for emotion, action, and memory. Brain Structure & Function. 2019;**224**(9):3001-3018. DOI: 10.1007/s00429-019-01945-2

[22] Chittka L, Giurfa M, Riffell JA. The mechanisms of insect cognition. Frontiers in Psychology. 2019;**10**:2751. DOI: 10.3389/fpsyg.2019.02751

[23] Davies R, Gagen MH, Bull JC, Pope EC. Maze learning and memory in a decapod crustacean. Biology Letters. 2019;**15**(10):20190407. DOI: 10.1098/rsbl.2019.0407

[24] Krichbaum S, Katz JS. Dogs demonstrate memory of incidentally encoded information. Learning & Behavior. 2021;**49**(2):173-174. DOI: 10.3758/s13420-020-00459-w

[25] Hampton RR, Engelberg JW, Brady RJ. Explicit memory and cognition in monkeys. Neuropsychologia. 2020;**138**:107326. DOI: 10.1016/j.neuropsychologia.2019.107326

[26] Klinzing JG, Niethard N, Born J. Mechanisms of systems memory consolidation during sleep. Nature Neuroscience. 2019;**22**(10):1598-1610. DOI: 10.1038/s41593-019-0467-3

[27] Acosta MT. Sleep, memory and learning. Medicina. 2019;**79**:29-32

[28] Mandal A, Prabhavalkar KS, Bhatt LK. Gastrointestinal hormones in regulation of memory. Peptides. 2018;**102**:16-25. DOI: 10.1016/j.peptides.2018.02.003

[29] Calabrese EJ, Rubio-Casillas A. Biphasic effects of THC in memory and cognition. European Journal of Clinical Investigation. 2018;**48**(5):e12920. DOI: 10.1111/eci.12920

[30] Abrahao KP, Salinas AG, Lovinger DM. Alcohol and the brain: Neuronal molecular targets, synapses, and circuits. Neuron. 2017;**96**(6):1223-1238. DOI: 10.1016/j.neuron.2017.10.032

[31] Voss JL, Bridge DJ, Cohen NJ, Walker JA. A closer look at the hippocampus and memory. Trends in Cognitive Sciences. 2017;**21**(8):577-588. DOI: 10.1016/j.tics.2017.05.008

[32] Josselyn SA, Tonegawa S. Memory engrams: Recalling the past and imagining the future. Science. 2020;**367**(6473):eaaw4325

[33] Della Sala S, Cubelli R. Forgetting. Cortex. 2021;**136**:157. DOI: 10.1016/j.cortex.2020.12.013

[34] Cherry KE, Brigman S, Burton-Chase AM, Baudoin KH. Perceptions of forgetfulness in adulthood. The Journal of Genetic Psychology.

2021;**182**(1):31-46. DOI: 10.1080/00221325.2020.1840330

[35] De Hoz L, Martin SJ, Morris RG. Forgetting, reminding, and remembering: The retrieval of lost spatial memory. PLoS Biology. 2004;**2**(8):E225. DOI: 10.1371/journal.pbio.0020225

[36] Brewin CR. Memory and forgetting. Current Psychiatry Reports. 2018;**20**(10):87. DOI: 10.1007/s11920-018-0950-7

[37] Cornell J, Salinas S, Huang HY, Zhou M. Microglia regulation of synaptic plasticity and learning and memory. Neural Regeneration Research. 2022;**17**(4):705-716. DOI: 10.4103/1673-5374.322423

[38] Poo MM, Pignatelli M, Ryan TJ, Tonegawa S, Bonhoeffer T, Martin KC, et al. What is memory? The present state of the engram. BMC Biology. 2016;**14**:40. DOI: 10.1186/s12915-016-0261-6

[39] Wang C, Yue H, Hu Z, Shen Y, Ma J, Li J, et al. Microglia mediate forgetting via complement-dependent synaptic elimination. Science. 2020;**367**(6478):688-694. DOI: 10.1126/science. aaz2288

[40] Mitchelmore C, Kjærulff KM, Pedersen HC, Nielsen JV, Rasmussen TE, Fisker MF, et al. Characterization of two novel nuclear BTB/POZ domain zinc finger isoforms: Association with differentiation of hippocampal neurons, cerebellar granule cells, and macroglia. The Journal of Biological Chemistry. 2002;**277**(9):7598-7609. DOI: 10.1074/jbc.M110023200

[41] Seifert G, Steinhäuser C. Heterogeneity and function of hippocampal macroglia. Cell and Tissue Research. 2018;**373**(3):653-670. DOI: 10.1007/s00441-017-2746-1

[42] Kol A, Adamsky A, Groysman M, Kreisel T, London M, Goshen I. Astrocytes contribute to remote memory formation by modulating hippocampal–cortical communication during learning. Nature Neuroscience. 2020;**23**(10):1229-1239

[43] Cubelli R, Della SS. Amnesia. Cortex. 2021;**136**:158. DOI: 10.1016/j.cortex.2020.12.014

[44] Custers EJ, Ten Cate OT. Very long-term retention of basic science knowledge in doctors after graduation. Medical Education. 2011;**45**(4):422-430. DOI: 10.1111/j.1365-2923.2010.03889. X

[45] Salvetti B, Morris RG, Wang SH. The role of rewarding and novel events in facilitating memory persistence in a separate spatial memory task. Learning & Memory. 2014;**21**(2):61-72. DOI: 10.1101/lm.032177.113

[46] Ismail FY, Fatemi A, Johnston MV. Cerebral plasticity: Windows of opportunity in the developing brain. European Journal of Paediatric Neurology. 2017;**21**(1):23-48. DOI: 10.1016/j.ejpn.2016.07.007. Epub 2016 Aug 9

[47] Nørby S. Varieties of graded forgetting. Consciousness and Cognition. 2020;**84**:102983. DOI: 10.1016/j.concog.2020.102983

[48] Chambers AM. The role of sleep-in cognitive processing: Focusing on memory consolidation. Wiley Interdisciplinary Reviews: Cognitive Science. 2017;**8**(3):e1433. DOI:10.1002/wcs.1433

[49] Ngo HV, Born J. Sleep and the balance between memory and forgetting. Cell. 2019;**179**(2):289-291. DOI: 10.1016/j.cell.2019.09.007

[50] Bergstein J, Erdelyi M. Recognition hypermnesia: How to

get it. Memory. 2008;**16**(7):689-702. DOI: 10.1080/09658210802169095

[51] Doolen AC, Radvansky GA. A novel study: Hypermnesia for books read years ago. Memory. 2021;**30**(2):1-12. DOI:10.1080/09658211.2021.1993262. Epubahead of print

[52] Hurlemann R. Noradrenergic control of emotion-induced amnesia and hypermnesia. Reviews in the Neurosciences. 2006;**17**(5):525-532. DOI: 10.1515/revneuro.2006.17.5.525

[53] Mulligan NW, Duke MD. Positive and negative generation effects, hypermnesia, and total recall time. Memory & Cognition. 2002;**30**(7): 1044-1053. DOI: 10.3758/bf03194322

[54] Zoellner LA, Sacks MB, Foa EB. Directed forgetting following mood induction in chronic posttraumatic stress disorder patients. Journal of Abnormal Psychology. 2003;**112**(3):508-514. DOI: 10.1037/0021-843x.112.3.508

[55] Sabandal JM, Berry JA, Davis RL. Dopamine-based mechanism for transient forgetting. Nature. 2021;**591**(7850):426-430. DOI: 10.1038/s41586-020-03154-y

[56] Nørby S. Forgetting and emotion regulation in mental health, anxiety and depression. Memory. 2018;**26**(3):342-363. DOI: 10.1080/09658211.2017.1346130

[57] Wang H, Zong Y, Han Y, Zhao J, Liu H, Liu Y. Compared of efficacy and safety of high-dose donepezil vs standard-dose donepezil among elderly patients with Alzheimer's disease: A systematic review and meta-analysis. Expert Opinion on Drug Safety. 2022;**31**:1-9. DOI: 10.1080/14740338.2022.2027905.87

[58] Fan F, Liu H, Shi X, Ai Y, Liu Q, Cheng Y. The efficacy and safety of Alzheimer's disease therapies: An updated umbrella review. Journal of Alzheimer's Disease. 2022;**85**(3):1195-1204. DOI: 10.2174/1567205019666220120112301

[59] Shi X, Ren G, Cui Y, Xu Z. Comparative efficacy and acceptability of cholinesterase inhibitors and memantine based on dosage in patients with vascular cognitive impairment: A network meta-analysis. Current Alzheimer Research. 2022;**19**(2):133-145. DOI: 10.2174/1567205019666220120112301

[60] Salloway S, Chalkias S, Barkhof F, Burkett P, Barakos J, Purcell D, et al. Amyloid-related imaging abnormalities in 2 phase 3 studies evaluating Aducanumab in patients with early Alzheimer disease. JAMA Neurology. 2022;**79**(1):13-21. DOI: 10.1001/jamaneurol.2021.4161

[61] Koch G, Spampinato D. Alzheimer disease and neuroplasticity. Handbook of Clinical Neurology. 2022;**184**:473-479. DOI: 10.1016/B978-0-12-819410-2.00027-8

[62] Jakobs M, Lee DJ, Lozano AM. Modifying the progression of Alzheimer's and Parkinson's disease with deep brain stimulation. Neuropharmacology. 2020;**171**:107860. DOI: 10.1016/j.neuropharm.2019.107860

[63] Luo Y, Sun Y, Tian X, Zheng X, Wang X, Li W, et al. Deep brain stimulation for Alzheimer's disease: Stimulation parameters and potential mechanisms of action. Frontiers in Aging Neuroscience. 2021;**13**:619543. DOI: 10.3389/fnagi.2021.619543

[64] Rahman SA, Nathan MD, Wiley A, Crawford S, Cohn AY, Harder JA, et al. A double-blind, randomized, placebo-controlled trial of suvorexant for the treatment of vasomotor

symptom-associated insomnia disorder in midlife women. Sleep. 2022;**45**(3):zsac007. DOI: 10.1093/sleep/zsac007

[65] Pell GS, Harmelech T, Zibman S, Roth Y, Tendler A, Zangen A. Efficacy of deep TMS with the H1 coil for anxious depression. Journal of Clinical Medicine. 2022;**11**(4):1015. DOI: 10.3390/jcm11041015

[66] Song P, Tong H, Zhang L, Lin H, Hu N, Zhao X, et al. Repetitive transcranial magnetic stimulation modulates frontal and temporal time-varying EEG network in generalized anxiety disorder: A pilot study. Frontiers in Psychiatry. 2022;**12**:1779201. DOI: 10.3389/fpsyt.2021.779201

Chapter 3

Neurophysiology of Emotions

Maurizio Oggiano

Abstract

Emotions are automatic and primary patterns of purposeful cognitive-behavioral organizations. They have three main functions: coordination, signaling, and information. First, emotions coordinate organs and tissues, thus predisposing the body to peculiar responses. Scholars have not reached a consensus on the plausibility of emotion-specific response patterns yet. Despite the limitations, data support the hypothesis of specific response patterns for distinct subtypes of emotions. Second, emotional episodes signal the current state of the individual. Humans display their state with verbal behaviors, nonverbal actions (e.g., facial movements), and neurovegetative signals. Third, emotions inform the brain for interpretative and evaluative purposes. Emotional experiences include mental representations of arousal, relations, and situations. Every emotional episode begins with exposure to stimuli with distinctive features (i.e., elicitor). These inputs can arise from learning, expressions, empathy, and be inherited, or rely on limited aspects of the environment (i.e., sign stimuli). The existence of the latter ones in humans is unclear; however, emotions influence several processes, such as perception, attention, learning, memory, decision-making, attitudes, and mental schemes. Overall, the literature suggests the nonlinearity of the emotional process. Each section outlines the neurophysiological basis of elements of emotion.

Keywords: emotion, emotion definition, emotion faces, facial expressions, emotional experience, elicitor, sign stimuli, reward system, James-Lange theory, Cannon-Bard theory

1. Introduction

The study of emotions has fascinated scholars from all over the world for millennia. Socrates and Plato dealt with it about two thousand five hundred years ago, and they probably were not even the first [1]. Although there have been considerable advances since then, our knowledge is far from complete.

In this chapter, the term *emotion* refers to "automatic and primary patterns of purposeful cognitive-behavioral organizations" [2]. Although there is no consensus, data suggest that emotions are "automatic models" since each subtype probably has specific neurophysiological layouts [3]. Furthermore, every emotional process is "primary" because, in certain situations, it coordinates the activity of the nervous system (e.g., perception, attention, and memory) [4]. The expression "cognitive-behavioral organizations" describes the coordinating nature of emotion and its ability to facilitate distinctive behavioral responses [5], while maintaining central control. In particular, brain processing makes it possible to learn (see Section 5.1), inhibit, and regulate emotional responses [6]. Emotions are then "purposeful" because they aim to prepare the body to respond to situations that have occurred repeatedly throughout evolution [4].

On the whole, these features reveal the essential functions of emotions, namely [7]:

- **Coordination:** Emotions coordinate organs and tissues, thus predisposing the body to peculiar responses (see Section 2).

- **Signaling:** Although the central nervous system (CNS) maintains the faculty of control (e.g., inhibition), coordination activities facilitate the production of distinctive behavioral responses and expressive signals of the individual's current emotional state (see Section 3).

- **Information:** The CNS interprets and evaluates emotional episodes. That allows individuals to partly consciously perceive emotions, learn from them, and direct behaviors (see Section 4).

2. Coordination

One of the first scientists to define the nature of emotions was probably William James. Until then, the prevalent idea was that situations evoke emotions that, in turn, trigger bodily changes. James instead claimed that "bodily changes follow directly the PERCEPTION of the exciting fact, and that our feeling of the same changes as they occur IS the emotion" (pp. 189–190) [8]. The Danish psychologist Lange developed similar concepts in the same period. Therefore, today, scholars refer to this idea as the *James-Lange theory*. Sensory systems send data about the current situation to the central nervous system. Subsequently, the CNS induces physiological changes (e.g., heartbeat and muscle tone). The following feeling of these changes is the emotion. In other words, there is no emotion without physiological changes. It seemed a counter-intuitive thesis even then. Nevertheless, several scholars accepted the James-Lange theory [9].

Some years later, Walter Bradford Cannon falsified James's idea. He considered that stopping sensory sensitivity would impair the central perception of physiological changes, thereby eliminating emotions. Thus, Cannon resected the animals' spinal cords. Results suggested that surgically operated individuals still felt emotions, though. Furthermore, it seemed the same physiological changes accompanied various emotions. Cannon concluded that emotions disturb the activity of the autonomic nervous system (ANS) [10], and Philip Bard enriched this view. In brief, this hypothesis (i.e., the *Cannon-Bard theory*) [9] is an "all-or-none" approach with only two autonomic patterns: *non-activated* versus *diffusely activated* [5].

Neither theory has disappeared over the years. Indeed, they have led to two contrasting approaches.

First, nonstate theories and general arousal models suggest the inexistence of specific internal states of emotions [11]. The two-factor theory [12], the component process model [13], and other hypotheses [14] reach the same conclusion. Empirical evidence does not support the idea that emotions have specific ANS patterns. Part of the data instead suggests that undifferentiated arousal accompanies emotional experiences [3]. Something in the body happens, but are the people who label it as an emotional experience (e.g., fear, joy, and sadness) [15]. Therefore, proponents of these theories (i.e., cognitive models) claim that emotions result from brain activity [5].

The second approach supports the existence of discrete emotions, each one characterized by specific neurophysiological and behavioral routines. In this case, scholars usually view emotions as an adaptive mechanism, a product of evolution [4]. Charles Darwin was probably the first to search for the cause of expressions [16]. Subsequently, several scientists focused on the link between emotions, autonomic activity, and behavioral responses (e.g., facial expressions) [17]. Proponents of this functional-evolutionary approach claim the existence of different emotions associated with bio-behavioral layouts [5].

There is a lively debate still today. In particular, scholars did not reach a consensus on the existence of emotion-specific response patterns. One explanation for this diatribe lies in the methodological challenges the study of emotions entails. Individual differences (e.g., emotion recognition skills), the choice of elicitor (e.g., there is no certainty that a given stimulus elicits a given emotion in people), indicators (e.g., a continuous recording of different physiological and behavioral measures) [5], and statistical methods are among these [18].

Despite the issues, data support the hypothesis of specific response patterns for distinct subtypes of emotions [3]. For example, in rodents, different types of fear correspond to independent neural substrates [19]. Indeed, emotional families are sets of states that share elicitors (see Section 5), autonomic patterns, expressions, and behavioral reactions [17]. The neural substrates of emotional subtypes facilitate different behavioral responses. As an illustration, consider the *Fight Flight Freeze System* [20]:

- The **flight** depends on norepinephrine activity from the locus coeruleus. Moreover, the amygdaloid complex (Amg) activates periaqueductal gray and brainstem autonomic nuclei.

- The **fight** has its neurobiological basis in the hypothalamic-pituitary-adrenal axis (HPA). In particular, the release of cortisol stimulates *gluconeogenesis* (i.e., the conversion of substrates into glucose) and *glycogenesis* (i.e., glycogen synthesis). That provides fuel for metabolism and activates the sympathetic division of the autonomic nervous system [21].

- **Freezing** also relies on a specific neural network. In brief, the central nucleus of the amygdala has connections to the lateral hypothalamus (i.e., which mediates autonomic sympathetic responses), medullar nuclei (i.e., that control parasympathetic response), and the ventrolateral periaqueductal gray [22].

3. Signaling

The bodily activity that occurs with emotions has a high signaling value. For example, organs and tissues, as well as the nervous system, can signal emotional states [20].

Furthermore, emotions and motor activation often correlate. That can affect the striated muscles of the neck, back, arms, or the smooth muscles of the blood vessels and alimentary tract. Similarly, facial muscles can also be part of the emotion [23]. All these activities can be *expressions*, namely, distinctive signals of emotional episodes [6].

However, emotions are not the only determinants of bodily signals. In particular, contextual and cognitive factors make it challenging to distinguish expressions from cues attributable to other causes. Individual differences (e.g., age, gender and learning) are often decisive in expression regulation. Indeed, the nervous system (e.g., premotor cortex and primary motor cortex) has the flexibility to adjust actions already planned to the current situation [9]. That means individuals can generally inhibit, mask, or even simulate expressions [6].

In brief, emotional signals can belong to three macro categories. First, *verbal behavior* refers to emotional expression through natural-historical languages. The second category, *nonverbal behavior* (*NVB*), concerns any type of action except the use of words. Gestures, gait, and posture are examples of NVB. Although facial expressions also fall into this category, they will be examined separately, given their significance to humans. Finally, autonomic activity can produce external manifestations (*neurovegetative signals*) interpretable as expressions (e.g., pupil diameter, heart rate, and breathing).

3.1 Verbal behavior

Voice and speech are the two components of the act of speaking. The voice features are pitch, volume, intensity, and rhythm. Instead, speech is the content of discourses. It includes vocabulary, grammar style, and structure [24].

Humans can use verbal behavior to express emotions [25]. Speaking is a faculty that recruits several anatomical structures: cerebral regions (e.g., the frontal lobe) [9] and the digestive and respiratory systems (e.g., lungs, larynx, sinus cavities of the vocal tract, palate, tongue, and teeth) [26]. Noteworthy, dysfunction of the frontal lobe is one of the determinants of alexithymia, a condition that involves among other things, difficulty or inability to verbalize emotions [27]. However, a vast cerebral network underpins verbal expression of emotion. The right inferior frontal cortex, the right posterior superior temporal cortex, the left mid-fusiform gyrus, the right inferior prefrontal and bilateral fusiform cortices, and the amygdaloid complex are part of this network [25].

3.2 Nonverbal behavior

Behavioral responses can be emotional clues. For example, gait (e.g., arm swing, length, and speed of stride) can reveal whether an individual is happy, sad, or angry [28]. Furthermore, the emotional state can influence posture (i.e., the position of the body or its parts) [29], produce acoustic signals, such as laughter [30], and alter the sound of voice (e.g., pitch, intensity, and tension) [31].

However, humans can voluntarily signal and fake emotional states through their bodies (e.g., facial expressions, gestures, and posture) [30]. Birds also have this ability. For example, the wild fork-tailed drongos (*Dicrurus adsimilis*) produce false mimicked alarm calls that scare meerkats (*Suricata suricatta*). Thereby, these birds steal meerkats' food [32].

It is unclear whether emotional expression management relies on a single neurocognitive system. The intentional inhibition of human motor responses depends, at least in part, on the activation of the right inferior frontal cortex (rIFC). Indeed, the activity of the rIFC is often associated with the deactivation of other brain regions important for emotions, such as the amygdaloid complex [33].

3.3 Facial expressions

The face is informative in several ways. For example, humans get clues about people's health through skin color [34]. Nonetheless, the main source of information is the activity of the facial muscles. Their contraction, in specific combinations, produces skin movements, namely, *facial expressions*. Moreover, they assume complex patterns according to the movement of the head and eyes.

The muscles of the face include two large groups. First, the *mastication muscles* (i.e., temporalis, masseter, and pterygoid muscles) have the primary task of moving the jaws and chewing. However, they can even participate in emotional expression. It is the trigeminal nerve that innervates these muscles (**Figure 1**).

The *expressive* or *mimetic* muscles are the second group. The facial nerve innervates these muscles. Indeed, their function is to configure the expression of the face. The temporofacial division of this nerve connects the muscles of the upper part of the face to both cerebral hemispheres. Instead, the cervicofacial facial nerve links the lower face only to the contralateral hemisphere (**Figure 2**).

The cerebral cortex controls voluntary movements through the corticospinal (or pyramidal) tract [23]. Two-thirds of this tract receives input from the motor cortex and the rest from the somatosensory areas, such as the parietal lobe [9]. For these reasons, emotional facial expressions seem to depend on the other trait, the extrapyramidal one.

The right side of the face could be dominant for emotional expressions. That is the idea of some scholars, based on some clinical evidence. For example, several people show a left bias during posed expressions [23]. Nevertheless, the empirical results are ambiguous, and academic speculations are divergent [35]. For instance, the approach-withdrawal model hypothesizes that emotions of "approach" (e.g., joy) coincide with more activity of the left frontal brain, and the "withdrawal" ones (e.g., fear) activate the right frontal brain to a superior extent [36].

Moreover, humans can exhibit brief, local contractions (i.e., *microexpressions*). Their duration varies from about 40 to 335 ms [37]. Microexpressions mainly involve

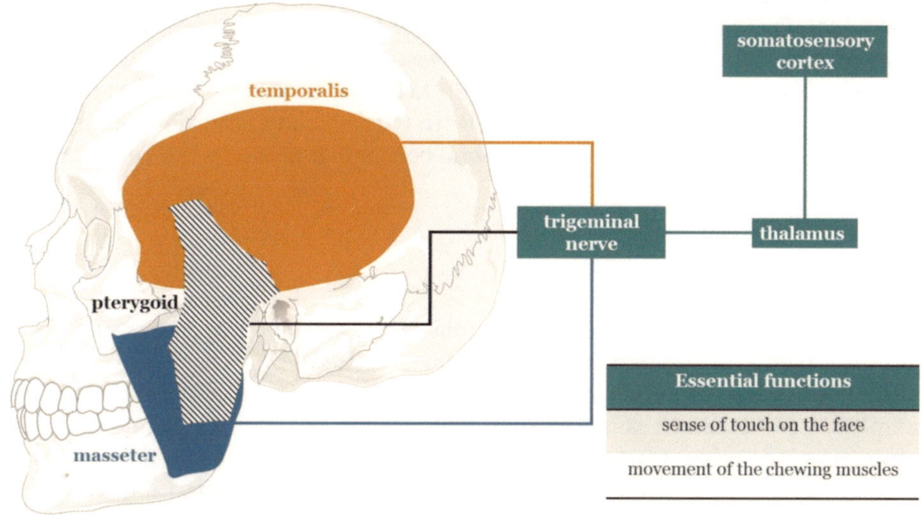

Figure 1.
Schematic representation of the motor pathways of mastication.

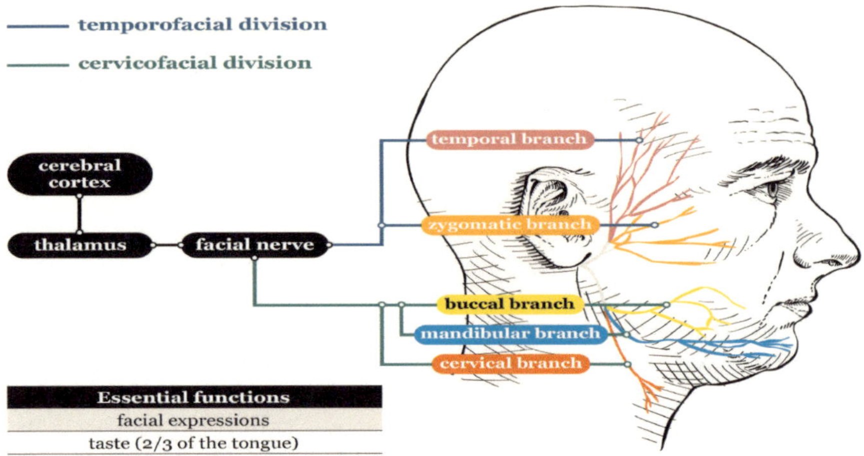

Figure 2.
Schematic representation of the pathways of human facial expressive muscles.

the upper face muscles (e.g., the frontalis) and occur unconsciously, at least in part. Indeed, it is the extrapyramidal tract that mediates their production [23]. However, their alleged unintentional nature has stretched their informative potential. In particular, several scholars believe that microexpressions are reliable signals of spontaneous emotions and lies. For example, law enforcement and airport security use microexpressions as lie-detecting clues. All this despite the experimental data being inconclusive and practical applications ineffective [38]. However, microexpressions could be functional in other fields (e.g., to survey the quality of the patient-therapist relationship) [39]. Noteworthy, only a few microexpressions seem unmanageable. For example, the eyebrow flash and contempt expression are more controllable [38].

Although the prototypical patterns are well known, there is a low coherence between facial displays and emotions. Specifically, the likelihood of a person showing an expression (e.g., the Duchenne smile) when feeling the corresponding emotion (e.g., joy) is often lower than chance, in the both laboratory [40] and naturalistic settings [41]. One of the determinants of this low emotion-expression coherence lies in *display rules*. In brief, they are laws of expression management based on various factors (e.g., context, roles, gender, and age). Learning these rules usually takes place in the first years of life. Thereby, humans learn to repeat, amplify, and inhibit the expression of emotions [42]. It is the cerebral cortex that mediates the voluntary inhibition of facial movements [23].

3.4 Neurovegetative signals

Despite limitations and still open questions, there are enough data to state that physiological changes accompany emotions (see Section 2).

Activities of the autonomic nervous system can induce appearance variations. For example, vasodilation can cause blood vessels to bulge and alter the color of the skin. Blushing (i.e., in embarrassment) and reddering (e.g., in anger) are two typical neurovegetative signals of an increase in the caliber of blood vessels. Conversely, vasoconstriction (e.g., in fear) produces blanching.

The body can also secrete various substances. For example, tear glands provoke crying [43] related to some types of sadness [44]. Similarly, the sweat glands produce sweat (e.g., in fear), and the salivary ones are responsible for the secretion of saliva, which is typical of certain emotional states, such as disgust and anger.

Other neurovegetative signals are piloerection and the change in pupil diameter. They can be cues to emotions (e.g., anger and fear) or other states (e.g., sexual appetite) [43].

At the central level, these ANS activities are outcomes of a neural network that involves the hypothalamus (Hy), which is essential for homeostasis. The Hy links with periaqueductal gray, the reticular formation, parabrachial nucleus, ventral tegmental area (VTA), and the raphe nuclei [45].

4. Information

Rather than describing the whole emotional process, the James-Lange theory focuses on the emotional experience, namely, what the individual perceives of emotion [11].

In particular, *emotional experiences* consist of mental representations that can relate to different aspects of emotions. First, the individual can perceive, even if only partially, **arousal** [46]. The central nervous system processes the information it retrieves from the body's activity. For example, reactions, such as wrinkles, blushing, and tearing are all signals, that potentially influence the perception of emotions [47]. Indeed, autonomic feedback (e.g., from sweating and respiration) is another essential feature of the emotional experience [48]. The second aspect is about the **relational content** (e.g., mental representations of dominance and submission). Third, the **situational content** is an integral part of the emotional experience. For example, people often link psychological situations to their emotions.

The elements of the emotional experience concern the *appraisal*, at least in part [46]. In brief, it consists of the cognitive assessments (e.g., of valence) accompanying emotions (e.g., positive or negative) [48]. According to the cognitive approach, the appraisal is necessary to elicit emotions, and it can also occur quickly and involuntarily [49].

The neurophysiological basis of emotional experience may rely on two distinct networks. A first circuit seems to allow value-based representation. The essential brain regions of this network are the basolateral amygdala (BLA), the anterior insula, and the orbitofrontal cortex (OFC). In particular, the BLA provides an initial assessment of a stimulus. The anterior insula instead allows the representation of interoceptive inputs. Finally, the OFC makes processing more flexible by including information regarding the context. BLA and OFC are interconnected with each other, as well as have connections with the cortical regions responsible for sensory representations. The second circuit of emotional experience seems to be a sort of affective working memory that holds and processes emotive information for short periods. Its neurophysiological basis lies in the amygdaloid complex and the reciprocal connections between the prefrontal cortex (PFC) and the anterior cingulate cortex [46].

5. Elicitors of emotions

An emotional episode begins through exposure to stimuli with specific features. For example, loss causes sadness [50]. In this sense, emotion is a process, and these stimuli (i.e., *elicitors*) are the inputs that initiate it.

Animals learn to feel certain emotions in specific situations. However, it is not just environmental cues that trigger an emotional episode. The state of the organism, behavior, and other complex faculties (e.g., thoughts and empathy) can be elicitors too.

5.1 Learning

Stimuli internal and external to the body can become elicitors of emotion through classical and operant conditioning. In particular, animals learn to associate a stimulus (S) with a specific emotional response (R). With an S-R association established, exposure to S may be sufficient to elicit the emotional response.

Yet, S-R associations can take complex forms. For instance, emotional reactions can even be self-reinforcing. An individual may experience a pleasant state that elicits behavior, which fuels repetition in a sort of loop [51]. However, the S-R associability is not absolute. For example, it seems impossible to teach a hungry pigeon to fly away by presenting it with food [52].

At the basis of emotional learning is a vast neural network that includes the reward system. Its architecture is complex and involves circuits for the cost/benefit assessment of specific reward values, reward expectations, and action selection (**Figure 3**). The amygdaloid complex and ventral striatum (vStr) underpin the appetitive

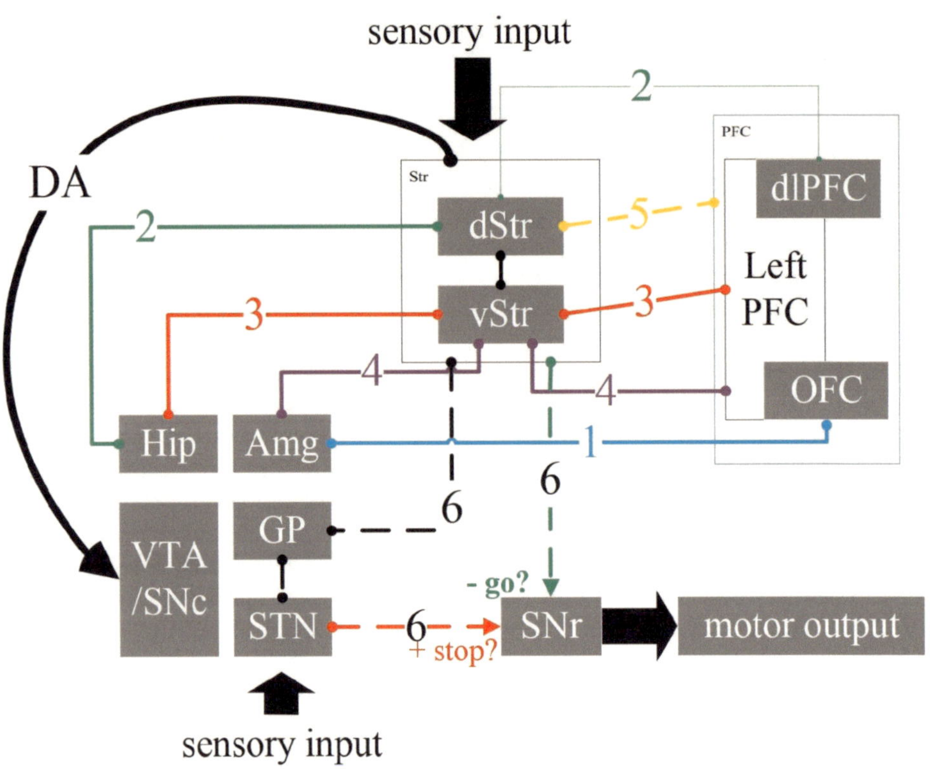

Figure 3.
Schematic representation of reward system. One: cost/benefit assessment of specific reward value. Two: cortical loop. Three: limbic loop. Four: reward expectation. Five: action selection. Six: go and stop processes. Abbreviations: Amg, Amygdaloid complex; DA, dopamine; dlPFC, dorsolateral prefrontal cortex; dStr, dorsal striatum; GP, globus pallidus; Hip, hippocampus; OFC, orbitofrontal cortex; SNc, substantia nigra pars compacta; SNr, substantia nigra pars reticulata; STN, subthalamic nucleus; vStr, ventral striatum; VTA, ventral tegmental area. This article was published in [53] Copyright Elsevier (2020).

processing of reward expectations. Moreover, dopamine (DA) pathways modulate motivation and behavior by connecting the ventral tegmental area, the substantia nigra pars compacta (SNc), and the striatum. The reward system is also capable of inhibiting the behavior. Specifically, five sub-circuits (or sub-loops) of the basal ganglia (BG) are essential for various functions, including action cancelation [53].

5.2 Expressions

Even emotional signals (see Section 3) can be elicitors. For example, breathing [54], vocalizations [55], and postures [56] may elicit emotions. Indeed, some experiments suggest that an expansive pose lowers cortisol and increases testosterone levels. That would be enough to produce feelings of power and increase risk tolerance [57]. However, attempts to replicate such data have failed [58].

In general, *peripheral feedback theories* propose the idea that emotional expressions can become elicitors [56]. For example, facial expressions could trigger emotions. According to this idea (i.e., the *facial feedback hypothesis*), movements of the face influence the release of some neurotransmitters, thus acting as an elicitor [59]. Therefore, feedback theories follow the path traced by the James-Lange theory (see Section 2).

Observing signals and behaviors in others can also initiate the emotional process. In this sense, indirect experience (e.g., imitation and emotional contagion) is a potential elicitor of emotion. The anterior insula and the rostral cingulate cortex are part of the neural network responsible for these mechanisms [60].

5.3 Empathy

Empathy consists of the emotional states produced by observation of individuals and situations. Thus, it is an elicitor of emotion *per se*. Scholars usually distinguish between *cognitive empathy* (CE) and *emotional empathy* (EE). The CE deals with mental perspective-taking, mentalizing, and the theory of mind. Instead, EE consists of the vicarious sharing of emotions [61].

The resulting emotion may be the same as that of the observed individual, but not necessarily. Indeed, the emotional experience can be so intense as to produce an *empathic overarousal*, which often induces disengagement [62]. That happens, for example, to paramedics who, being exposed to traumatic events, adopt coping strategies, such as emotional detachment [63]. In other cases, emotions felt may be diverse from those observed. That is the case of *Schadenfreude* (i.e., the pleasure caused by the misfortune of others) [64].

Given the complexity of empathy as a faculty, it seems clear that its neurobiological basis is equally complex. For example, its neural substrate includes the insula, cingulate cortex, and the interoceptive network [65]. Furthermore, empathic responses probably depend on various processes (e.g., emotional contagion and affective mentalizing) that underpin distinct neural mechanisms. The temporoparietal junction, medial temporal lobe, prefrontal cortex, and dopamine pathways are part of the circuits of cognitive empathy. Instead, the neural substrates of emotional empathy include the frontal gyrus, insula, anterior cingulated cortex, and oxytocin paths [66].

5.4 Sign stimuli

Animals can feel emotions even when exposed to stimuli they have never encountered. For instance, the smell of predators they have never seen before scares rats [67].

Some scholars claim that humans also have innate fears (e.g., snakes) [68]. However, experimental results are controversial [69].

Several animals can respond to limited aspects of the environment (i.e., *sign stimuli*), ignoring the rest [70]. The common toad (*Bufo bufo*) produces defensive responses (e.g., stiff-legged) when faced with relatively simple perceptual patterns (dummies) with specific configurational features reminiscent of their predators (i.e., snakes) [71]. Moreover, newborn babies prefer and imitate human face schematizations (i.e., *facelike patterns*) [72].

From a neurobiological point of view, it is unlikely that there are cells in the nervous system specialized in innately identifying specific stimuli. The determinants of any sign stimuli as elicitors could lie, then, in the biological predisposition [52].

6. Outcomes

Emotions influence several processes, including [4]:

- **Perception**: Emotional states can magnify some perceptual aspects to the detriment of others. For instance, afraid-of-falling people overestimate the steepness of hills [73].

- **Attention:** According to a theoretical approach, the central nervous system allocates more cognitive resources to selective attention and vigilance under the influence of negative emotions. Instead, positive ones spread these resources. That hypothesis (i.e., *broaden-and-build theory*) [74] is subject to debate, however [75]; in this sense, each emotion may have selective effects on attentional performance [76].

- **Learning:** Emotions heighten some mnemonic aspects at the expense of others. In particular, several factors (e.g., personality and age) define the enhancement and impairment of learning due to emotional influences [77].

- **Memory:** Considering their relationship with attention and learning, it seems logical that emotions influence memory. The *flashbulb memory* (i.e., the vivid remembering of details of when a person learned about a specific dramatic event) is an excellent example of this [78].

- **Decision making:** Emotions impact the evaluation and interpretation processes. For example, sadness may lead people to overestimate difficulties [2].

- **Attitudes and mental schemes:** Some emotional episodes can shape attitudes, mentalities, and values. For instance, awe produces overwhelming and elevating experiences that strengthen the sense of unity. In particular, those who experience emotional experiences of this type usually develop a new vision of life and the universe [79].

The neurophysiological basis of the relationship between emotions and cognitive processes is not yet fully understood. In brief, neurophysiological superimposition of emotions and other mechanisms may underpin emotional outcomes. For example, amygdala, hippocampus, and orbitofrontal cortex have a role in several neural functions [80].

7. Emotional and mood disorders

Some emotion-based symptoms may appear in various conditions, such as schizophrenia (e.g., anhedonia), borderline personality disorder (e.g., emotional instability), addiction (e.g., euphoria and dysphoria), and so on. These conditions can be related to several features of emotions, such as intensity, frequency, adaptivity, physiology, expression, and experience [81]. However, some scholars disagree with placing these disorders in the same category. Indeed there is the risk of generalizing qualitatively diverse states (e.g., emotions, moods, and affect) [7].

Several scholars use the term "*emotional disturbance*" to refer to psychopathologies that include emotion-related symptoms, such as regulation problems, phobias, specific deficits (e.g., lack of empathy), and so on [81]. It is challenging to briefly delineate the etiology of these emotion-related psychopathologies. In brief, there are hereditary, epigenetic, developmental, environmental, and behavioral determinants. From a neurophysiological point of view, emotion disturbances usually result from cerebral peculiar functioning. Indeed, the bases of such conditions often include inefficient reuptake [82], irregularities in synaptic proteins, and neuronal density [83]. Even social activities (e.g., play) can influence the development of these brain mechanisms and shape cerebral maturation [84]. These factors can influence the functioning of a vast neural network that includes the prefrontal cortex, limbic system, striatum, thalamus, and brainstem [83].

8. Conclusions

Albeit limitations [5], the literature suggests that emotions predispose the body to timely recognition and response to specific circumstances. Situations identify ancestral problems, and the responses illustrate the solutions that have proved most profitable for evolutionary success [4]. In this sense, the emotional process has as its central themes the body's coordination [5], the signaling of the individual's state [17], and the processing by the central nervous system of both endogenous and exogenous information [46].

The literature does not allow a conclusive illustration of the neurophysiology of emotions. Nevertheless, each emotional subtype likely has its patterns [3]. It seems then better to speak of families rather than single emotions [17].

Due to several factors, the emotional process affects performance in different domains (e.g., perception and attention). These factors include the partial overlap of the neural basis of emotion and other faculties and the numerous brain interconnections. Furthermore, elicitors are heterogeneous and even include the essential elements of emotion (e.g., expressions). That suggests the nonlinearity of the emotional process. In other words, emotions could have stochastic or aleatory progress: In probabilistic terms, each element can initiate, be a part of, or be their outcome [7].

Conflict of interest

The authors declare no conflict of interest.

Thanks

Thanks to every enthusiast, scholar, and researcher who came before me. Thanks to those who deal with emotions today, and to those who will do so in the future.

Everyone's contribution is a step forward in the path of knowledge. Thanks to every reader, without whom the effort of writing this chapter would have been futile. Ideologies separate us, and emotions bring us together.

Abbreviations

Amg	amygdaloid complex
ANS	autonomic nervous system
BG	basal ganglia
BLA	basolateral amygdala
CE	cognitive empathy
CNS	central nervous system
DA	dopamine
dlPFC	dorsolateral prefrontal cortex
dStr	dorsal striatum
EE	emotional empathy
GP	globus pallidus
Hip	hippocampus
Hy	hypothalamus
HPA	hypothalamic-pituitary-adrenal axis
NVB	nonverbal behavior
OFC	orbitofrontal cortex
PFC	prefrontal cortex
rIFC	right inferior frontal cortex
SNc	substania nigra pars compacta
STN	subthalamic nucleus
vStr	ventral striatum
VTA	ventral tegmental area

Author details

Maurizio Oggiano
European Mind and Metabolism Association, Via Valtellina, Rome, Italy

*Address all correspondence to: maurizio.oggiano.79@gmail.com

References

[1] Solomon RC. The philososphy of emotions. In: Lewis M, Haviland-Jones JM, Barrett LF, editors. Handbook of Emotions. New York, NY, US: The Guilford Press; 2008. pp. 3-16

[2] Caprara GV. Motivare è riuscire: Le ragioni del successo. In: To Motivate is to Achieve: The Reasons for Success. Bologna, Italia: Il Mulino; 2013. p. 267

[3] Kreibig SD. Autonomic nervous system activity in emotion: A review. Biological Psychology. 2010;**84**:394-421. DOI: 10.1016/j.biopsycho.2010.03.010

[4] Tooby J, Cosmides L. The evolutionary psychology of the emotions and their relationship to internal regulatory variables. In: Lewis M, Haviland-Jones JM, Barrett LF, editors. Handbook of Emotions. New York, NY, US: The Guilford Press; 2008. pp. 114-137

[5] Levenson RW. The autonomic nervous system and emotion. Emotion Review. 2014;**6**:100-112. DOI: 10.1177/1754073913512003

[6] Ekman P. Expression and the nature of emotion. In: Scherer K, Ekman P, editors. Approaches to Emotion. Hillsdale, NJ: Lawrence Erlbaum; 1984. pp. 319-344

[7] Oggiano M. Origins of emotion in humans and other animals. Sign stimuli as elicitors of emotion families. Sign stimuli as elicitors of emotion families [thesis] Rome, Italy: Uninettuno University; 2020. DOI: 10.13140/RG.2.2.14919.65443

[8] James W. What is an emotion? Mind. 1884;**9**:188-205

[9] Bear MF, Connors BW, Paradiso MA, editors. Neuroscience: Exploring the Brain. 3rd ed. Philadelphia, PA, US: Lippincott Williams and Wilkins Publishers; 2006. p. 857

[10] Cannon WB. The James-Lange theory of emotions: A critical examination and an alternative theory. The American Journal of Psychology. 1927;**39**:106-124. DOI: 10.2307/1415404

[11] Lewis M. The emergence of human emotions. In: Lewis M, Haviland-Jones JM, Barrett LF, editors. Handbook of Emotions. New York, NY, US: The Guilford Press; 2008. pp. 304-319

[12] Schachter S, Singer J. Cognitive, social, and physiological determinants of emotional state. Psychological Review. 1962;**69**:379-399. DOI: 10.1037/h0046234

[13] Scherer KR. Emotion as a process: Function, origin and regulation. Social Science Information. 1982;**21**:555-570. DOI: 10.1177/053901882021004004

[14] Barrett LF. Are emotions natural kinds? Perspectives on Psychological Science. 2006;**1**:28-58. DOI: 10.1111/j.1745-6916.2006.00003.x

[15] Friedman BH. Feelings and the body: The Jamesian perspective on autonomic specificity of emotion. Biological Psychology. 2010;**84**:383-393. DOI: 10.1016/j.biopsycho.2009.10.006

[16] Darwin C. The Expression of the Emotions in Man and Animals. 3rd ed. London, UK: John Murray; 1872. p. 400. DOI: 10.1037/10001-000

[17] Ekman P. An argument for basic emotions. Cognition and Emotion. 1992;**6**:169-200. DOI: 10.1080/02699939208411068

[18] Halsey LG. The reign of the p-value is over: What alternative analyses could we employ to fill the power vacuum? Biology Letters. 2019;**15**:20190174. DOI: 10.1098/rsbl.2019.0174

[19] Gross CT, Canteras NS. The many paths to fear. Nature Reviews Neuroscience. 2012;**13**:651-658. DOI: 10.1038/nrn3301

[20] Balconi M. Neuropsychology of Communication. Milan, Italy: Springer-Verlag Italia srl; 2010. p. 223. DOI: 10.1007/978-88-470-1584-5

[21] Kemeny ME, Shestyuk A. Emotions, the neuroendocrine and immune systems, and health. In: Lewis M, Haviland-Jones JM, Barrett LF, editors. Handbook of Emotions. New York, NY, US: The Guilford Press; 2008. pp. 661-675

[22] Roelofs K. Freeze for action: Neurobiological mechanisms in animal and human freezing. Philosophical Transactions of the Royal Society B: Biological Sciences. 2017;**372**:20160206. DOI: 10.1098/rstb.2016.0206

[23] Rinn WE. The neuropsychology of facial expression: A review of the neurological and psychological mechanisms for producing facial expressions. Psychological Bulletin. 1984;**95**:52-77. DOI: 10.1037/0033-2909.95.1.52

[24] Allport GW, Cantril H. Judging personality from voice. The Journal of Social Psychology. 1934;**5**:37-55. DOI: 10.1080/00224545.1934.9921582

[25] Kotz SA, Paulmann S. Emotion, language, and the brain. Language and Linguistics Compass. 2011;**5**:108-125. DOI: 10.1111/j.1749-818X.2010.00267.x

[26] Aglioti SM, Fabbro F. Neuropsicologia del linguaggio [Neuropsychology of Language]. Bologna, Italia: Società editrice il Mulino; 2006. p. 240

[27] Larsen JK, Brand N, Bermond B, Hijman R. Cognitive and emotional characteristics of alexithymia: A review of neurobiological studies. Journal of Psychosomatic Research. 2003;**54**: 533-541. DOI: 10.1016/S0022-3999(02) 00466-X

[28] Keating CF. The life and times of nonverbal communication theory and research past, present, future. In: Matsumoto D, Hwang HC, Frank MG, editors. APA Handbook of Nonverbal Communication. Washington, DC, US: American Psychological Association; 2016. pp. 17-42. DOI: 10.1037/14669-002

[29] Matsumoto D, Hwang HC, Frank MG. The body: Postures, gait, proxemics, and haptics. In: Matsumoto D, Hwang HC, Frank MG, editors. APA Handbook of Nonverbal Communication. Washington, DC, US: American Psychological Association; 2016. pp. 387-400. DOI: 10.1037/14669-015

[30] Frank MG, Shaw AZ. Evolution and Nonverbal Communication. In: Matsumoto D, Hwang HC, Frank MG, editors. APA Handbook of Nonverbal Communication. Washington, DC, US: American Psychological Association; 2016. pp. 45-76. DOI: 10.1037/14669-003

[31] Scott S, McGettigan C. The voice: From identity to interactions. In: Matsumoto D, Hwang HC, Frank MG, editors. APA Handbook of Nonverbal Communication. Washington, DC, US: American Psychological Association; 2016. pp. 289-305. DOI: 10.1037/14669-011

[32] Flower T. Fork-tailed drongos use deceptive mimicked alarm calls to steal

food. Proceedings of the Royal Society B: Biological Sciences. 2011;**278**:1548-1555. DOI: 10.1098/rspb.2010.1932

[33] Berkman ET, Burklund L, Lieberman MD. Inhibitory spillover: Intentional motor inhibition produces incidental limbic inhibition via right inferior frontal cortex. NeuroImage. 2009;**47**:705-712. DOI: 10.1016/j.neuroimage.2009.04.084

[34] Re DE, Rule NO. Appearance and physiognomy. In: Matsumoto D, Hwang HC, Frank MG, editors. APA Handbook of Nonverbal Communication. Washington, DC, US: American Psychological Association; 2016. pp. 221-256. DOI: 10.1037/14669-009

[35] Thompson JK. Right brain, left brain; left face, right face: Hemisphericity and the expression of facial emotion. Cortex. 1985;**21**:281-299. DOI: 10.1016/s0010-9452(85)80033-2

[36] Coan JA, Allen JJ, Harmon-Jones E. Voluntary facial expression and hemispheric asymmetry over the frontal cortex. Psychophysiology. 2001;**38**: 912-925. DOI: 1 0.1111/1469-8986.3860912

[37] Pfister T, Li X, Zhao G, Pietikäinen M. Recognising spontaneous facial micro-expressions. In: International Conference on Computer Vision. Barcelona: IEEE; 2011. pp. 1449-1456. DOI: 10.1109/ICCV.2011.6126401

[38] Burgoon JK. Microexpressions are not the best way to catch a liar. Frontiers in Psychology. 2018;**9**:1672. DOI: 10.3389/fpsyg.2018.01672

[39] Datz F, Wong G, Löffler-Stastka H. Interpretation and working through contemptuous facial micro-expressions benefits the patient-therapist relationship. International Journal of Environmental Research and Public Health. 2019;**16**:4901. DOI: 10.3390/ijerph16244901

[40] Reisenzein R, Studtmann M, Horstmann G. Coherence between emotion and facial expression: Evidence from laboratory experiments. Emotion Review. 2013;**5**:16-23. DOI: 10.1177/1754073912457228

[41] Fernández-Dols JM, Crivelli C. Emotion and expression: Naturalistic studies. Emotion Review. 2013;**5**:24-29. DOI: 10.1177/1754073912457229

[42] Ekman P, Friesen WV. Nonverbal leakage and clues to deception. Psychiatry. 1969;**32**:88-106. DOI: 10.1080/00332747.1969.11023575

[43] Levenson RW. Blood, sweat, and fears: The autonomic architecture of emotion. Annals of the New York Academy of Sciences. 2003;**1000**: 348-366. DOI: 10.1196/annals.1280.016

[44] Arias JA, Williams C, Raghvani R, Aghajani M, Baez S, Belzung C, et al. The neuroscience of sadness: A multidisciplinary synthesis and collaborative review. Neuroscience and Biobehavioral Reviews. 2020;**111**: 199-228. DOI: 10.1016/j.neubiorev.2020.01.006

[45] Wager TD, Barrett LF, Bliss-Moreau E, Lindquist KA, Duncan S, Kober H, et al. The neuroimaging of emotion. In: Lewis M, Haviland-Jones JM, Barrett LF, editors. Handbook of Emotions. New York, NY, US: The Guilford Press; 2008. pp. 249-271

[46] Barrett LF, Mesquita B, Ochsner KN, Gross JJ. The experience of emotion. Annual Review of Psychology. 2007;**58**: 373-403. DOI: 10.1146/annurev.psych.58.110405.085709

[47] de Melo CM, Kenny P, Gratch J. Influence of autonomic signals on perception of emotions in embodied agents. Applied Artificial Intelligence. 2010;**24**:494-509. DOI: 10.1080/08839514.2010.492159

[48] Kirkland T, Cunningham WA. Neural basis of affect and emotion. Wiley Interdisciplinary Reviews: Cognitive Science. 2011;**2**:656-665. DOI: 10.1002/wcs.145

[49] Arnold MB. Emotion and Personality. Vol. 1. New York, US: Columbia University Press; 1960. p. 288

[50] Joaquim RM, de Oliveira FCS, Fajardo RS, Caramaschi S. Psychobiology of sadness: Functional aspects in human evolution. EC Psychology and Psychiatry. 2018;**7**:1015-1022

[51] Cervone D, Pervin LA. Personality, Theory and Research. 13th ed. John Wiley and Sons, Inc: Hoboken, NJ; 2016. p. 560

[52] Seligman ME. On the generality of the laws of learning. Psychological Review. 1970;**77**:406-418. DOI: 10.1037/h0029790

[53] Oggiano M, Zoratto F, Palombelli G, Festucci F, Laviola G, Curcio G, et al. Striatal dynamics as determinants of reduced gambling vulnerability in the NHE rat model of ADHD. Progress in Neuro-Psychopharmacology and Biological Psychiatry. 2020;**100**:109886. DOI: 10.1016/j.pnpbp.2020.109886

[54] Philippot P, Chapelle G, Blairy S. Respiratory feedback in the generation of emotion. Cognition and Emotion. 2002;**16**:605-627. DOI: 10.1080/02699930143000392

[55] Hatfield E, Hsee CK. The impact of vocal feedback on emotional experience and expression. Journal of Social Behavior and Personality. 1995;**10**: 293-313

[56] Duclos SE, Laird JD, Schneider E, Sexter M, Stern L, Van Lighten O. Emotion-specific effects of facial expressions and postures on emotional experience. Journal of Personality and Social Psychology. 1989;**57**:100-108. DOI: 10.1037/0022-3514.57.1.100

[57] Carney DR, Cuddy AJ, Yap AJ. Power posing: Brief nonverbal displays affect neuroendocrine levels and risk tolerance. Psychological Science. 2010; **21**:1363-1368. DOI: 10.1177/0956797610383437

[58] Ranehill E, Dreber A, Johannesson M, Leiberg S, Sul S, Weber RA. Assessing the robustness of power posing: No effect on hormones and risk tolerance in a large sample of men and women. Psychological Science. 2015;**26**:653-656. DOI: 10.1177/0956797614553946

[59] Adelmann PK, Zajonc RB. Facial efference and the experience of emotion. Annual Review of Psychology. 1989;**40**: 249-280. DOI: 10.1146/annurev.ps.40.020189.001341

[60] Scheber MH, Baker JF. Descending control of movement. In: Squire LR, Bloom FE, Spitzer NC, du Lac S, Ghosh A, Berg D, editors. Fundamental Neuroscience. 3rd ed. Burlington, MA, US: Academic Press; 2008. pp. 987-1016

[61] Smith A. Cognitive empathy and emotional empathy in human behavior and evolution. The Psychological Record. 2006;**56**:3-21. DOI: 10.1007/BF03395534

[62] Hoffman ML. Empathy and prosocial behavior. In: Lewis M, Haviland-Jones JM, Barrett LF, editors. Handbook of Emotions. New York, NY,

US: The Guilford Press; 2008. pp. 440-455

[63] Regehr C, Goldberg G, Hughes J. Exposure to human tragedy, empathy, and trauma in ambulance paramedics. American Journal of Orthopsychiatry. 2002;**72**:505-513. DOI: 10.1037/0002-9432.72.4.505

[64] Smith RH, Powell CA, Combs DJ, Schurtz DR. Exploring the when and why of schadenfreude. Social and Personality Psychology Compass. 2009;**3**:530-546. DOI: 10.1111/j.1751-9004.2009.00181.x

[65] Bernhardt BC, Singer T. The neural basis of empathy. Annual Review of Neuroscience. 2012;**35**:1-23. DOI: 10.1146/annurev-neuro-062111-150536

[66] Shamay-Tsoory SG. The neural bases for empathy. The Neuroscientist. 2011;**17**:18-24. DOI: 10.1177/10738584103 79268

[67] Panksepp J, Biven L. The Archaeology of Mind: Neuroevolutionary Origins of Human Emotion. New York, NY, US: W. W. Norton and Company; 2012. p. 592

[68] Ohman A. Fear and anxiety: Overlaps and dissociations. In: Lewis M, Haviland-Jones JM, Barrett LF, editors. Handbook of Emotions. New York, NY, US: The Guilford Press; 2008. pp. 709-728

[69] Brosch T, Sharma D. The role of fear-relevant stimuli in visual search: A comparison of phylogenetic and ontogenetic stimuli. Emotion. 2005;**5**:360-364. DOI: 10.1037/1528-3542.5.3.360

[70] Manning A, Dawkins MS. An Introduction to Animal Behaviour. 5th ed. Cambridge, UK: Cambridge University Press; 1998. p. 460

[71] Ewert JP. Neural mechanisms of prey-catching and avoidance behavior in the toad (*Bufo bufo* L.). Brain Behavior and Evolution. 1970;**3**:36-56. DOI: 10.1159/000125462

[72] Valenza E, Simion F, Cassia VM, Umiltà C. Face preference at birth. Journal of Experimental Psychology: Human Perception and Performance. 1996;**22**:892-903. DOI: 10.1037/0096-1523.22.4.892

[73] Zadra JR, Clore GL. Emotion and perception: The role of affective information. Wiley Interdisciplinary Reviews: Cognitive Science. 2011;**2**:676-685. DOI: 10.1002/wcs.147

[74] Uddenberg S, Shim WM. Seeing the world through target-tinted glasses: Positive mood broadens perceptual tuning. Emotion. 2015;**15**:319-328. DOI: 10.1037/emo0000029

[75] Gable P, Harmon-Jones E. The blues broaden, but the nasty narrows: Attentional consequences of negative affects low and high in motivational intensity. Psychological Science. 2010;**21**:211-215. DOI: 10.1177/0956797609359622

[76] Taylor AJ, Bendall RCA, Thompson C. Positive emotion expands visual attention ... Or maybe not. The Cognitive Psychology Bulletin. 2017;**2**:33-37

[77] Tyng CM, Amin HU, Saad MN, Malik AS. The influences of emotion on learning and memory. Frontiers in Psychology. 2017;**8**:1454. DOI: 10.3389/fpsyg.2017.01454

[78] Finkenauer C, Luminet O, Gisle L, El-Ahmadi A, Van Der Linden M, Philippot P. Flashbulb memories and the underlying mechanisms of their formation: Toward an emotional-integrative model. Memory and

Cognition. 1998;**26**:516-531.
DOI: 10.3758/BF03201160

[79] Yaden DB, Iwry J, Slack KJ, Eichstaedt JC, Zhao Y, Vaillant GE, et al. The overview effect: Awe and self-transcendent experience in space flight. Psychology of Consciousness: Theory, Research, and Practice. 2016;**3**:1-11. DOI: 10.1037/cns0000086

[80] Rolls ET. A theory of emotion, and its application to understanding the neural basis of emotion. Cognition and Emotion. 1990;**4**:161-190. DOI: 10.1080/02699939008410795

[81] Kring AM. Emotion disturbances as transdiagnostic processes in psychopathology. In: Lewis M, Haviland-Jones JM, Barrett LF, editors. Handbook of Emotions. New York, NY, US: The Guilford Press; 2008. pp. 691-705

[82] Levinson DF. Genetics of major depression. In: Gotlib IM, Hammen CL, editors. Handbook of Depression. New York, NY, US: The Guilford Press; 2009. pp. 165-186

[83] Price JL, Drevets WC. Neural circuits underlying the pathophysiology of mood disorders. Trends in Cognitive Sciences. 2012;**16**:61-71. DOI: 10.1016/j.tics.2011.12.011

[84] Parvopassu A, Oggiano M, Festucci F, Curcio G, Alleva E, Adriani W. Altering the development of the dopaminergic system through social play in rats: Implications for anxiety, depression, hyperactivity, and compulsivity. Neuroscience Letters. 2021;**760**:136090. DOI: 10.1016/j.neulet.2021.136090

Chapter 4

Neuronal Architecture and Functional Organization of Olfactory Glomeruli

Thomas Heinbockel

Abstract

In the antennal lobes of insects and olfactory bulbs of vertebrates, the primary processing of olfactory information occurs within specialized units, called glomeruli. Glomeruli are discrete areas of densely packed, fine neuropil, usually ensheathed in glia cells. Glomeruli are the sites of synaptic interaction between axons of olfactory receptor cells and dendrites of central olfactory neurons. This chapter reviews the functional significance of this neuronal architecture, the glomerulus, with particular emphasis on results obtained in the sphinx moth, Manduca sexta. How is neuronal circuitry of olfactory glomeruli functionally organized, what attributes of olfactory stimuli are analyzed in glomeruli and how are these attributes processed and encoded in them? Glomeruli have been found in different invertebrate groups, such as crustaceans and insects with the glomeruli in the antennal lobes and the deutocerebrum, and molluscs with subepithelial glomeruli in the tentacle, as well as in different vertebrate groups such as amphibians, birds, fish, and mammals with glomeruli in the olfactory bulb. The organization of primary olfactory centers into glomeruli in diverse species suggests that glomeruli have a common and fundamental function in the processing of information about chemosensory stimuli and that glomeruli across taxa may share similar means of processing olfactory input.

Keywords: antenna, behavior, brain module, CNS, insect, *Manduca sexta*, neural coding, olfaction, orientation, pheromone, smell, synaptic integration

1. Introduction

Glomeruli in the brains of insects and vertebrates are the morphological and physiological structures where the primary processing of olfactory information takes place [1]. Glomeruli are housed in the olfactory centers of insects, the antennal lobes, and in the olfactory bulbs of vertebrates. Their widespread presence in different taxa has been interpreted to suggest common functionality. Experimental evidence based on recordings from principal output neurons in olfactory glomeruli of vertebrates and invertebrates supports this notion [2, 3]. The striking structural similarity, as well as the similarity of the responses to odor stimulation between neurons in the insect antennal lobe and vertebrate olfactory bulb, suggests that glomerular microcircuits across taxa may share similar means of processing olfactory input [1, 4, 5]. Studies on glomerular circuitry address the central question of the functional organization of olfactory glomeruli.

The antennal lobes of the sphinx moth *M. sexta* have emerged as a model system to determine mechanisms underlying olfactory information processing in early olfactory centers such as the glomerular microcircuits. (1) In *M. sexta*, the antennal lobes house a male-specific olfactory subsystem. This subsystem is specialized to process information about the female sex pheromone [6]. Input and output relationships in this experimentally advantageous model system can be precisely defined. (2) Other glomeruli in the antennal lobes of *M. sexta* are clearly different in both function and morphology compared to the male-specific subsystem that comprises the macroglomerular complex (MGC). The MGC consists of three glomeruli, the toroid-1, toroid-2, and the cumulus [7, 8]. (3) The MGC receives input from antennal sensory neurons [9] that are specifically tuned to one of the two essential components of the female sex pheromone [10].

The glomeruli of the MGC process information about the two essential pheromone components of the female sex pheromone. The components of the odor stimulus released by the female have been determined in terms of the concentration and ratio of the pheromone components. The number of neurons projecting from the MGC to higher brain centers is relatively limited. About 30 to 40 projection neurons (PNs) innervate the MGC, and about 860 PNs innervate all the glomeruli in the AL [11]. Many local interneurons (LNs) and PNs in the ALs have been described both morphologically and physiologically [3, 8, 12, 13].

The goal of research on olfactory glomeruli is to understand the role(s) of individual glomeruli, for example, the glomeruli that constitute the MGC in olfaction, namely, the toroid-1, toroid-2, and the cumulus, by analyzing how the neural circuits associated with these glomeruli process pheromonal information. The functional organization of the MGC can be studied by means of single-unit intracellular recording, staining and laser scanning confocal microscopy, and more recently, imaging techniques, multi-unit recordings, and computational models [14–22]. This line of research attempts to address several topics: How do features of the stimulus determine pheromone-evoked response characteristics of MGC interneurons? How do MGC interneurons discern pheromone components in a complex odor blend? Can MGC–PNs resolve and encode the naturally intermittent temporal structure of pheromonal stimuli? Do the two essential pheromone components serve specific and different roles? Answers to these questions will help define the functional role of glomeruli in olfaction and will aid our understanding of how different features of an odor stimulus are processed in the brain.

2. The chemical senses

The chemical senses are the oldest senses. The earliest living organisms monitored their environment with chemoreception in order to sense the availability of nutrients [23, 24] and thus to respond to different chemicals. Higher organisms face the challenge of reacting to various internal and external chemicals, for example, hormones, neurotransmitters, neural recognition molecules, and intra- and interspecific olfactory, and gustatory signals [24, 25].

The olfactory pathway starts with peripheral structures. In the case of vertebrates, olfactory receptor cells are located in the nasal cavity in the olfactory epithelium. In insects, sensilla on the antennae of insects houses olfactory receptor cells [26]. Two areas in olfactory research have been under heavy investigation: (a) the transduction mechanisms taking place at the olfactory receptor cells and (b) the synaptic mechanisms acting at the first synaptic relay in the olfactory pathway, including synaptic plasticity and learning, that is, in the olfactory bulb (OB) of vertebrates and the antennal lobes (ALs) of insects [1, 27–30].

3. The glomerulus in olfaction

The structural unit of organization in the AL or OB is the glomerulus [24, 31–37], that is, the neuropil is arranged into discrete areas ensheathed by a glial envelope [38, 39]. In *M. sexta*, glial cells play an important role in the sculpturing of glomeruli, since early removal of glial cells results in an absence of these subunits [38]. Glomeruli are the sites of synaptic interaction between primary olfactory axons and dendritic arborizations of central olfactory neurons [40]. Unlike that observed in vertebrates, evidence for moths and cockroaches suggests that in insects, few or no synaptic interactions take place in the neuropil outside the glomeruli [41, 42].

The brain and nervous system can be considered as arrangements of modular structures. Glomeruli are a prime example of such modular structures that are repeated in a specific brain region. It was Camillo Golgi (1874, cited in [34]) who first noted glomeruli. Since his early discovery, other modular structures have been described. Examples include columns, barrels, barreloids, and blobs [33, 34]. Considerable variation has been described for these modular structures in different species. In closely related species, one of them can lacks such an iterated module of brain organization but still achieves the same behavioral functions as the species that has them [34].

Glomeruli have also been found in the cerebellar cortex and the thalamic regions of vertebrates [25, 43]. Olfactory glomeruli have a long evolutionary history as they have been described in phylogenetically old animal groups. These groups include marine crustacea, fishes, onychophora, myriapoda, and mollusca. Glomeruli appeared before animals transitioned from marine to terrestrial life forms [25].

Glomeruli are not only structural modules but also functional units [33, 44]. 2-deoxyglucose (2-DG) studies in neonatal rat pups established a focal point in the dorsal part of the olfactory bulb, the modified glomerular complex. This is a small group of glomeruli involved in processing of suckling odor cues. In *Drosophila melanogaster*, 2-DG mapping of odor-induced neuronal activity in the ALs labeled distinct and histologically identified glomeruli [45, 46]. In insects, the macroglomerular complex has been established as the first central site where information about the female sex pheromone is processed [6, 47]. During odor stimulation of the rat olfactory epithelium, neighboring mitral/tufted cells, that is, the output neurons of the olfactory bulb that innervate the same glomerulus in the olfactory bulb, were frequently simultaneously excited or inhibited compared to cells that innervated different glomeruli [48].

The existing data indicates that glomeruli are functional units such that information about odorants is represented in a spatial manner among glomeruli. When the olfactory epithelium is stimulated with most odorants, the resulting responses in the

AL or olfactory bulb are spatial gradients or patterns of activity in more than one glomerulus [23, 45, 46, 49–51]. Three measures of neural activation (voltage-sensitive dyes, the 2-DG method, and *c-fos* expression) have revealed that in mammals, different odors elicit overlapping but distinctly different patterns of glomerular activity [51–54]. In the cockroach *Periplaneta americana*, stimulation with the female sex pheromone evokes responses in a very limited number of neurons and glomeruli, whereas general odorants result in responses in different output neurons representing more than 10 out of 130 glomeruli [55]. In *D. melanogaster*, stimulation with complex odors as well as with individual odors results in a spatial pattern of 2-DG activity in different specific subsets of antennal lobe glomeruli [45, 46].

A synthesis of the diffuse as well as specific aspects of the primary olfactory projections to central sites came from Ken Mori et al. [56, 57]. They characterized individual mitral/tufted cells based on the range of odor molecules effective in activating each cell. Individual mitral/tufted cells showed excitatory responses to groups of molecules with similar chemical structure [57]. Imamura et al. [56] developed a model for the activation of individual mitral/tufted cells by a range of odor molecules. In the model that takes into account work in different research groups, an olfactory sensory neuron expresses one or, at most, a few different types of receptor proteins. Subsequently, a neuron is activated by odor molecules with similar structure. The olfactory pathway is thought to work with a one cell-one receptor rule [58] such that a sensory cell expresses only one among hundreds of possible molecular receptors [59]. Neurons with the same or similar receptor proteins send one axon each to one or a few glomeruli and thus define glomerular function [60, 61]. The tuning specificity of the mitral/tufted cells thus reflects the specificity of the receptor protein [54, 56]. Recent studies have indicated that individual receptor probes hybridize to a small number of olfactory glomeruli. This suggests that axons of sensory neurons expressing the same olfactory receptor protein converge on only a small number of glomeruli [60, 61]. Together with the notion that individual mitral/tufted cells arborizing in single glomeruli have similar response specificities, the resulting picture is that each glomerulus appears to have a unique mixture of inputs [52]. This input, in turn, limits its odor specificity, also known as its molecular receptive range.

4. The antennal lobes of the Sphinx Moth *M. sexta*

The insect antenna consists of three segments, namely, the scape, pedicel, and flagellum. The entire length of the antenna has hairs or sensilla on its surface. On the first two segments, the sensilla houses mechanosensitive neurons. These project to mechanosensory centers in the deutocerebrum [62]. In the sphinx moth *M. sexta*, the long flagellum, divided into 85–90 annuli, is equipped with about $4x10^5$ sensilla. These represent several modalities, such as mechanosensation, hygroreception, and olfaction [10, 63–65]. The sensory neurons in olfactory and possibly other antennal sensilla send their axons to the antennal lobes (**Figure 1**). The sensory neurons converge onto central interneurons. In *P. americana*, the convergence ratio between olfactory sensory neurons and projection neurons can be as high as 5000 to 1, and in rabbits, the ratio between sensory neurons and mitral cells is about 1000 to1 [40, 67]. The antennal lobe of *M. sexta* contains about 64 spheroidal glomeruli [68, 69]. In male *M. sexta*, a macroglomerular complex located near the entrance of the antennal nerve into the antennal lobes has been identified (**Figure 1**) [6].

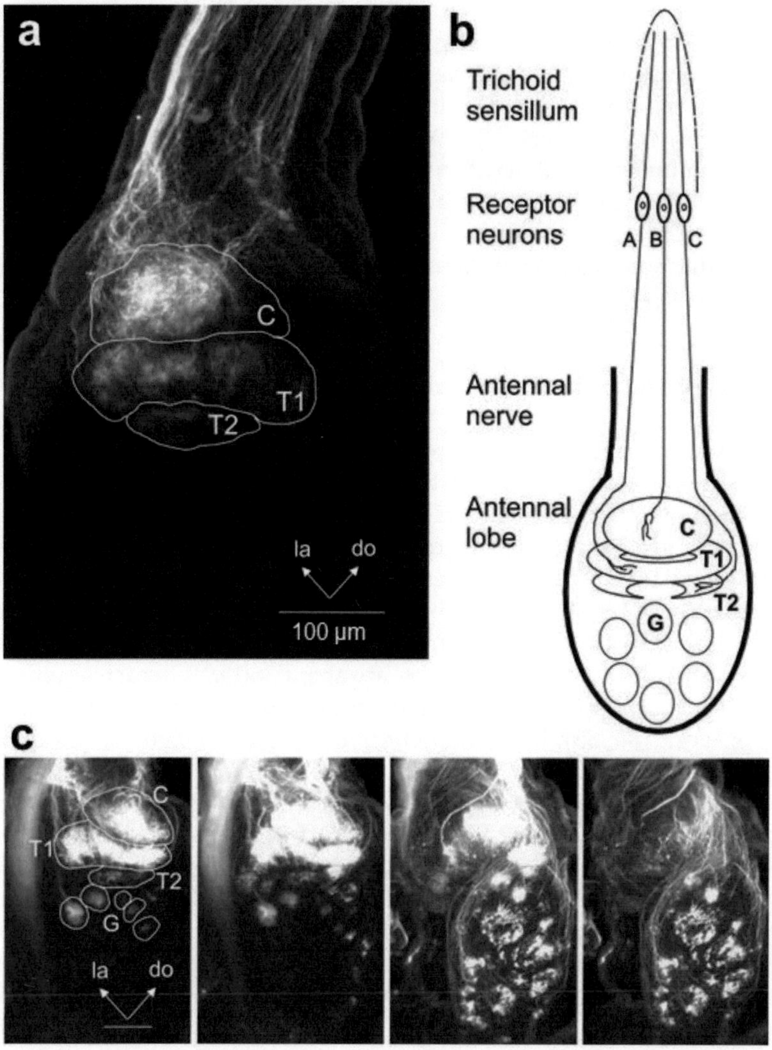

Figure 1.
This figure shows how antennal sensory neurons project to glomeruli in the antennal lobes of Manduca sexta. Sensory neurons were anterogradely labeled with rhodamine dextran. The diagram and images show frontal views. (a) Olfactory receptor neurons in long antennal trichoid sensilla project to the three glomeruli of the macroglomerular complex (MGC: C-cumulus, T1-toroid-1, and T2-toroid-2). (b) a schematic diagram illustrates how receptor neurons project to the antennal lobe. Receptor neurons from a trichoid sensillum project to the three glomeruli of the macroglomerular complex. (c) if receptor neurons in long trichoid sensilla and other antennal sensilla were labeled, axonal projections would go to the macroglomerular complex and ordinary glomeruli in the antennal lobes. Optical sections were taken at different depths in anterior to posterior direction through the antennal lobes shown from left to right. C – Cumulus, do – Dorsal, la – Lateral, T1 – Toroid-1, T2 – Toroid-2. Scale bar: 100 μm. From [66].

The first glomeruli in insects were described in the deutocerebrum of the bee by Kenyon [25, 70, 71]. In *M. sexta*, a closer anatomical analysis of glomeruli revealed a complex substructure of discrete domains and laminae within individual glomeruli [7, 72]. In bees, however, glomeruli have a relatively simple organization [73].

In contrast to the large differences in the number of glomeruli among different animal species, insect antennal systems present highly invariant glomerular organizations with regard to shape, size, location, and number within a species [74]. This has been shown for a variety of species including the fruit fly *Drosophila melanogaster* [75, 76], sphinx moth *M. sexta* [68], moth *Mamestra brassicae* [31], bee *Apis mellifera* [77], and cockroach *Blaberus craniifer* [78]. This invariance was also found to be true for the iulid *Cylindroiulus punctatus* (Diplopoda) [79] and in a vertebrate, the zebrafish (*Brachydanio rerio*) [80]. The number of glomeruli in all of these species is relatively small (18 for *C. punctatus* to 174 in worker bees). It is more difficult to verify numerical invariance in vertebrates with several thousand glomeruli [55]. The only identified vertebrate glomerulus is the modified glomerular complex for detection of the maternal suckling pheromone in rats [81].

5. Morphology and immunocytochemistry of neurons in the antennal lobe

Three classes of interneurons are present in the antennal lobes [11, 82]: (1) local, amacrine interneurons (LNs), with arborizations limited to the antennal lobe; (2) projection neurons (PNs) that send axons to higher order brain centers; and (3) centrifugal neurons that send axons from higher order brain centers into the antennal lobe (**Figures 2** and **3**). Sensory neurons from the antenna send their axon into one glomerulus only [9, 62] where it forms synapses with LNs, presumably mediated by acetylcholine [83]. The somata of antennal lobe LNs and PNs form three groups (lateral, medial, and anterior) [82]. There are about 360 LNs in each antennal lobe of *M. sexta*. They can innervate many, and perhaps all, glomeruli and appear to be mostly GABAergic [84, 85]. Different neurophysiological categories of local interneurons have been observed with respect to patterns of postsynaptic activity [13]. Evidence for unidirectional synaptic interactions between local interneurons and projection neurons as well as for disinhibitory pathways between these two types of neurons was found [13]. About 860 PNs project axons out of the antennal lobe through various antenno-cerebral tracts to different parts of the protocerebrum, for example, the calyces of the mushroom body and the lateral horn of the protocerebrum [11]. The third group of neurons, centrifugal neurons, is small in number and consists of a variety of cell types with unique morphologies, some of which innervate all glomeruli of one or both antennal lobes [82, 86]. The antennal lobe possesses a single serotonin-immunoreactive neuron [86]. This neuron has its soma in one antennal lobe, which innervates all glomeruli in the contralateral antennal lobe where it forms and receives synapses and has arborizations in the ipsilateral and contralateral protocerebrum [87].

Acetylcholine and GABA are the most prominent neurotransmitters in the antennal lobe [83]. Evidence that acetylcholine may serve as a transmitter has been reported for antennal sensory neurons [88] and some classes of projection neurons [89]. Acetylcholine may be released by primary afferent axons synapsing onto AL neurons [88, 90–93]. GABA is prominent in local interneurons and is also present in a subset of PNs [84]. GABA has an important role in the synaptic inhibition of PNs [85]. An IPSP is evoked when the antenna is stimulated with an odor. The IPSP is mediated by a chloride conductance and is sensitive to reversible blockade by picrotoxin and bicuculline. GABA hyperpolarizes neurons and inhibits their spontaneous nerve impulse firing. Several antennal lobe neurons are immunoreactive for biogenic amines. These neurons have wide dendritic arborizations and are thought to have widespread effects. Possibly, these neurons mediate central modulation of synaptic activity or threshold levels within the antennal lobe [86].

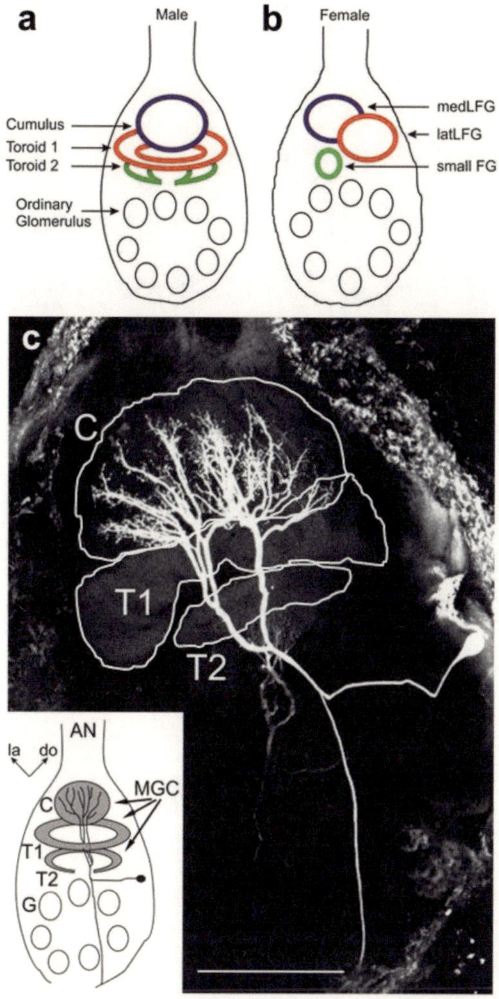

Figure 2.
Diagrammatic representation of sexually isomorphic glomeruli (G) and sexually dimorphic glomeruli in (a) male and (b) female Manduca sexta. (c) Laser-scanning confocal micrograph of an antennal lobe projection neuron in the moth antennal lobe of M. sexta. Image of a projection neuron in the macroglomerular complex (MGC-PN) with arborizations confined to the cumulus. The inset illustrates the organization of the antennal lobe with the macroglomerular complex (MGC) and other glomeruli (G). latLFG – Lateral large female glomerulus, medLFG – Medial large female glomerulus, smallFG – Small female glomerulus, C - cumulus, T1 - toroid-1, T2 - toroid-2, G – glomerulus, MGC – Macroglomerular complex, la – Lateral, do - dorsal. Scale bar: 100 μm. Modified from [66].

In LNs and PNs, several putative neuropeptides appear to be colocalized with classical transmitters [89].

6. The male-specific macroglomerular complex

In male *M. sexta*, the approximately 42,000 long trichoid sensilla commonly each contain two bipolar olfactory-sensory neurons that project to the macroglomerular complex (MGC) in the AL [9, 63–65]. Each of these two neurons is very sensitive

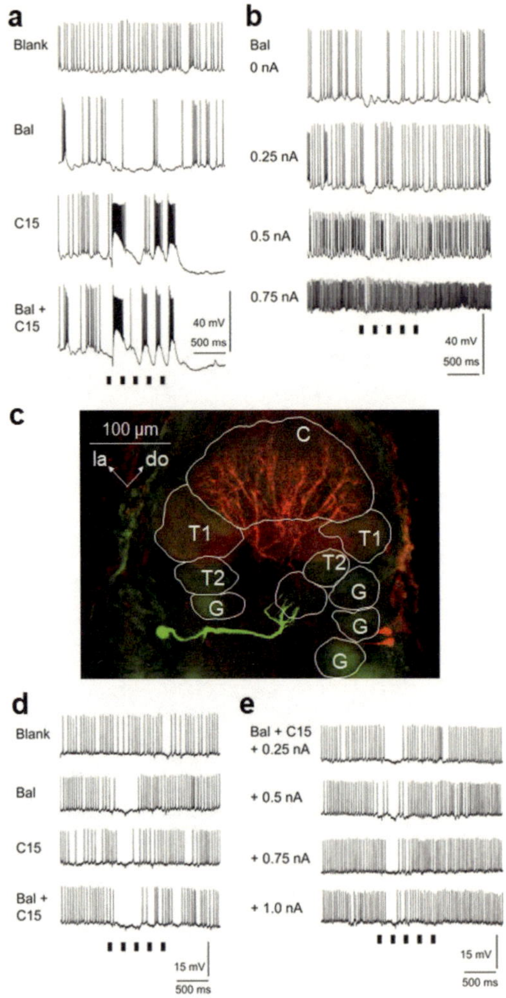

Figure 3.

The figure shows the intracellularly recorded physiological responses of labeled projection neurons in the antennal lobes of the sphinx moth Manduca sexta during antennal stimulation with pheromone. (a) this C15-specialist neuron responded with membrane potential inhibition to a stimulus that contained C15. This was followed by strong depolarization and again inhibition (third and fourth trace). Likewise, antennal stimulation with the pheromone blend of C15 + BAL evoked a mixed response. In contrast, stimulation with bombykal (BAL) evoked an inhibitory response. The antenna was stimulated with five 50-ms stimulus pulses at 5 Hz. The stimulus markers are depicted as black boxes beneath the records. (b) the membrane potential of the neuron was depolarized by injecting current through the recording electrode. As a result, the nerve impulse firing frequency increased, whereas the first stimulus pulse of BAL induced a membrane hyperpolarization and reduction in firing (inhibition). (c) the laser scanning confocal micrograph shows the morphology of two projection neurons in the antennal lobe. The neuron labeled in red, stained with biocytin, is described in figure panels (a) and (b), whereas panels (d) and (e) show the responses of the green, uniglomerular projection neuron, which is in Lucifer yellow (frontal view). C15-specialist neuron, the red neuron, has dendritic branches in the cumulus glomerulus of the macroglomerular complex and not in the toroid-1 or any other glomerulus. The green neuron sent dendritic branches into only one ordinary glomerulus. (d) Electrophysiological recordings from an antennal lobe projection neuron that innervated one of the ordinary glomeruli. When the antenna was stimulated with bomybkal, C15, or a blend of both pheromone components, the neuron responded with a reduction of the firing rate. As in panels (a) and (b), five identical stimulus pulses were delivered to the ipsilateral antenna at a frequency of 5 Hz. (e) Antennal stimulation of the same neuron with the pheromone blend resulted in inhibition, even when the membrane potential of the neuron was depolarized through current injection. The first antennal pheromone stimulus resulted in membrane hyperpolarization. C - cumulus; do - dorsal; G - ordinary glomerulus; la - lateral; me - medial; T1 - toroid-1; T2 - toroid-2. Scale bar = 100 µm. From [66].

to stimulation with one of the two major female sex-pheromone components, bombykal ((E,Z)-10,12-hexadecadienal) and a hexadecatrienal ((E,E,Z)-10,12,14-hexadecatrienal) [10]; that is, they have narrow molecular receptive ranges and constitute highly specific input channels.

In addition to the 64 spheroidal, ordinary glomeruli, the antennal lobe of *M. sexta* houses the sexually dimorphic MGC [69]. The MGC consists of at least three glomeruli (**Figures 2** and **3**) [7, 8, 16, 66]. One is donut-shaped (the "toroid-1"), and the other has a more globular structure (the "cumulus"). The third one (the toroid-2) is of unknown function and appears to have a donut shape as well. The cumulus is situated on the toroid and closer to the entrance of the antennal nerve. Projection neurons (PNs) with arborizations in the toroid-1 respond preferentially to antennal stimulation with the principal pheromone component bombykal (Bal-specialist MGC-PNs), whereas PNs arborizing in the cumulus respond preferentially to the second key

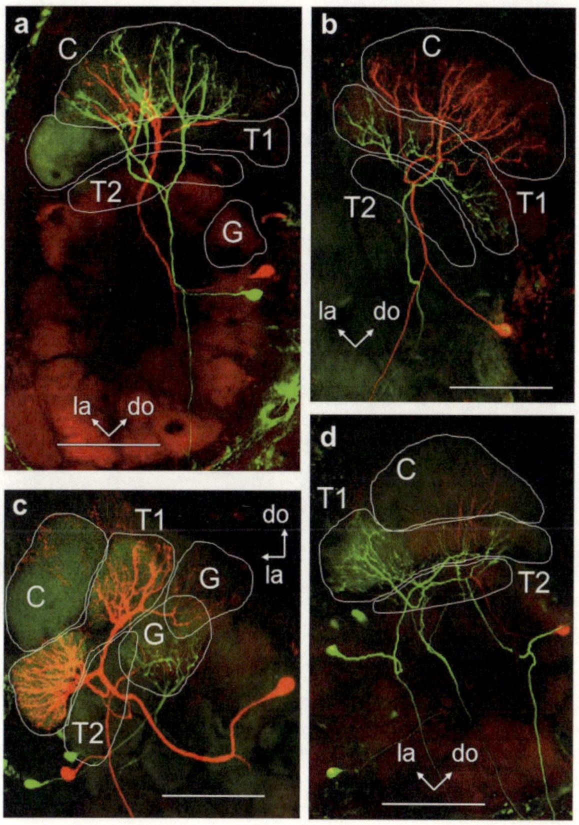

Figure 4.
Laser scanning confocal images illustrating the morphological diversity of projection neurons in the antennal lobes of the sphinx moth Manduca sexta. (a) Two C15 (E,Z-11,13-pentadecadienal) -specialist MGC-PNs (projection neurons of the macroglomerular complex) with arborizations confined to the cumulus. While the branches of the two neurons apparently overlapped in certain parts of the cumulus (indicated in yellow), other parts were innervated by just one of the two neurons. (b) One C15 specialist MGC-PN arborizing in the cumulus (green), and one bombykal (Bal) -specialist MGC-PN arborizing in the toroid-1 (red). (c) One bombykal (Bal) -specialist MGC-PN arborizing in the toroid-1 (red), and two projection neurons (red and green) innervating ordinary glomeruli adjacent to the MGC. (d) Several MGC-PNs innervating either the cumulus (red) or the toroid-1 (green). C – Cumulus, T1 – Toroid-1, T2 – Toroid-2, G – Ordinary glomerulus, do – Dorsal, la - lateral. Scale bar: 100 μm. Modified from [16].

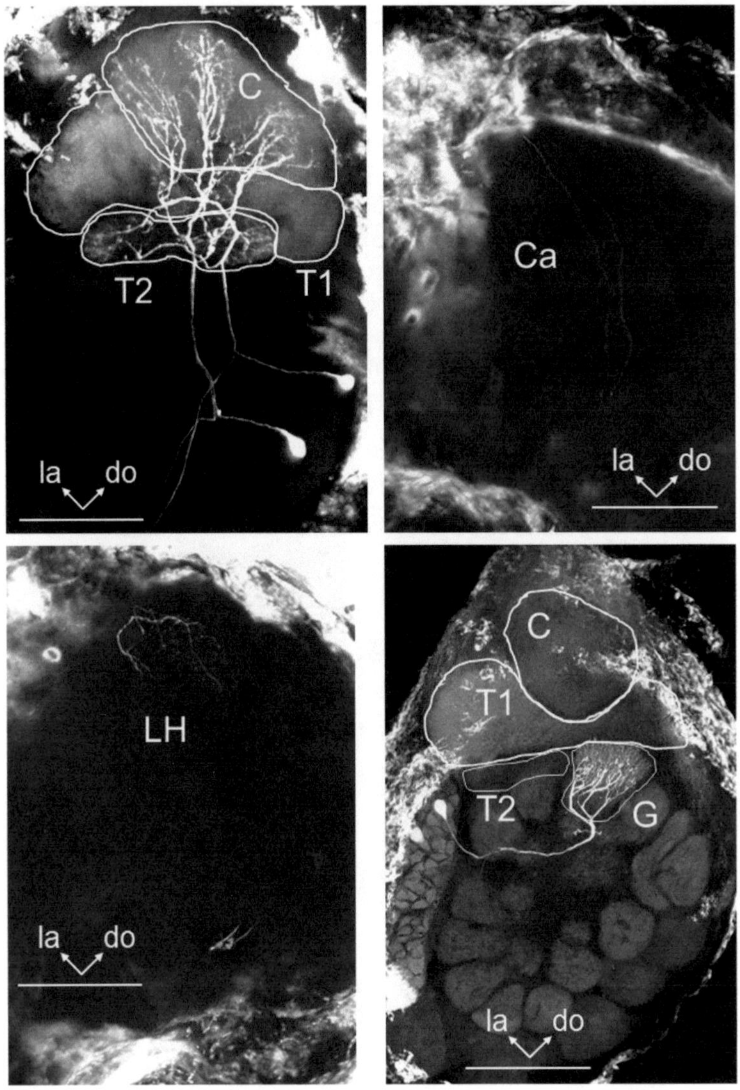

Figure 5.
Morphology (frontal view) of antennal lobe projection neurons. (a) a C15-specialist MGC-PN sent dendrites into the cumulus, while the branches of another MGC-PN were confined to the toroid-2. (b) the axons of the neurons shown in (a) left the antennal lobe and projected via the inner antenno-cerebral tract to the ipsilateral protocerebrum where they sent collaterals into the calyces (Ca) of the mushroom body and (c) terminated in the lateral horn (LH). (d) another PN showed branches in an ordinary glomerulus adjacent to the MGC. Do - dorsal, la - lateral, C – Cumulus, G – Ordinary glomerulus, T1 – Toroid-1, T2 -toroid-2. Scale bar: 100 μm.

component of the pheromone, a hexadecatrienal [8]. These neurons also respond to E,Z-11,13-pentadecadienal (C15), a chemically more stable mimic of the second key component of the sex pheromone E,E,Z-10,12,14-hexadecatrienal [10, 94] and are referred to as C15-specialist MGC-PNs.

Many AL neurons in *M. sexta* have been characterized morphologically and physiologically (**Figures 3–5**) [3, 4, 6, 11, 12, 85]. Neurophysiological studies of the pheromone-specific olfactory subsystem in male moths have focused on three properties of the sex-pheromone stimulus and on how these properties affect the central processing of sex-pheromone information [4, 12, 83]. The properties are the quality or chemical composition of the pheromone blend, the quantity or concentrations of individual pheromone components, and the temporal structure of the stimulus or its intermittency. Odor stimuli such as pheromones exist in wind plumes in the form of filaments and blobs of different concentration.

7. Conclusions

An important issue in the organization and operation of the insect olfactory system is the functional significance of glomeruli in the antennal lobes. Research on the sphinx moth *Manduca sexta* has provided a firm foundation of technical experience and knowledge about an experimentally favorable model system that allows the study of glomerular structure and function with greater precision than has been possible in other species [95–97]. Specifically, glomeruli in the olfactory subsystem of male *M. sexta*, which are designated for pheromone processing with its anatomically and functionally identified, male-specific neuropil, contribute to our understanding of the neuronal architecture and functional organization of olfactory glomeruli.

Acknowledgements

This publication resulted in part from research support from the National Science Foundation [NSF IOS-1355034], Howard University College of Medicine and the District of Columbia Center for AIDS Research, an NIH-funded program [P30AI117970], which is supported by the following NIH co-funding and participating institutes and centers: NIAID, NCI, NICHD, NHLBI, NIDA, NIMH, NIA, NIDDK, NIMHD, NIDCR, NINR, FIC, and OAR. The content is solely the responsibility of the authors and does not necessarily represent the official views of the NIH.

Conflict of interest

The author declares no conflict of interest.

Author details

Thomas Heinbockel
Department of Anatomy, Howard University College of Medicine, Washington, DC, USA

*Address all correspondence to: theinbockel@howard.edu

References

[1] Shepherd GM. Neurobiology, modules for molecules. Nature. 1992;**358**(6386):457-458

[2] Hamilton KA, Kauer JS. Intracellular potentials of salamander mitral/tufted neurons in response to odor stimulation. Brain Research. 1985;**338**(1):181-185

[3] Christensen TA, Hildebrand JG. Male-specific, sex pheromone-selective projection neurons in the antennal lobes of the moth *Manduca sexta*. Journal of Comparative Physiology. A. 1987;**160**(5):553-569

[4] Christensen TA, Heinbockel T, Hildebrand JG. Olfactory information processing in the brain: Encoding chemical and temporal features of odors. Journal of Neurobiology. 1996;**30**(1):82-91

[5] Hildebrand JG, Shepherd GM. Mechanisms of olfactory discrimination: Converging evidence for common principles across phyla. Annual Review of Neuroscience. 1997;**20**:595-631

[6] Matsumoto SG, Hildebrand JG. Olfactory mechanisms in the moth *Manduca sexta*: Response characteristics and morphology of central neurons in the antennal lobes. Proceedings of the Royal Society of London B. 1981;**213**:249-277

[7] Strausfeld NJ. Cellular organization in male-specific olfactory neuropil in the moth *Manduca sexta*. In: Elsner N, Singer W, editors. Dynamics and Plasticity in Neuronal Systems. Stuttgart, Germany: Thieme; 1989. p. 79

[8] Hansson BS, Christensen TA, Hildebrand JG. Functionally distinct subdivisons of the macroglomerular complex in the antennal lobe of the male sphinx moth *Manduca sexta*. The Journal of Comparative Neurology. 1991;**312**:264-278

[9] Christensen TA, Harrow ID, Cuzzocrea C, Randolph PW, Hildebrand JG. Distinct projections of two populations of olfactory receptor neurons in the antennal lobe of the sphinx moth *Manduca sexta*. Chemical Senses. 1995;**20**:313-323

[10] Kaissling K-E, Hildebrand JG, Tumlinson JH. Pheromone receptor cells in the male moth *Manduca sexta*. Archives of Insect Biochemistry and Physiology. 1989;**10**:273-279

[11] Homberg U, Christensen TA, Hildebrand TA. Structure and function of the deutocerebrum in insects. Annual Review of Entomology. 1989;**34**:477-501

[12] Kanzaki R, Arbas EA, Strausfeld NJ, Hildebrand JG. Physiology and morphology of projection neurons in the antennal lobe of the male moth *Manduca sexta*. Journal of Comparative Physiology. A. 1989;**165**:427±453

[13] Christensen TA, Waldrop B, Harrow ID, Hildebrand JG. Local interneurons and information processing in the olfactory glomeruli in the moth *Manduca sexta*. Journal of Comparative Physiology. A. 1993;**173**:385-399

[14] Heinbockel T, Hildebrand JG. Antennal receptive fields of pheromone-responsive projection neurons in the antennal lobes of the male sphinx moth *Manduca sexta*. Journal of Comparative Physiology. A. 1998;**183**(2):121-133

[15] Heinbockel T, Kloppenburg P, Hildebrand JG. Pheromone-evoked

potentials and oscillations in the antennal lobes of the sphinx moth *Manduca sexta*. Journal of Comparative Physiology. A. 1998;**182**(6):703-714

[16] Heinbockel T, Christensen TA, Hildebrand JG. Temporal tuning of odor responses in pheromone-responsive projection neurons in the brain of the sphinx moth *Manduca sexta*. The Journal of Comparative Neurology. 1999;**409**(1):1-12

[17] Heinbockel T, Christensen TA, Hildebrand JG. Representation of binary pheromone blends by glomerulus-specific olfactory projection neurons. Journal of Comparative Physiology. A, Neuroethology, Sensory, Neural, and Behavioral Physiology. 2004;**190**(12):1023-1037

[18] Reisenman CE, Heinbockel T, Hildebrand JG. Inhibitory interactions among olfactory glomeruli do not necessarily reflect spatial proximity. Journal of Neurophysiology. 2008;**100**(2):554-564

[19] Christensen TA, Pawlowski VM, Lei H, Hildebrand JG. Multi-unit recordings reveal context-dependent modulation of synchrony in odor-specific neural ensembles. Nature Neuroscience. 2000;**3**(9):927-931

[20] Delahunt CB, Riffell JA, Kutz JN. Biological mechanisms for learning: A computational model of olfactory learning in the *Manduca sexta* moth, with applications to neural nets. Frontiers in Computational Neuroscience. 2018;**19**:102

[21] Martin JP, Lei H, Riffell JA, Hildebrand JG. Synchronous firing of antennal-lobe projection neurons encodes the behaviorally effective ratio of sex-pheromone components in male *Manduca sexta*. Journal of Compound Physiology A. 2013;**199**(11):963-979

[22] Lei H, Reisenman CE, Wilson CH, Gabbur P, Hildebrand JG. Spiking patterns and their functional implications in the antennal lobe of the tobacco hornworm *Manduca sexta*. PLoS One. 2011;**6**(8):e23382

[23] Shepherd GM. Principles of specificity and redundancy underlying the organization of the olfactory system. Microscopy Research and Technique. 1993;**24**(2):106-112

[24] Hildebrand JG. Analysis of chemical signals by nervous systems. Proceedings of the National Academy of Sciences of the United States of America. 1995;**92**(1):67-74

[25] Dethier VG. Five hundred million years of olfaction. In: Colbow K, Allison F, Linvilles RH, editors. Wright Lectures on Olfactory Research. Burnaby, B.C., Canada: Simon Fraser University; 1990. pp. 1-37

[26] Carr WE, Gleeson RA, Trapido-Rosenthal HG. The role of perireceptor events in chemosensory processes. Trends in Neurosciences. 1990;**13**(6):212-215

[27] Anton S, Rössler W. Plasticity and modulation of olfactory circuits in insects. Cell and Tissue Research. 2021;**383**:149-164

[28] Braubach O, Croll RP. The glomerular network of the zebrafish olfactory bulb. Cell and Tissue Research. 2021;**383**:255-271

[29] Marachlian E, Klappenbach M, Locatelli F. Learning-dependent plasticity in the antennal lobe improves discrimination and recognition of odors in the honeybee. Cell and Tissue Research. 2021;**383**:165-175

[30] Mori K, Sakano H. Olfactory circuitry and Behavioral decisions.

Annual Review of Physiology. 2021;**83**:231-256

[31] Rospars JP. Invariance and sex-specific variations of the glomerular organization in the antennal lobes of a moth, Mamestra brassicae, and a butterfly, Pieris brassicae. Journal of Compound Neurology. 1983;**220**(1):80-96

[32] Strausfeld NJ. Insect vision and olfaction: Common design principles of neuronal organization. In: Singh RN, Strausfeld NJ, editors. Neurobiology of Sensory Systems. Boston, MA: Springer; 1989a

[33] Shepherd GM. Contribution toward a theory of olfaction. In: Colbow K, Allison F, Linvilles RH, editors. Wright Lectures on Olfactory Research. Simon Fraser University; 1990. pp. 61-109

[34] Purves D. Neural Activity and the Growth of the Brain. Cambridge: Cambridge University Press; 1994

[35] Fuscà D, Kloppenburg P. Odor processing in the cockroach antennal lobe—The network components. Cell and Tissue Research. 2021;**383**:59-73

[36] Paoli M, Galizia GC. Olfactory coding in honeybees. Cell and Tissue Research. 2021;**383**:35-58

[37] Wheelwright M, Whittle CR, Riabinina O. Olfactory systems across mosquito species. Cell and Tissue Research. 2021;**383**:75-90

[38] Tolbert LP, Oland LA. A role for glia in the development of organized neuropilar structures. Trends in Neurosciences. 1989;**12**(2):70-75

[39] Tolbert LP. Intercellular interactions in the constructions of olfactory glomeruli in an insect. In: Døving KB, editor. Proc Xth International Symposium on Olfaction and Taste (ISOT X). Oslo, Norway: University of Oslo; 1990. pp. 236-245

[40] Shepherd GW, Chen WR, Greer CA. Olfactory bulb. In: Shepherd GM, editor. The Synaptic Organization of the Brain. Oxford: New York, NY; 2004. pp. 165-216

[41] Tolbert LP, Hildebrand JG. Organization and synaptic ultrastructure of glomeruli in the antennal lobes of the moth*Manduca sexta*: A study using thin sections and freeze-fracture. Proceedings of the Royal Society London. 1981;**24**(3):213-279

[42] Boeckh J, Tolbert LP. Synaptic organization and development of the antennal lobe in insects. Microscopy Research and Technique. 1993;**24**(3):260-280

[43] Heimer L. The Human Brain and Spinal Cord. New York: Springer; 1983. p. 402

[44] Lodovichi C. Topographic organization in the olfactory bulb. Cell Tissue Research. 2021;**383**(1):457-472

[45] Rodrigues V, Buchner E. [3H]2-deoxyglucose mapping of odor-induced neuronal activity in the antennal lobes of Drosophila melanogaster. Brain Research. 1984;**324**(2):374-378

[46] Rodrigues V. Spatial coding of olfactory information in the antennal lobe of Drosophila melanogaster. Brain Research. 1988;**453**(1-2):299-307

[47] Boeckh J, Boeckh V. Threshold and odor specificity of pheromone-sensitive neurons in the deutocerebrum ofAntheraea pernyi andA. polyphemus (Saturnidae). Journal of Comparative Physiology. 1979;**132**:235-242

[48] Buonviso N, Chaput MA. Response similarity to odors in olfactory bulb output cells presumed to be connected to the same glomerulus: Electrophysiological study using simultaneous single-unit recordings. Journal of Neurophysiology. 1990;**63**(3):447-454

[49] Kauer JS. Contributions of topography and parallel processing to odor coding in the vertebrate olfactory pathway. Trends in Neurosciences. 1991;**14**(2):79-85

[50] Scott JW. Central processing of olfaction. The Journal of Steroid Biochemistry and Molecular Biology. 1991;**39**(4B):593-600

[51] Kauer JS, Cinelli AR. Are there structural and functional modules in the vertebrate olfactory bulb? Microscopy Research and Technique. 1993;**24**(2):157-167

[52] Shepherd GM, Firestein S. Making scents of olfactory transduction. Current Biology. 1991;**1**(4):204-206

[53] Guthrie KM, Anderson AJ, Leon M, Gall C. Odor-induced increases in c-fos mRNA expression reveal an anatomical "unit" for odor processing in olfactory bulb. Proceedings of the National Academy of Sciences of the United States of America. 1993;**90**(8):3329-3333

[54] Mori K, Shepherd GM. Emerging principles of molecular signal processing by mitral/tufted cells in the olfactory bulb. Seminars in Cell Biology. 1994;**5**(1):65-74

[55] Boeckh J, Distler P, Ernst KD, Hösl M, Malun D. Olfactory bulb and antennal lobe. In: Schild D, editor. Chemosensory Information Processing. Berlin, Heidelberg: Springer; 1990. pp. 201-228

[56] Imamura K, Mataga N, Mori K. Coding of odor molecules by mitral/tufted cells in rabbit olfactory bulb. I. Aliphatic compounds. Journal of Neurophysiology. 1992;**68**(6):1986-2002

[57] Katoh K, Koshimoto H, Tani A, Mori K. Coding of odor molecules by mitral/tufted cells in rabbit olfactory bulb.II. Aromatic compounds. Journal of Neurophysiology. 1993;**70**(5):2161-2175

[58] Lancet D. Olfaction. Exclusive Receptors. Nature. 1994;**372**(6504): 321-322

[59] Chess A, Simon I, Cedar H, Axel R. Allelic inactivation regulates olfactory receptor gene expression. Cell. 1994;**78**(5):823-834

[60] Ressler KJ, Sullivan SL, Buck LB. Information coding in the olfactory system: Evidence for a stereotyped and highly organized epitope map in the olfactory bulb. Cell. 1994;**79**(7):1245-1255

[61] Vassar R, Chao SK, Sitcheran R, Nuñez JM, Vosshall LB, Axel R. Topographic organization of sensory projections to the olfactory bulb. Cell. 1994;**79**(6):981-991

[62] Kloppenburg P, Camazine SM, Sun XJ, Randolph P, Hildebrand JG. Organization of the antennal motor system in the sphinx moth *Manduca sexta*. Cell and Tissue Research. 1997;**287**(2):425-433

[63] Sanes JR, Hildebrand JG. Structure and development of antennae in a moth,*Manduca sexta*. Developmental Biology. 1976;**51**(2):280-299

[64] Keil TA. Fine structure of the pheromone-sensitive sensilla on the antenna of the hawkmoth, Manduca sexta. Tissue Cell. 1989;**21**(1):139-151

[65] Lee JK, Strausfeld NJ. Structure, distribution and number of surface sensilla and their receptor cells on the olfactory appendage of the male moth *Manduca sexta*. Journal of Neurocytology. 1990;**19**(4):519-538

[66] Heinbockel T, Shields VD, Reisenman CE. Glomerular interactions in olfactory processing channels of the antennal lobes. Journal of Compound Physiology A. 2013;**99**(11):929-946

[67] Boeckh J, Ernst KD, Sass H, Waldow U. Anatomical and physiological characteristics of individual neurones in the central antennal pathway of insects. Journal of Insect Physiology. 1984;**30**(1):15-26

[68] Rospars JP, Hildebrand JG. Anatomical identification of glomeruli in the antennal lobes of the male sphinx moth *Manduca sexta*. Cell and Tissue Research. 1992;**270**(2):205-227

[69] Rospars JP, Hildebrand JG. Sexually dimorphic and isomorphic glomeruli in the antennal lobes of the sphinx moth *Manduca sexta*. Chemical Senses. 2000;**25**(2):119-129

[70] Kenyon FC. The brain of the bee. A preliminary contribution to the morphology of the nervous system of the arthropoda. The Journal of Comparative Neurology. 1896;**6**:133-210

[71] Ernst KD, Boeckh J, Boeckh V. A neuroanatomical study on the organization of the central antennal pathways in insects. Cell and Tissue Research. 1977;**176**(3):285-306

[72] Strausfeld NJ. Selective staining reveals complex microstructure within antennal lobe glomeruli of *Manduca sexta*.Cellular organization in male-specific olfactory neuropil in the moth *Manduca sexta*. In: Elsner N, Barth FG, editors. Sense organs: interface between environment and behavior. Stuttgart, Germany: Thieme; 1988. p. 67

[73] Masson C, Mustaparta H. Chemical information processing in the olfactory system of insects. Physiological Reviews. 1990;**70**(1):199-245

[74] Rospars JP, Chambille I. Deutocerebrum of the cockroach Blaberus craniifer Burm. Quantitative study and automated identification of the glomeruli. Journal of Neurobiology. 1981;**12**(3):221-247

[75] Stocker RF. The organization of the chemosensory system in Drosophila melanogaster: A review. Cell and Tissue Research. 1994;**275**(1):3-26

[76] Stocker RF, Lienhard MC, Borst A, Fischbach KF. Neuronal architecture of the antennal lobe in Drosophila melanogaster. Cell and Tissue Research. 1990;**262**(1):9-34

[77] Arnold G, Masson C, Budharugsa S. Comparative study of the antennal lobes and their afferent pathway in the worker bee and the drone (Apis mellifera). Cell and Tissue Research. 1985;**242**:593-605

[78] Chambille I, Masson C, Rospars JP. The deutocerebrum of the cockroach Blaberus craniifer Burm. Spatial organization of the sensory glomeruli. Journal of Neurobiology. 1980;**11**(2):135-157

[79] Duy-Jacquemin MN, Arnold G. Spatial organization of the antennal lobe in Cylindroiulus punctatus (leach) (Myriapoda: Diplopoda). International Journal of Insect Morphology and Embryology. 1991;**20**(4-5):205-214

[80] Baier H, Korsching S. Olfactory glomeruli in the zebrafish form an invariant pattern and are identifiable

across animals. The Journal of Neuroscience. 1994;**14**(1):219-230

[81] Teicher MH, Stewart WB, Kauer JS, Shepherd GM. Suckling pheromone stimulation of a modified glomerular region in the developing rat olfactory bulb revealed by the 2-deoxyglucose method. Brain Research. 1980;**194**(2):530-535

[82] Homberg U, Montague RA, Hildebrand JG. Anatomy of antenno-cerebral pathways in the brain of the sphinx moth *Manduca sexta*. Cell and Tissue Research. 1988;**254**(2):255-281

[83] Hildebrand JG, Christensen TA, Arbas EA, Hayashi JH, Homberg U, Kanzaki R, et al. Olfaction in *Manduca sexta*: Cellular mechanisms of responses to sex pheromone. In: Duce IR, editor. Proc NEUROTOX 91 – Molecular Basis of Drug & Pesticide Action. London: Elsevier Applied Science; 1992. pp. 323-338

[84] Hoskins SG, Homberg U, Kingan TG, Christensen TA, Hildebrand JG. Immunocytochemistry of GABA in the antennal lobes of the sphinx moth *Manduca sexta*. Cell and Tissue Research. 1986;**244**(2):243-252

[85] Waldrop B, Christensen TA, Hildebrand JG. GABA-mediated synaptic inhibition of projection neurons in the antennal lobes of the sphinx moth, Manduca sexta. Journal of Compound Physiology A. 1987;**161**(1):23-32

[86] Kent KS, Hoskins SG, Hildebrand JG. A novel serotonin-immunoreactive neuron in the antennal lobe of the sphinx moth *Manduca sexta* persists throughout postembryonic life. Journal of Neurobiology. 1987;**18**(5):451-465

[87] Sun XJ, Tolbert LP, Hildebrand JG. Ramification pattern and ultrastructural characteristics of the serotonin-immunoreactive neuron in the antennal lobe of the moth *Manduca sexta*: A laser scanning confocal and electron microscopic study. The Journal of Comparative Neurology. 1993;**338**(1):5-16

[88] Sanes JR, Hildebrand JG. Acetylcholine and its metabolic enzymes in developing antennae of the moth,*Manduca sexta*. Developmentaal Biology. 1976;**52**(1):105-120

[89] Homberg U. Immunocytochemical demonstration of transmitter candidates in the central olfactory pathways in the sphinx moth *Manduca sexta*. In: Døving KB, editor. Proc Xth International Symposium on Olfaction and Taste (ISOT X). Oslo, Norway: University of Oslo; 1990. pp. 151-158

[90] Sanes JR, Prescott DJ, Hildebrand JG. Cholinergic neurochemical development of normal and deafferented antennal lobes during metamorphosis of the moth. Brain Research. 1977;**119**(2):389-402

[91] Maxwell GD, Tait JF, Hildebrand JG. Regional synthesis of neurotransmitter candidates in the CNS of the moth *Manduca sexta*. Compound Biochemistry Physiology C. 1978;**61C**(1):109-119

[92] Hildebrand JG, Hall LM, Osmond BC. Distribution of binding sites for 125I-labeled alpha-bungarotoxin in normal and deafferented antennal lobes of *Manduca sexta*. Proceedings of the National Academy of Sciences of the United States of America. 1979;**76**(1):499-503

[93] Waldrop B, Hildebrand JG. Physiology and pharmacology of acetylcholinergic responses of interneurons in the antennal lobes of the moth*Manduca sexta*. Journal of Comparative Physiology. 1989;**164**:433-441

[94] Tumlinson JH, Brennan MM, Doolittle RE, Mitchell ER, Brabham A,

Mazomenos BE, et al. Identification of a pheromone blend attractive to *Manduca sexta* (L.) males in a wind tunnel. Archives of Insect Biochemistry and Physiology. 1989;**10**(4):255-271

[95] Martin JP, Beyerlein A, Dacks AM, Reisenman CE, Riffell JA, Lei H, et al. The neurobiology of insect olfaction: Sensory processing in a comparative context. Progress in Neurobiology. 2011;**95**(3):427-447

[96] Riffell JA, Hildebrand JG Adaptive processing in the insect olfactory system. In: von der Emde G, Warrant E (eds) The Ecology of Animal Senses. Springer; 2016. pp. 3-24

[97] Lei H, Oland LA, Riffell JA, Beyerlein A, Hildebrand JG. Implications from microcircuits of a moth antennal lobe for olfactory information processing [updated version]. In: Shepherd G, Grillner S, editors. Handbook of Brain Microcircuits. Oxford University Press; 2018. pp. 333-344

Chapter 5

Quantitative Electroencephalography for Probing Cognitive and Behavioral Functions of the Human Brain

Richard M. Millis, Merin Chandanathil, Ayoola Awosika, Fidelis Nwachukwu, Ravindrasingh Rajput, Sheetal Naik and Kishan Kadur

Abstract

Previous studies have shown that quantitative electroencephalography (qEEG) provides measures of brain wave voltage and symmetry within each of the standard bandwidths. These qEEG measures are neurophysiological correlates of brain wave signatures for various aspects of cognition and behavior and are susceptible to neurofeedback training for improving human performance. Using exam scores and an individualized self-inventory (ISI) of psychosocial interactions, we provide unique data for probing behavioral and cognitive performance of medical students. Increments in voltage within the standard theta (4–7 Hz) and beta (15–20 Hz) frequencies and decrements in the theta–beta ratio (TBR) suggest improvements in attentional control. Associations between right-sided frontal alpha asymmetry (fAA) and ISI scores for negative self-perceptions suggest a novel qEEG signature for emotional balance. These findings suggest that changes in qEEG voltages and asymmetries may be predictive of improvements in attentional control, cognitive performance, and psychosocial skills, as well as serving as surrogate markers for neurofeedback training-related changes in neuroplasticity.

Keywords: electroencephalography, academic performance, psychosocial interactions, theta–beta ratio, frontal alpha asymmetry

1. Introduction

Academic learning requires a person to interact with other individuals [1]. Depression and anxiety may be good predictors of learning disabilities and academic underachievement [2] probably because the emotional state is a marker for an individual's ability to manage their psychosocial interactions [3]. There are numerous examples of brilliant persons with high intelligence quotients who do not do anything meaningful with their intelligence because of difficulties with psychosocial interactions. Such persons may be deficient in emotional intelligence or balance [4]. For medical students, overall health and wellness appear to be correlated with their academic performance [5]. Academic achievement in medical school is a good example of a situation wherein stress can unmask latent mental or emotional imbalances which, in turn, lead to academic underachievement or failure [6]. Previous studies from our laboratory suggest that quantitative electroencephalographic (qEEG) measures of theta–beta ratio (TBR) and frontal alpha asymmetry (fAA) may be useful neurophysiological correlates of academic achievement and negative perceptions of psychosocial interactions in medical students [7–10]. TBR is reported to be a qEEG marker for a person's capacity to focus their attention on salient information [11]. Between a control human structure–function course introductory exam and a comparison structure–function exam on different topics, we have reported significant increments in voltage within the standard theta and beta frequencies combined with a significant decrement in TBR [7]. These findings were associated with no significant changes in the magnitude of voltages in the standard delta and alpha bandwidths and, therefore, suggest an overall increase in attention control for our pilot study cohort. Frontal (F8–F7) alpha asymmetry (fAA) is reported to be a qEEG marker for negative emotions. In our pilot study cohort, we previously reported a significant negative correlation between the magnitude of F8–F7 frontal alpha asymmetry and "depressed" score on an individualized self-inventory (ISI) of the cohort's perceptions of their psychosocial interactions, which purported to be indicative of negative perceptions of themselves [10].

This chapter is intended as a primer for neurophysiological evaluation of qEEG brain maps and demonstrates how qEEG technology is becoming a useful tool for probing the human brain. The relatively inexpensive qEEG brain maps showing electrical activity are analogous to the much more expensive functional magnetic resonance imaging (fMRI) maps; therefore, validating qEEG as the "poor man's" fMRI. We will demonstrate the utility of qEEG by interpreting the qEEG maps of individual medical students exhibiting the highest and lowest TBRs and fAAs.

2. Methodology

The methods and protocols described herein were approved by the American University of Antigua College of Medicine Research Council which served as the University's Institutional Review Board (IRB). The study subjects provided their informed consent. A total of 10 male subjects were recruited; 1 subject discontinued the study due to his ill health. Females were excluded because of the potential confounding influence of hormonal changes associated with their menstrual cycles. Nine individuals underwent 5–10 minutes of eyes-closed (EC) qEEG measurements at 19 standard electrode sites [12–20]. The Brain Master Discovery System (Brain Master Technologies, Inc., Bedford, OH, USA) was used to take the qEEG readings 3 days before each of the first two summative block examinations covering standard first-semester integrated basic science courses. For the purposes of this chapter, the qEEG voltage brain maps were selected for the subjects with the highest and lowest exam scores to demonstrate the spectrum of changes found to be associated with academic achievement.

The qEEG measurements were performed with subjects seated comfortably with their eyes-closed, in a dimly lit room. After manual editing with the New Mind Maps online editing tool (New Mind Technologies, Roswell, GA, USA), the mean ± SD of the voltage amplitudes, stated in μV, and the mode frequency in each bandwidth, expressed in Hz, were measured for the following standard qEEG frequencies: delta, theta, alpha, and beta. The average theta-beta ratio of the voltages recorded at each of the 19 standard electrode sites was computed. All qEEG recordings were performed at the same time in the morning, after overnight fasting to avoid confounding factors related to food ingestion and metabolism. The subjects refrained from recreational drugs such as alcohol, marijuana, nicotine, and caffeine for the 24 hours before the qEEG recording session. None of the patients reported using prescription or recreational drugs within the previous month, according to self-report. Preliminary results suggest that eye opening, which is known to inhibit alpha brainwave voltage amplitude, resulted in changes in alpha voltage that were not uniformly reproducible. The qEEG measurements were therefore only interpreted for the closed-eye condition. Frontal alpha asymmetry (fAA) was computed from the mean alpha voltages at the frontal recording sites F7 (left) and F8 (right) as follows: ([F8 – F7]/F8 × 100). The inferior frontal gyrus is an area where "mirror neurons," hypothesized to process information about psychosocial interactions have been identified [21]. Negative asymmetry values are indicators of right-sided, nondominant hemispheric alpha asymmetry resulting from greater activation of the right frontal cortex. Negative fAA was defined as a qEEG recording where the average alpha voltage at the left frontal F7 scalp electrode was greater than the voltage at the symmetrical right frontal F8 scalp electrode. Positive asymmetry values are indicators of left-dominant hemisphere alpha asymmetry. Positive fAA was defined as qEEG where the mean alpha voltage at the right frontal F8 scalp electrode was greater than the voltage at the symmetrical left frontal F7 scalp electrode.

Within 8 hours of qEEG measurement, each study subject completed an Interactive Self-Report Inventory (ISI, New Mind Technologies, Roswell GA, USA) online at the New Mind Maps website (https://www.newmindmaps.com). For the purposes of this chapter, the ISI scoring of "depressed" psychosocial interactions was used to demonstrate the correlation and potential utility of the fAA measurement. Depressed individuals are expected to have had negative thoughts about themselves based on a correlation coefficient > 0.8 between relevant inventory items and the Beck's Depression Inventory (New Mind Technologies, Roswell GA, USA). For the purposes of this chapter, the qEEG asymmetry brain maps were selected for the subjects with the highest and lowest ISI "depressed" scores to demonstrate the spectrum of changes found to be associated with a subject's negative perceptions of their psychosocial interactions.

3. Results

3.1 Theta–beta ratio

Theta–beta ratio (TBR) is reported to be a qEEG marker for a person's capacity to focus their attention on salient information. For this cohort, between the control block 1 and the comparison block 2 human structure–function course examinations, we have reported significant increments in voltage within the standard theta and beta EEG frequencies combined with a significant decrement in the theta–beta ratio. (TBR) These findings were associated with no significant changes in the magnitude of voltages in the standard delta and alpha bandwidths and, therefore, suggest an overall increase in attention control for the cohort.

Figure 1 presents qEEG maps comparing the control exam (upper panel) and comparison exam (lower panel) measures of the voltages from the subject earning the highest block 2 exam score of 90%.

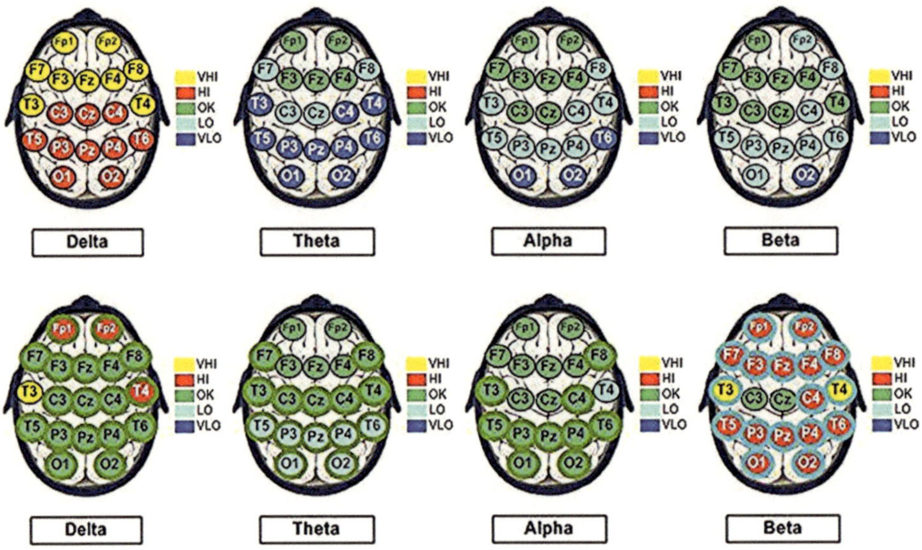

Figure 1.
qEEG voltage maps for the subject with the highest exam score. Green circles indicate the scalp recording sites where the voltages are in the normal reference range (OK = normal) for each of the standard delta (1–3 Hz), theta (4–7 Hz), alpha (8–12 Hz), and beta (15–20 Hz) EEG bandwidths. Light blue (LO = low) and dark blue (VLO = very low) circles indicate voltages 1 and 2 standard deviations below the normal reference range; red (HI = high) and yellow (VHI = very high) circles are voltages 1 and 2 standard deviations above the normal reference range, respectively. Upper panel: eyes-closed control map recorded 3 days before the first control exam covering introductory material taught in the subject's first structure–function course exam. Lower panel: eyes-closed comparison map recorded 3 days before the comparison exam covering material taught in the subject's second human structure–function course exam, 5 weeks after the control recording. Theta and beta voltages are used to compute theta–beta ratio, a qEEG marker for attentional control.

3.2 Theta bandwidth changes for the subject with the highest exam score

Five recording sites exhibit no changes from normal theta frequencies coded green, four sites show theta increases from 1 standard deviation below normal (light blue) to normal (green), seven sites show theta increases from 2 standard deviations below normal (dark blue) to 1 standard deviation below normal (light blue), and three sites show theta increases from 2 standard deviations below normal (dark blue) to normal (green). These findings are indicative of an overall increase in theta voltage for the subject with the highest exam score. The predominant change in theta is observed at 14 of 19 sites wherein increases in theta voltage are found.

3.3 Beta bandwidth changes for the subject with the highest exam score

Two recording sites exhibit no changes from normal beta frequencies coded green, five sites show beta increases from normal (green) to 1 standard deviation above normal (red), nine sites show beta increases from 1 standard deviation below normal (light blue) to 1 standard deviation above normal (red), one site shows a beta increase from 2 standard deviations below normal (dark blue) to 1 standard deviation above normal (red), and two sites shows beta increases from normal (green) to 2 standard deviations above normal (yellow). These findings are indicative of an overall increase in beta voltage for the subject with the highest exam score. The predominant change in beta is observed at 12 of 19 sites wherein increments in beta voltage are found.

3.4 Theta-beta ratio for the subject with the highest exam score

The theta and beta voltage changes resulted in larger increases in beta than in theta voltage, thereby decreasing the theta-beta ratio (TBR) by 11.7% from 0.93 to 0.84. This range of TBR indicates that the subject with the highest exam score is functioning with 7%–16% more beta than theta bandwidth voltage.

Figure 2 shows representative qEEG maps comparing the control block 1 (upper panel) and comparison block 2 (lower panel) voltage measurements from the subject earning the lowest block 2 exam score of 48%.

3.5 Theta bandwidth changes for the subject with the lowest exam score

One recording site shows no change in theta voltage from normal (green), five sites show no changes in theta voltage from 1 standard deviation below normal (light blue), nine sites show no changes in theta voltage from 2 standard deviations below normal (dark blue), and four sites show decreases in theta voltage from normal (green) to 1 standard deviation below normal (light blue). These findings indicate minimal changes in theta voltage for the subject with the lowest exam score. The predominant change in theta is observed at 4 of 19 sites wherein decrements from normal to 1 standard deviation below normal are found.

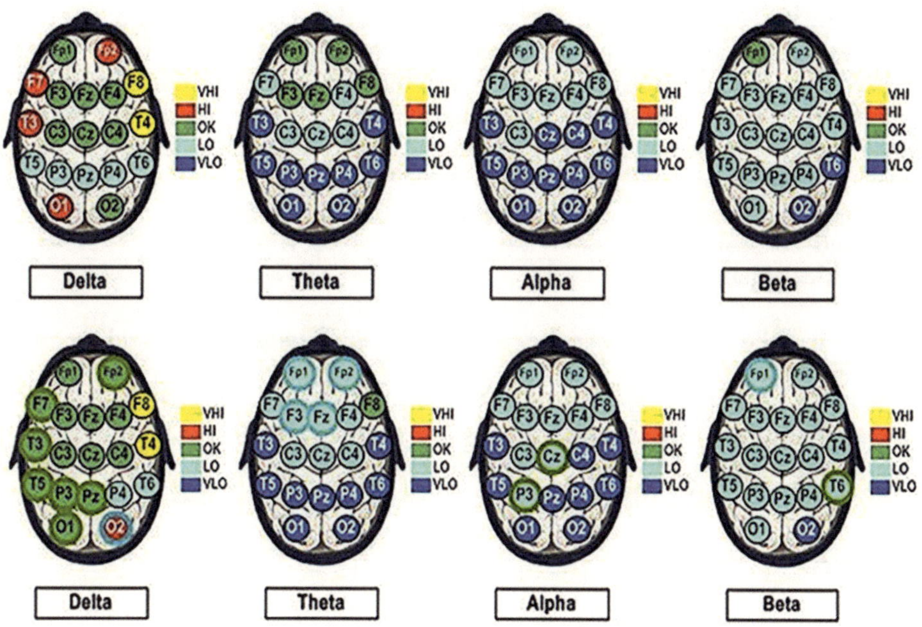

Figure 2.
qEEG brain maps for the subject with the lowest exam score. Green circles indicate the scalp recording sites where the voltages are in the normal reference range (OK = normal) for each of the standard delta (1–3 Hz), theta (4–7 Hz), alpha (8–12 Hz), and beta (15–20 Hz) EEG bandwidths. Light blue (LO = low) and dark blue (VLO = very low) circles indicate voltages 1 and 2 standard deviations below the normal reference range; red (HI = high) and yellow (VHI = very high) circles are voltages 1 and 2 standard deviations above the normal reference range, respectively. Upper panel: eyes-closed control map recorded 3 days before the first control exam covering introductory material taught in the subject's first structure–function course exam. Lower panel: eyes-closed comparison map recorded 3 days before the comparison exam covering material taught in the subject's second human structure–function course exam, 5 weeks after the control recording.

3.6 Beta bandwidth changes for the subject with the lowest exam score

A total of 16 recording sites show no change in beta voltage from 1 standard deviation below normal (light blue), one site shows no change in beta voltage from 2 standard deviations below normal (dark blue), one site shows a decrease in beta voltage from normal (green) to 1 standard deviation below normal (light blue), and one site shows an increase in beta voltage from 2 standard deviations below normal to 1 standard deviation below normal. These findings also indicate minimal changes in beta voltage for the subject with the lowest exam score. The predominant change in beta is observed at 16 of 19 sites wherein no changes from 1 standard deviation below normal are found.

3.7 Theta-beta ratio for the subject with the lowest exam score

As for the subject with the highest exam score, the theta and beta voltages in this subject with the lowest exam score also resulted in larger increases in beta than in theta voltage, thereby decreasing the theta-beta ratio (TBR) by 16.1%, from 1.55 to 1.30. This range of TBR indicate that the subject with the lowest exam score is functioning with 30%–55% more theta than beta bandwidth voltage.

3.8 Frontal alpha asymmetry

Frontal (F8–F7) alpha asymmetry (fAA) is reported to be a qEEG marker for negative emotions. In the same cohort, we previously reported significant negative correlation between the magnitude of F8–F7 frontal alpha asymmetry and "depressed" score on an individualized self-inventory (ISI), purported to be indicative of negative perceptions of a person's psychosocial interactions. No significant correlations were found between the ISI and the exam scores.

Figure 3 shows representative qEEG maps from the subject with the highest "depressed" ISI score (upper panel) and from the subject with the lowest "depressed" ISI score (lower panel).

3.9 Frontal alpha asymmetry in the subject with the highest "depressed" score

F8–F7 fAA in the subject with the highest "depressed" score shows F8 voltage 1 standard deviation above normal (coded red) combined with F7 voltage 1 standard deviation below normal (coded blue). There is no normal asymmetry in this map for any of the qEEG bandwidths because none of the symmetrical right–left recording sites are less than 1 standard deviation from each other, considered equal, even or no asymmetry.

3.10 Frontal alpha asymmetry in the subject with the lowest "depressed" score

F8–F7 fAA in the subject with the lowest "depressed" score shows F8 voltage 1 standard deviation below normal (coded blue) combined with F7 voltage 1 standard deviation below normal (coded red). Normal in this map is shown for the theta voltage at recording sites T4–T3 wherein both voltages are less than 1 standard deviation from each other, considered equal or even asymmetry. It is noteworthy that the F8–F7 voltages for this subject with the lowest "depressed" ISI score are reverse of those in the subject with the highest "depressed" score. This pattern is indicative of right-sided

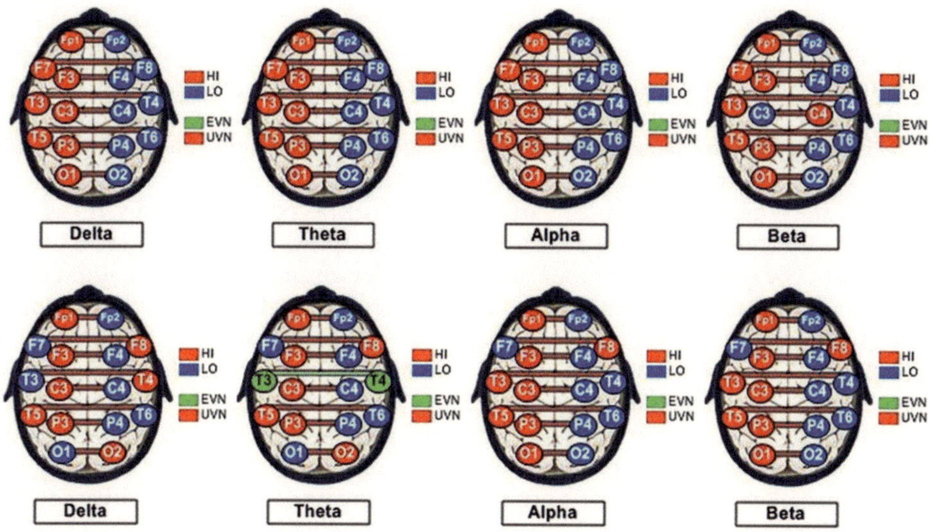

Figure 3.
qEEG symmetry maps for the subjects with the highest and lowest "depressed" ISI scores. Green circles joined by horizontal lines indicate the symmetrical homologous right and left scalp recording sites where the voltages are ≤0.5 μV (EVN = even) for each of the standard delta (1–3 Hz), theta (4–7 Hz), alpha (8–12 Hz), and beta (15–20 Hz) EEG bandwidths. Blue (LO = low) and red (HI = high) circles joined by horizontal red lines (UVN = uneven) indicate the recording sites where the voltages are >0.5 μV lower and higher than the symmetrical homologous site, respectively. Upper panel: eyes-closed map recorded on the same day as completion of the Individualized Self Inventory (ISI) for the subject with the highest "depressed" ISI score. Lower panel: eyes-closed map recorded on the same day as completion of the ISI for the subject with the lowest "depressed" score. The highest and lowest "depressed" ISI scores are indicative of increased and decreased negative perceptions of a person's psychosocial interactions, respectively.

fAA in the subject with the highest "depressed" score and left-sided fAA in the subject with the lowest "depressed" score.

4. Discussion

4.1 Neurophysiological markers for academic underachievement

Most people believe that academic underachievement is caused by low intelligence, motivation, or social background [22–24]. This view fails to explain the substantial proportion of extremely bright, driven, and affluent individuals who display academic underachievement. Academic underachievement has long been hypothesized to be caused by inter-individual variations in preferred learning styles (such as auditory vs. visual learning). For instance, it has been demonstrated that in a first-year chemistry course at an Australian institution, introverted Myers-Briggs Personality Type students outperform extroverted students [25]. Students with reflective personalities and visual learning style are reported to have the best academic performance in an ophthalmology course for fifth-year students in a Chilean medical school [26]. Although learning style preferences are hard to measure, most curricula demand that students successfully use a variety of learning methods [25].

To date, research on optimizing academic performance has encountered an inability to translate what is known about learning style preferences to how effectively students use the critical nodes and hubs in their cerebral cortex for learning. This barrier to learning outcomes research has been effectively overcome by the advent of computer-based technologies for measuring electrical and metabolic functions of the cerebral cortex such as qEEG, functional MRI (fMRI), and psychometric testing. While fMRI provides information about metabolic activity in the brain, quantitative qEEG is primarily a measure of electrical activity. Due to the tight coupling between metabolic and electrical signals in the brain [27], each can be thought of as a surrogate for the other. In that context, an inexpensive qEEG record can be useful as a surrogate for a very expensive fMRI recording [28]. qEEG has been validated by the US Food & Drug Administration as a medical device and diagnostic method for identifying children diagnosed with attention deficit disorder (ADD) and attention deficit hyperactivity disorder (ADHD) syndromes (FDA News Release July 15, 2013). qEEG has also been shown to be useful in selecting children and adults who are likely to respond to psychostimulant treatment [11–13]. qEEG is also increasingly used for neurofeedback training in sports and academia. qEEG profiles indicative of improved athletic performance in major league baseball players [29], and Olympic athletes with performance decline following injury have been reported [14]. qEEG-based neurofeedback training has also been shown to be effective in improving the neurosurgery skills of ophthalmic microsurgeons [15]. A relatively specific qEEG signature indicates working memory deficits in low-achieving high school students, compared with high-achieving students [16]. Specific changes in qEEG, indicating increased brain performance, also as a result of certain yoga practices [17, 18] suggest the ability of yoga training to produce improvements in academic performance.

4.2 The problem of academic underachievement in medical school

Concerns about academic under-preparedness associated with the transition from the preprofessional phase to the professional phase of undergraduate medical education are likely to be the same as those associated with transition from the preclinical to the clinical phase [30]. A key feature of low academic achievement in the first two preclinical years of medical school appears to be an underappreciation of the volume and complexity of the information needed to be learned. Students are often hampered by not having developed a systematic method for handling complex information during their preprofessional undergraduate training. A "tried-and-true," albeit highly individualized, method of mastering complex information is usually needed to provide the impetus for developing confidence in test taking skills and for meeting the challenges of translating complex basic science knowledge into evaluation, problem solving and differential diagnosis. The outcome of this difficult transition is twofold: (i) a large number of students exhibit low academic achievement on formative and summative examinations designed to test medical information processing skills and substandard performance on standardized (NBME, USMLE Step 1) examinations designed to evaluate readiness for entry into the clinical phase of undergraduate medical training; and (ii) many students exhibit high academic performance on their formative and summative course examinations but low performance on the standardized NBME and USMLE. We have previously reported that TBR, a measure of attentional control, is negatively correlated with academic achievement in medical students [7–9]. In this chapter, we depict and interpret the qEEG brain maps demonstrating the spectrum of deviations from the normative reference values for qEEG theta and beta voltages and higher versus lower TBR seemed to identify students exhibiting neurophysiological and cognitive deviations from the norms, which could explain their underachievement. To our knowledge, this is the first report depicting qEEG brain maps of individuals at two ends of a spectrum of academic performance on a medical school exam. The significance of the qEEG maps is quite obvious from the global changes in color-coded range of voltages shown in **Figure 1** and **2**. The map of the subject with the highest exam score shows colors indicative of substantial increases in both their average theta frequency and their average beta frequency voltages **Figure 1**. The map of the subject with the lowest exam score shows colors indicative of very little change in their theta and beta voltages (**Figure 2**). We speculate that these color-coded voltage changes suggest that the subject with the highest exam score might have improved neural plasticity, compared to the subject with the lowest exam score. We envisage such usage of qEEG as a putative marker for identifying individuals who might benefit from neurofeedback or other types of brain training and educational counseling. Such targeted interventions might be expected to change the qEEG brain maps and neural plasticity associated with studying for a high-stakes medical school examination, as described in this chapter, or preparing for other rigorous academic challenges.

4.3 Evaluating the brain networks involved in facilitating complex learning

The advent of qEEG and fMRI technologies have led to the discovery of important correlations in the electrical and metabolic profiles between areas of the cerebral cortex considered to be the main nodes and hubs for learning and memory. The key networks are the cingulate, arcuate and uncinate. In this study, the qEEG data were interpreted by measuring, within each of the EEG brain wave bandwidths (beta, alpha, theta, delta), magnitude (electrical power), dominant frequency and coherence between recording sites. We have successfully used alpha and beta coherences to demonstrate correlations between the amount of communication between the cortical tissue in the vicinity of each interhemispheric recording site, a measure of network development, integrity, and function in this same cohort of medical students with apparent attentional dysregulation and academic underachievement [9]. Our study is corroborated by research demonstrating correlations between the brain's executive functions and the frontal theta, alpha and beta interhemispheric coherences of 168 Iranian university students in their twenties [31], as well as by a metanalysis suggesting that the brain's executive function is a positive predictor of academic performance in primary school children [32].

4.4 The cingulate network in complex learning

The cingulate network is also known as the dorsal pathway for cognition, and activity in this network represents the cerebral cortex's neutral or idling gear. The cingulate network may be the equivalent of a computer's default mode network, involving areas that are operating when a person's eyes are closed and in a state of "day-dreaming" or not involved in deep thought and problem-solving [33]. Linkage of electrical magnitude (voltage), dominant frequency, and coherence in the prefrontal (Fp1, Fp2), frontal (F3, F4, F7, F8), central (C3, Cz, C4), parietal (P3, Pz, P4) and occipital (O1, O2) electrodes is thought to reflect activity of the cingulate default mode network. The cingulate network is named for the anterior and posterior cingulate cortical areas, the anterior involved in high-level information processing of socialization, empathy, outcome/error monitoring and action planning. Error-related negative emotional responses are found to be diminished in individuals with anterior cingulate lesions [34]. The posterior cingulate is linked to interpreting emotional salience and both the anterior and posterior cingulate have strong connectivity to the insular cortex, the main area for integrating and interpreting interoceptive responses [35]. The cingulate network appears to contain critical nodes and hubs for motivating emotional aspects of learning and memory [36]. Mastery of a lengthy, rigorous medical curriculum may require normal range of electrical activity in the cingulate network to support empathy-motivated and self-corrective learning paradigms. qEEG deviation from the norm in the cingulate network could, therefore, provide a key signature for academic underachievement in a medical curriculum. We have previously reported that right-sided fAA, a measurement of negatively valenced emotions, is positively correlated with negative self-perceptions of the psychosocial interactions among medical students [10]. In this chapter, we show and interpret the qEEG brain maps demonstrating a spectrum of fAA correlating with negative self-perceptions. We also provide the first depiction in the scientific literature of qEEG brain maps from subjects at both ends of an emotional scale from high to low "depressed" ISI score and from high to low nondominant hemispheric, right-sided fAA. These findings should be interpreted cautiously because of a report that there was no correlation between these variables in a robust multiverse analysis of five "clinically depressed" populations [37].

4.5 The hippocampus in complex learning

The hippocampus is a gray structure in the center of the brain and is necessary for normal learning, memory, mood, and emotion [38]. Learning and memory are also highly dependent on neurogenesis. The adult brain generates new brain cells at the rate of 700 neurons per day in the hippocampus and by the age of 50, humans are thought to exchange the entire population of neurons with which they were born [39]. In adult laboratory animals, blocking brain neurogenesis limits the animal's ability to navigate the environment, a function highly dependent on working memory [40] and blocking neurogenesis also results in depression and the inability of antidepressant medications to work [41]. Brain-derived neurotrophic factor (BDNF) is thought to be the main stimulator of neurogenesis in the human brain [42]. Aerobic exercise, learning and sexual activities, 20%–30% calorie restriction, diets high in flavonoids (e.g., curcumin), resveratrol (e.g., blueberries, grapes, dark chocolate, and red wine), omega-3 fatty acids (e.g., walnuts, fatty fish such as salmon), folic acid and zinc are known to increase BDNF and neurogenesis, whereas diets high in saturated fats, sugars and ethanol, vitamins A, B and E deficiencies, sleep deprivation, stress, aging, inflammation, as well as exposure to high plasma levels of cortisol decrease BDNF and neurogenesis [43]. We have previously shown robust correlations between hippocampal neurogenesis and physical activity, maze learning and environmental enrichment in normal healthy rats, as well as in rats recovering from kainite-induced epileptic seizures [44–46]. The results of these studies support the notion that interventions which increase hippocampal neurogenesis are also likely to increase the cognitive learning and memory functions of the human brain [47]. Studying the effects of hippocampal neurogenesis was beyond the scope of the qEEG studies reviewed herein. However, modafinil, a drug known to stimulate neurogenesis, is reported to decrease qEEG voltage, across all the standard EEG frequencies [48]. This finding of a limitation on qEEG voltage suggests that in the presence of active neurogenesis, we may not be able to observe the large global changes in theta and beta power indicative of the putative improvement in neural plasticity depicted in the (**Figure 1**) brain map from the subject with the highest exam score.

5. Conclusion

This chapter introduces qEEG brain map interpretation in newly matriculated medical students transitioning from undergraduate science to the preclinical basic medical science phase of their medical training exhibiting a spectrum from high to low theta-beta ratio (TBR), a putative measure of attentional control, and in students showing a spectrum of right-sided and left-sided frontal alpha asymmetry (fAA), a putative measure of negative emotions. The qEEG changes in TBR are highly correlated with academic performance on a first-semester human structure–function (anatomy-physiology) exam and the qEEG changes in fAA are highly correlated with "depressed" scores on an individualized self-inventory of their psychosocial interactions. These brain maps suggest that changes in qEEG voltages and asymmetries may be predictive of changes in attentional control, cognitive performance, and psychosocial skills and may serve as surrogate markers for neurofeedback training related changes in neuroplasticity and in cognitive learning and memory functions of the human brain.

Acknowledgements

This research is supported by a grant from the American University of Antigua College of Medicine. The authors are grateful for the expert technical assistance of Dr. Vasavi R. Gorantla for performing the qEEG recordings.

Author details

Richard M. Millis*, Merin Chandanathil, Ayoola Awosika, Fidelis Nwachukwu, Ravindrasingh Rajput, Sheetal Naik and Kishan Kadur
Department of Pathophysiology, American University of Antigua College of Medicine, Antigua, Antigua & Barbuda, West Indies

*Address all correspondence to: rmillis@auamed.net

References

[1] Hurst B, Wallace R, Nixon SB. The impact of social interaction on student learning. Reading Horizons: A Journey of Literacy and Language Arts. 2013;**52**(3):375-398. Available from: https://scholarworks.wmich.edu/reading_horizons/vol52/iss4/5

[2] Humensky J, Kuwabara SA, Fogel J, Wells C, Goodwin B, Van Voorhees BW. Adolescents with depressive symptoms and their challenges with learning in school. The Journal of School Nursing. 2010;**26**:377-392

[3] Lane RD, Smith R. Levels of emotional awareness: Theory and measurement of a socio-emotional skill. Journal of Intelligence. 2021;**9**(3):42. DOI: 10.3390/jintelligence9030042

[4] Hunt J, Fitzgerald M. The relationship between emotional intelligence and transformational leadership: An investigation and review of competing claims in the literature. American International Journal of Social Science. 2013;**2**:30-38

[5] Haight SJ, Chibnall JT, Schindler DL, Slavin SJ. Associations of medical student personality and health/wellness characteristics with their medical school performance across the curriculum. Academic Medicine. 2012;**87**:476-485

[6] van Batenburg-Eddes T, Jolles J. How does emotional wellbeing relate to underachievement in a general population sample of young adolescents: A neurocognitive perspective. Frontiers in Psychology. 2013;**4**:673. DOI: 10.3389/fpsyg.2013.00673

[7] Gorantla VR, Parsons G, Sayed E, Fadel S, Olukoga C, Volkova YA, et al. Electroencephalographic correlates of brain adaptations to medical school academic challenges. Journal of Clinical and Diagnostic Research. 2018;**12**(4):11388. DOI: 10.7860/JCDR/2018/32222.11388

[8] Gorantla VR, Bond V Jr, Dorsey J, Tedesco S, Kaur T, Simpson M, et al. qEEG measures of attentional and memory network functions in medical students: Novel targets for pharmacopuncture to improve cognition and academic performance. Journal of Pharmacopuncture. 2019;**22**(3):166-170. DOI: 10.3831/KPI.2019.22.022

[9] Gorantla VR, Tedesco S, Chandanathil M, Maity S, Bond V Jr, Lewis C, et al. Associations of alpha and beta interhemispheric EEG coherences with indices of attentional control and academic performance. Behavioural Neurology. 2020;**2020**:4672340. DOI: 10.1155/2020/4672340

[10] Millis RM, Arcaro J, Palacios A, Millis GL. Electroencephalographic signature of negative self perceptions in medical students. The Cureus Journal of Medical Science. 2022;**14**(2):e22675. DOI: 10.7759/cureus.22675

[11] Nuwer MR, Buchhalter J, Shepard KM. Quantitative EEG in attention-deficit/hyperactivity disorder: A companion payment policy review for clinicians and payers. Neurology Clinical Practice. 2016;**6**(6):543-548. DOI: 10.1212/CPJ.0000000000000308

[12] White JN, Hutchens TA, Lubar JF. Quantitative EEG assessment during neuropsychological task performance in adults with attention deficit hyperactivity disorder. Journal of Adult Development. 2005;**12**(2):113-121

[13] Sangal RB, Sangal JM. Use of EEG beta-1 power and theta/beta ratio over Broca's area to confirm diagnosis of attention deficit/hyperactivity disorder in children. Clinical EEG and Neuroscience. 2015;**46**(3):177-182

[14] Snyder SM, Rugino TA, Hornig M, Stein MA. Integration of an EEG biomarker with a clinician's ADHD evaluation. Brain and Behavior: A Cognitive Neuroscience Perspective. 2015;**5**:e00330. DOI: 10.1002/brb3.330

[15] Graczyk M, Pachalska M, Ziółkowski A, Manko G, Łukaszewska B, Kochanowicz K, et al. Neurofeedback training for peak performance. Annals of Agricultural and Environmental Medicine. 2014;**21**(4):871-875

[16] Ros T, Moseley MJ, Bloom PA, Benjamin L, Parkinson LA, Gruzelier JH. Optimizing microsurgical skills with EEG neurofeedback. BMC Neuroscience. 2009;**10**:87. DOI: 10.1186/1471-2202-10-87

[17] Aguirre-Pérez DM, Otero-Ojeda GA, Pliego-Rivero FB, Ferreira-Martínez AA. Relationship of working memory and EEG to academic performance: A study among high school students. International Journal of Neuroscience. 2007;**117**(6):869-882

[18] Field T, Diego M, Hernandez-Reif M. Tai chi/yoga effects on anxiety, heart rate, EEG and math computations. Complementary Therapies in Clinical Practice. 2010;**16**(4):235-238

[19] Trakroo M, Bhavanani AB, Pal GK, Udupa K, Krishnamurthy N. A comparative study of the effects of asan, pranayama and asan-pranayama training on neurological and neuromuscular functions of Pondicherry police trainees. International Journal of Yoga. 2013;**6**:96-103

[20] Putman P, Verkuil B, Arias-Garcia E, Pantazi I, van Schie C. EEG theta/beta ratio as a potential biomarker for attentional control and resilience against deleterious effects of stress on attention. Cognitive, Affective, & Behavioral Neuroscience. 2014;**14**(2):782-791

[21] Rajmohan V, Mohandas E. Mirror neuron system. Indian Journal of Psychiatry. 2007;**49**:66-69. DOI: 10.4103/0019-5545.31522

[22] Sirin SR. Academic achievement and socioeconomic status: A meta-analytic review of research. Review of Educational Research. 2005;**75**:417-453

[23] Caro DH. Socio-economic status and academic achievement trajectories from childhood to adolescence. Canadian Journal of Education. 2009;**32**:558-590

[24] Kieffer J. Socioeconomic status, english proficiency, and late-emerging reading difficulties. Educational Researcher. 2010;**39**:484-486

[25] Yeung A, Read J, Schmid S. Students' learning style, and academic performance in first year chemistry. UniServe Science Blended Learning Symposium Proceedings 2005; 137-142.

[26] Pellon M, Nome S, Aran A. Relationship between learning styles and academic performance in fifth graders enrolled in a medical course. Revista Brasileira de Oftalmologia. 2013;**72**:181-184

[27] Jolivet R, Coggan JS, Allaman I, Magistretti PJ. Multi-timescale modeling of activity-dependent metabolic coupling in the neuron-glia-vasculature ensemble. PLoS Computational Biology. 2015;**11**:e1004036

[28] Duyn JH. EEG-fMRI methods for the study of brain networks during sleep.

Frontiers in Neurology. 2012;**3**:100. DOI: 10.3389/fneur.2012.00100

[29] Sherlin LH, Larson NC, Sherlin RM. Developing a performance brain training™ approach for baseball: A process analysis with descriptive data. Applied Psychophysiology and Biofeedback. 2013;**38**(1):29-44. DOI: 10.1007/s10484-012-9205-2

[30] Malau-Aduli BS, Roche P, Adu M, et al. Perceptions and processes influencing the transition of medical students from pre-clinical to clinical training. BMC Medical Education. 2020;**20**:279. DOI: 10.1186/s12909-02002186-2

[31] Basharpoor S, Heidari F, Molavi P. EEG coherence in theta, alpha, and beta bands in frontal regions and executive functions. Applied Neuropsychology. Adult. 2019;**28**(3):310-317. DOI: 10.1080/23279095.2019.1632860

[32] Pascual AC, Muhoz NM, Robres AQ. The relationship between executive functions and academic performance in primary education: Review and meta-analysis. Frontiers in Psychology. 2019;**10**:1582. DOI: 10.3389/fpsyg.2019.01582

[33] Tu W, Ma Z, Ma Y, Dopfel D, Zhang N. Suppressing anterior cingulate cortex modulates default mode network and behavior in awake rats. Cerebral Cortex. 2021;**31**(1):312-323. DOI: 10.1093/cercor/bhaa227

[34] Lavin C, Melis C, Mikulan E, Gelormini C, Huepe D, Ibañez A. The anterior cingulate cortex: An integrative hub for human socially driven interactions. Frontiers in Neuroscience. 2013;7:64. DOI: 10.3389/fnins.2013.00064

[35] Eng GK, Collins KA, Brown C, Ludlow M, Tobe RH, Iosifescu DV, et al. Relationships between interoceptive sensibility and resting-state functional connectivity of the insula in obsessive–compulsive disorder. Cerebral Cortex. 8 Mar 2022;**14**. DOI: 10.1093/cercor/bhac014

[36] Rolls ET. The cingulate cortex and limbic systems for emotion, action and memory. Brain Structure and Function. 2019;**224**(9):3001-3018. DOI: 10.1007/s00429-019-01945-2

[37] Kołodziej A, Magnuski M, Ruban A, Brzezicka A. No relationship between frontal alpha asymmetry and depressive disorders in a multiverse analysis of five studies. eLife 2021;**10**:e60595. https://doi.org/10.7554/eLife.60595

[38] Anand KS, Dhikav V. Hippocampus in health and disease: An overview. Annals of Indian Academic Neurology. 2012;**15**(4):239-246. DOI: 10.4103/0972-2327.104323

[39] Spalding KL, Bergmann O, Alkass K, Bernard S, Salehpour M, Huttner HB, et al. Dynamics of hippocampal neurogenesis in adult humans. Cell. 2013;**153**:1219-1227

[40] Yau S, So K-F. Involvement of adult hippocampal neurogenesis in learning and forgetting. Neural Plasticity. 2015;**2015**:717958. DOI: 10.1155/2015/717958

[41] Hanson ND, Owens MJ, Nemeroff CB. Depression, antidepressants, and neurogenesis: A critical reappraisal. Neuropsychopharmacology. 2011;**36**: 2589-2602

[42] Jeon YK, Ha CH. Expression of brain-derived neurotrophic factor, IGF-1 and cortisol elicited by regular aerobic exercise in adolescents. Journal of Physical Therapy Science. 2015;**27**:737-741

[43] Numakawa T, Nakajima S, Adachi N, Furuta M, Odaka H, Kunugi H et al,

The role of brain-derived neurotrophic factor (BDNF) in comorbid depression: Possible linkage with steroid hormones, cytokines, and nutrition. Frontiers in Psychiatry. 2014;**5**:136. DOI: 10.3389/fpsyt.2014.00136

[44] Gorantla VR, Sirigiri A, Volkova YA, Millis RM. Effects of swimming exercise on limbic and motor cortex neurogenesis in the Kainate-lesion model of temporal lobe epilepsy. Cardiovascular Psychiatry and Neurology. 2016;**2016**:3915767. DOI: 10.1155/2016/3915767

[45] Gorantla VR, Thomas SE, Millis RM. Environmental enrichment and brain neuroplasticity in the Kainate rat model of temporal lobe epilepsy. Journal of Epilepsy Research. 2019;**9**(1):51-64. DOI: 10.14581/jer.19006

[46] Gorantla VR, Maity S, Millis RM. Pyramid exploration intervention, environmental enrichment, aerobic swimming exercise and brain neuroplasticity in the kainate rat model of temporal lobe epilepsy. In: Misciagna S, editor. Epilepsy - Update on Classification, Etiologies, Instrumental Diagnosis and Treatment. London: IntechOpen; 2020:73453. DOI: 10.5772/intechopen.93914

[47] Babcock KR, Page JS, Fallon JR, Webb AE. Adult hippocampal neurogenesis in aging and Alzheimer's disease. Stem Cell Reports. 2021;**16**(4):681-693. DOI: 10.1016/j.stemcr.2021.01.019

[48] Vas S, Casey JM, Schneider WT, Kalmar L, Morton AJ. Wake-promoting and EEG spectral effects of modafinil after acute or chronic administration in the R6/2 mouse model of Huntington's disease. Neurotherapeutics. 2020;**17**(3):1075-1086. DOI: 10.1007/s13311-020-00849-y

Chapter 6

Resting-State Brain Network Analysis Methods and Applications

Yunxiang Ge and Weibei Dou

Abstract

Resting-state fMRI has been widely applied in clinical research. Brain networks constructed by functional connectivity can reveal alterations related to disease and treatment. One of the major concerns of brain network application under clinical situations is how to analyze groups of data to find the potential biomarkers that can aid in diagnosis. In this paper, we briefly review common methods to construct brain networks from resting-state fMRI data, including different ways of the node definition and edge calculation. We focus on using a brain atlas to define nodes and estimate edges by static and dynamic functional connectivity. The directed connectivity method is also mentioned. We then discuss the challenges and pitfalls when analyzing groups of brain networks, including functional connectivity alterations, graph theory attributes analysis, and network-based statistics. Finally, we review the clinical application of resting-state fMRI in neurorehabilitation of spinal cord injury patients and stroke patients, the research on the mechanism and early diagnosis of neurodegenerative diseases, such as multiple system atrophy, as well as the research on brain functional network alteration of glioma patients.

Keywords: resting-state fMRI, brain networks, graph theory attributes, dynamic functional connectivity, network-based statistics, neurorehabilitation, multiple system atrophy, glioma

1. Introduction

Magnetic resonance imaging (MRI) is a multimodal technique that can noninvasively reflect the structure and function of the human brain. Structural MRI (sMRI), including longitudinal (spin-lattice) relaxation time T1-weighted and transverse (spin-spin) relaxation time T2-weighted imaging, has been applied to investigate the structural features of the brain. Based on the different relaxation times of different tissue, T1-weighted and T2-weighted images can be used to reflect the volume of grey matter, white matter, as well as lesions caused by infarction or hemorrhage. Diffusion MRI (dMRI), such as diffusion-weighted imaging (DWI) and diffusion tensor imaging (DTI), can be used to measure water diffusion along different directions and tract neural fiber counts and orientation. Functional MRI (fMRI) reflects neural activity during a period of time by measuring the relative amount of deoxygenated hemoglobin and oxygenated hemoglobin in the blood flow, which is also called the blood-oxygen-level-dependent (BOLD) signal. The fMRI is becoming popular in clinical situations to investigate the functional alterations following disease or treatment.

The fMRI experiment can be categorized into task fMRI and resting-state fMRI (rs-fMRI). For task fMRI, subjects need to perform a specific task, such as finger tapping or receive external stimulation like heat or sound during the scanning session. Resting-state fMRI, on the other hand, is collected when the subject lies still in the scanner, without doing any movement or thinking anything particular, and keeping awake at all time. Researchers focus on the spontaneous neural activity reflected by the BOLD signal under resting conditions. The correlation of signals related to spatially distinct regions is commonly defined as functional connectivity (FC) [1].

In the recent two decades, several methods have been developed to analyze functional connectivity in the resting state, including seed-based analysis, independent component analysis (ICA) [2], and resting-state network (RSN) method [3, 4]. The network method characterizes brain spontaneous activity as a graph, where nodes are defined as brain regions and edges are represented as connectivity between regions. There are different ways to calculate the connectivity, including static and dynamic functional connectivity and directed connectivity. Furthermore, features proposed in network science can be adopted to characterize the brain network topology, such as graph theory attributes [5].

Resting-state fMRI has been applied to clinical research and applications [6, 7]. In clinical situations, a common research paradigm is performing group comparison and searching for inter-group significant different features. Researchers are interested in whether a group of patients is significantly different from a group of healthy controls, or whether the same group of patients shows significant recovery after treatment. The identified significant different features may be the potential biomarker to aid in diagnosis as well as treatment. More importantly, the location of the significant different feature is of great interest, since each brain region has its unique function. As a result, this requires comparing groups of brain networks and other extracted network features. In clinical research, there are two key techniques of brain network analysis, the method of network construction and significant difference analysis of groups of brain networks.

In the following sections, we first describe how to construct brain networks from resting-state fMRI data, including different node definitions and a range of connectivity measurements. Then, we present common group analysis methods of brain networks. The clinical application of brain network analysis is also reported. We also propose several future directions in brain network research and end the chapter with a conclusion.

2. Constructing brain networks

Unlike structural and diffusion MRI, the fMRI scanning captures the BOLD signal in a period of time that typically lasts for several minutes. The collected data are a time series, and the "sampling period" is called repetition time (TR). That is, whole-brain data are collected every TR seconds. Before constructing brain networks, the data need to be preprocessed to clean out non-neural artifacts, including physiological signals like breath and heartbeat, head movements, and scanner noise. Then the nodes of the network are defined and connectivity between each pair of nodes is calculated. The whole data processing pipeline is shown in **Figure 1**.

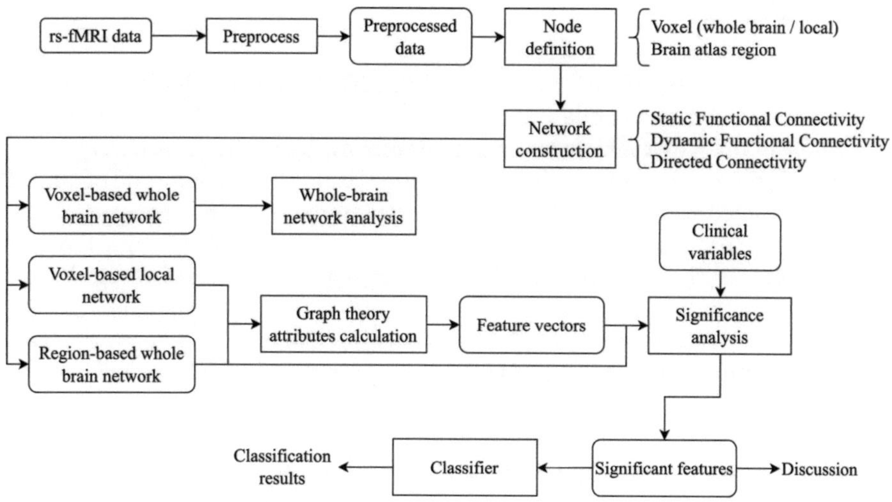

Figure 1.
Resting-state fMRI data process pipeline.

2.1 fMRI preprocessing

The preprocessing of fMRI data is necessary since there are non-neural noises in the signal. There are openly available toolboxes to carry out preprocessing, such as Statistical Parametric Mapping (SPM), FMRIB Software Library (FSL), and Data Processing Assistant for Resting-State fMRI (DPARSF) [8]. Common preprocessing procedures begin by removing the first 10 time points to let the subject be familiar with the scanning environment. Since the scanning of fMRI data within a repetition period (2s) is done in a slice-by-slice manner, the exact collection time of the first slice and the last slice has a time difference. To correct this difference, a procedure called slice timing correction needs to be performed. Then the head motion is corrected so that each voxel corresponds to the same brain location in the scanning series.

For group analysis, the data of different subjects need to be co-registered or normalized to the Montreal Neurological Institute (MNI) standard space. The data then undergoes smoothing using a Gaussian filter with a specified full-width-half-maximum (FWHM) value. After that, the linear trend in the signal is removed and nuisance covariates, such as white matter, cerebral spinal fluid (CSF), and global signal, are regressed out. At last, the data are filtered to keep signals within 0.01-0.08 Hz, since signals within this frequency range are reported to reflect spontaneous neural activities.

Although numerous preprocessing steps have been developed, there is still no consensus on the standard fMRI data preprocessing pipeline. The controversy is centered on the nuisance covariates regression, especially global signal regression (GSR) [9] and white matter signal regression [10]. Other researchers tried to optimize the preprocessing across multiple outcome measures [11], for low-frequency fluctuation analysis [12] and specific patients, such as stroke patients [13]. We have also investigated how the choices of preprocessing parameters and steps influence statistical analysis results [14]. The preprocessing of fMRI data remains to be a complex but important research topic.

2.2 Node definition

The most basic node definition is the voxel in a 3D fMRI image. Each voxel within the brain can be treated as a node and the constructed voxel-based network covers the whole brain. However, since the spatial resolution of fMRI is relatively high (2mm–4mm), the number of voxels is rather large (around the magnitude of 100,000) and the constructed network requires huge computation power for further analysis. Researchers have proposed specialized methods, such as the Parallel Graph-theoretical Analysis (PAGANI) toolkit to accelerate the processing of voxel-based whole-brain networks [15].

On the other hand, the nodes of the brain network can be defined as regions in the brain. The preprocessed data of voxels within a region are averaged spatially as the signal related to this node. The region can be specified manually by drawing regions of interest (ROI). Independent component analysis (ICA) can also reveal the component region but requires specifying parameters, such as the number of components. Both methods require human intervention and depend heavily on expert knowledge.

We proposed a fuzzy node definition method in Ref. [16] for tumor-brain, named "Spatial-Neighborhood and Functional-Correlation (SNFC)" based on fuzzy connectedness. It is a self-adapting method where the network was divided into functional connection and spatial adjacency. In the SNFC method, fuzzy connectedness between two voxels acts as a measurement to decide if they belong to the same node. Each voxel in the brain could be mapped into two feature spaces—structure feature space S and correlation feature space C. Let $s_{i,k}$ represent the spatial relationship between voxel v_i and voxel v_k, acting as a judgment of the neighboring relationship. $c_{i,k}$ is the correlation coefficient between the BOLD signal of v_i and v_k. The features of structural space S guarantee the principle of the spatial neighborhood and the features of correlation space C ensure the principle of consistency. Fuzzy connectedness between two voxels could be defined as the following:

$$FC_{i,k} = s_{i,k} \cdot c_{i,k} \tag{1}$$

If $FC_{i,k} > T$, then v_i and v_k belong to the same node, where T is the correlation threshold determining whether the correlation of two voxels is strong enough to be in the same node.

The nodes can also be defined using regions in the brain atlas to avoid the subjective error caused by human intervention and enable automatic processing for large cohorts of data. The most known brain atlas is the Brodmann atlas, created by the German anatomist Korbinian Brodmann based on cytoarchitecture [17]. Another popular brain parcellation is the Automated Anatomical Labeling (AAL) atlas [18]. The AAL atlas focuses on brain structure and the finer partition of certain cortices was proposed in AAL2 [19] and AAL3 [20]. Apart from structure, the brain atlas derived from diffusion and functional data is getting more attention. The Brainnetome Atlas was proposed based on DTI data with fine-grained parcellation [21]. Researchers also developed functional atlas, such as the Atlas of Intrinsic Connectivity of Homotopic Areas (AICHA) that considered the homolog of regions in both hemispheres [22]. The above-defined network is called a region-based whole-brain network. We can also construct networks within a region. In this scenario, the voxels are defined as nodes, and the network only consists of voxels within a region. The constructed network is called a voxel-based local network, representing the topology within certain regions. We proposed a multilevel brain network joint analysis method on voxel-based whole-brain networks, voxel-based local networks as well as region-based whole-brain networks (**Figure 2**) [23].

Node definition has a fundamental influence on the topology of the brain network. Different atlas parcels the cerebrum and cerebellum based on different information, and it plays a key role in linking physiological regions to abstract brain network nodes. However, similar to the preprocessing of fMRI data, there is no gold standard for the node definition. Several researches have been carried out to investigate the effect of node definition on network analysis [24], resting-state networks [25], and the topology of both functional networks [26] and structural networks [27]. It is still an open question and needs more thorough research.

2.3 Static and dynamic functional connectivity

Edges in brain networks are represented by the connectivity between nodes. One of the most common connectivity measures is functional connectivity (FC). In 1995, Biswal et al firstly reported the correlation of intrinsic low-frequency BOLD signal fluctuation under resting-state and since then, multiple efforts have been devoted to FC analysis [1, 3]. Functional connectivity is commonly defined as the Pearson correlation between the BOLD signal of spatially distant regions. In recent years, researchers realized that FC ignores the dynamics of neural activity and developed dynamic functional connectivity (DFC) or Chronnectome [28–30]. The research on DFC is becoming popular and has attracted lots of attention.

Technically speaking, FC or static functional connectivity (SFC) is calculated using the whole time series, whereas DFC utilizes a sliding time window and the correlation of signals within the window is calculated. The window then moves from the beginning of the BOLD signal to the end, with a pre-defined step size. As a result, the connectivity shows dynamic fluctuations as the window slides, and each scanning session is associated with a series of brain networks, or a dynamic brain network. In

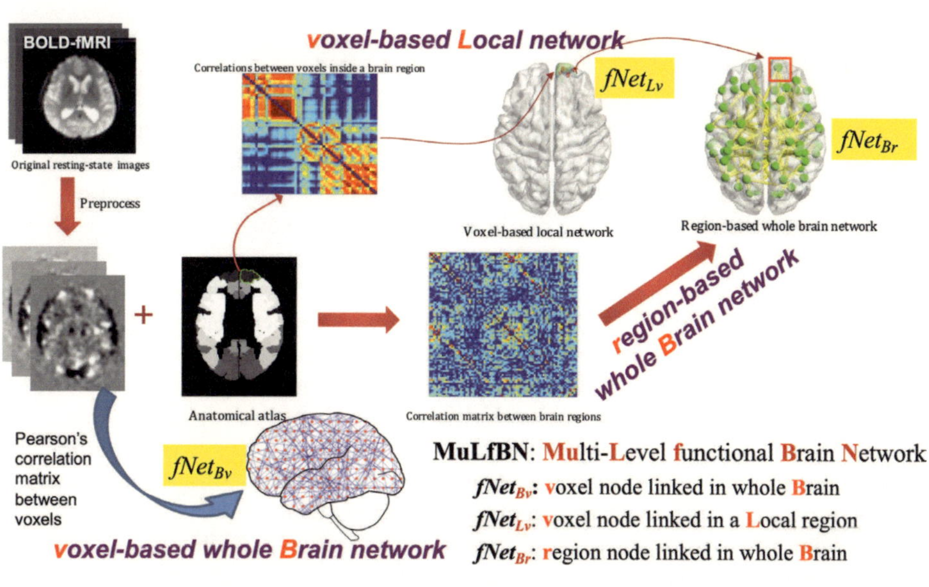

Figure 2.
Construction of multilevel functional brain networks.

contrast, there is only one static network related to the scanning session. The network is usually represented by a graph adjacency matrix, which is a square symmetric matrix and the (i,j) value equals the connection of node i to node j. For a dynamic network, there is a time axis along with the adjacency matrix.

There are two major parameters regarding DFC calculation—the window length and the sliding step size. With a longer window length, the dynamics of neural activity might be averaged out while a shorter window length can capture transient signal changes. The step size controls the temporal resolution of DFC. Normally it is specified as several TRs. We investigated the optimal window width by using the small-world property as criteria [31]. Node degree distribution has exponential truncated power-law in the small-world network, and the normal human brain network shows a strong small-world property. The reasonable window width range was verified on both SNFC-based and voxel-based whole-brain networks. Results show that the smallest window width is 200 seconds and 260 seconds for normal subjects and brain tumor patients, respectively. Leonardi et al also studied the theory between window length and filter cut-off frequency during preprocessing [32]. Apart from the two window parameters, the shape of the sliding window is another concern. The rectangular window is the simplest solution, but other choices such as tapered window exist. Mokhtari et al also proposed a modulated rectangular (mRect) window to reduce spectral modulations [33].

We also proposed a dynamic network analysis method for enlarging the training samples required by an unsupervised learning classification algorithm [34], such as a classical backpropagation neural network classifier containing a hidden layer. It reached the optimal accuracy of 100% for classifying glioma patients and normal subjects.

Despite controversies, DFC has been used to investigate diseases, such as schizophrenia [35], post-traumatic stress disorder (PTSD) [36], Parkinson's Disease [37], and autism [38]. It has also been applied to lifespan studies [39] and cognitive research [40]. From either a methodological or application view, the research on DFC is still insufficient.

2.4 Directed connectivity

As the definition implies, both SFC and DFC contain no directional information. Effective connectivity (EC) can measure the directional influence of one region toward another area by calculating the causal relationships between time series. Commonly adopted EC estimation methods are structural equation modeling (SEM) [41], dynamic causal modeling (DCM) [42], and Granger causality analysis (GCA) [43, 44]. The computation cost becomes unacceptable for SEM and DCM as the number of nodes increases [43]. Several amendments have been proposed to reduce the computation requirement of DCM recently [45, 46], but the model complexity is still challenging for clinical applications. We proposed a method based on convergent cross-mapping (CCM) that can reflect the interactions between regions in a dynamic, nonlinear, and deterministic way, which is not covered by GCA [47]. The method overview, together with the extended network-based statistic, is shown in **Figure 3**.

CCM was originally developed to detect causality in complex ecosystems [48]. It acts as a complement to GCA as CCM assumes the system to be deterministic and dynamical, while GCA works for a stochastic system and requires separability. In GCA, if removing X decreases the predictability of Y, it can be deduced that X causes

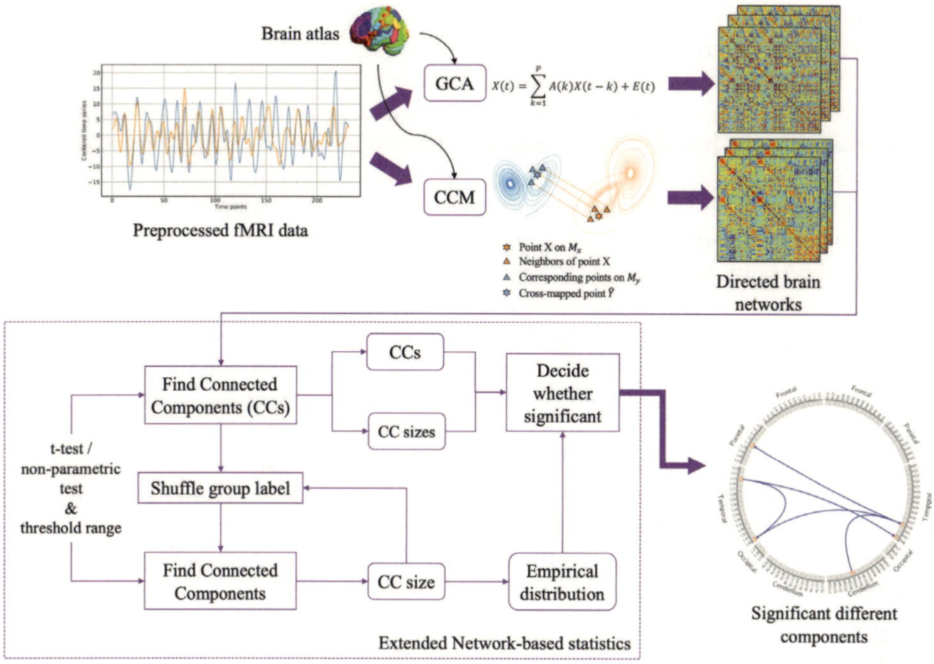

Figure 3.
CCM-based directed connectivity estimation and extended network-based statistic method.

Y, and in a brain network scenario, there is a directed connection from X to Y. On the other hand, in deterministic dynamic systems where CCM was developed, we can measure how well Y can estimate X to determine the causal relationship from X to Y, which then determines the directed connectivity strength from X to Y. The procedure of estimating X using Y is called cross-mapping. CCM is also applicable under situations where separability is not guaranteed. GCA, on the other hand, may produce erroneous results [49]. As for the brain, it is a dynamic system whose functional organization is poorly understood [50]. Utilizing CCM to estimate directed connectivity between regions could facilitate the investigation of brain activity as well as enable novel clinical applications.

3. Analyzing group differences in brain networks

After brain network construction, for each scanning of each subject, the preprocessed fMRI data were converted to a brain network represented by a graph adjacency matrix. The next question is how to find the difference between groups of brain networks. Here we summarize two popular methods to further analyze brain networks.

3.1 Significance analysis

The most basic method is analyzing functional connectivity directly. Specifically, suppose we are comparing two groups of networks. Each connectivity value is extracted from every network, forming two sets of values. Statistical hypothesis testing can be adopted to decide whether this connection shows a significant difference as well as which group is higher. After performing a comparison on every connection in the network, the group difference network consisting of significant different connections is obtained. All edges with a significant difference were stored in a network for further discussion. We can also select several regions based on prior knowledge, such as the sensorimotor area or visual area, to further filter the set of significant different connections.

Another method is calculating graph theory attributes. Graph theory characterizes the topology of the network by nodal and global attributes. Common node level graph theory attributes are betweenness centrality, clustering coefficient, local efficiency, modularity, and weighted degree, while the network level graph theory attributes include global efficiency and characteristic path length. Small-worldness is also a common index used in brain network analysis. For multilevel brain networks, we define intra-region features as the attributes calculated at voxel-based local networks, and the attributes calculated at region-based whole-brain networks are called inter-region features. We can calculate the global feature of the voxel-based local network (intra-region features), and the nodal feature of the region-based whole-brain network (inter-region features). As a result, for each graph attribute, we obtain a feature vector whose length equals the number of nodes in the network, representing the whole-brain network feature.

After obtaining feature vectors of graph theory attributes, we can perform a statistical comparison on each region similar to FC analysis. The feature at each region is extracted, forming two sets of values; and statistical testing is used to find significant regions or significant different features. Moreover, the clinical relevance of the features can be evaluated by assessing the correlation of features and clinical scores, which produces features with significant correlation. The intersection of significant different and significant correlated features is selected for further discussion and following analysis.

We also investigated methods to analyze dynamic graph theory attributes [51]. For dynamic brain networks, at each sliding window location, the obtained brain network is static, and graph theory attributes can be calculated. As the window slides, graph theory attributes at each window location are estimated, forming the dynamic graph theory attributes of the dynamic network. To combine static and dynamic attributes together with clinical scores, we proposed an analysis framework [51]. The strength and stability of dynamic graph attributes were calculated. We found significant different and correlated features for both static and dynamic networks, as well as their intersection. The resulting features were further analyzed using receiver-operating curves (ROC) to test their ability in classification.

A controversy regarding the above analysis method is the multiple comparison problem. For each single statistical comparison with a 0.05 significance level, there is a 0.05 chance of obtaining a false positive. However, when performing multiple statistical comparisons at the same time, the chance of getting at least one false positive would become higher as the number of comparisons increases. To tackle this problem, correction methods, such as Bonferroni correction and false discovery rate (FDR) correction, were proposed. The basic idea behind these correction methods is to decrease the single comparison significance level according to the number of comparisons. However, since the amount of comparison is related to the number of nodes in the network, and certain features show high within group variance, directly applying correction might result in no significant result. We argue that statistical comparison can be seen as a feature selection procedure. The significant or selected features are then fed into the next module, such as a classifier. During feature selection, we should keep as much useful information as possible. The uncorrected significant features are preliminary scanning results and taking the intersection of significant different and correlated features further select clinically relevant information. Searching for intersected significant features might be an alternative method to multiple comparison correction.

3.2 Network-based statistics

For brain networks, to overcome the multiple comparison issue, network-based statistics (NBS) was proposed, enabling direct comparison of groups of brain networks [52]. NBS assumes that the effect or the group difference forms a certain structure instead of distributed single connections. The edge-wise comparison is performed first and the links are thresholded according to the test statistics or p-values obtained from the edge-wise comparison, producing a binarized difference network. It then searches for structures or connected components in the binarized difference network. The size of the component, defined as the number of edges or nodes, is used to determine if the component is significant by a permutation test, where group labels of samples are randomly shuffled and the same procedure is performed to search for the maximum component size. The permutation is repeated 5000 times and the empirical distribution of the component size is obtained. An empirical p-value can be assigned to the original connected component by calculating the ratio of the number of permutations, where the maximal size is larger than the original size, to the total permutation number.

Compared with edge-wise comparison and direct edge-wise correction, NBS provides higher statistical power at the cost of coarser spatial resolution in detecting differences [52]. In other words, NBS can only declare the connected component as a whole to be significant. It draws no conclusion on the significance of each single connection within the component. However, the original NBS only works for symmetric adjacency matrices, which corresponds to functional connectivity.

Based on directed connectivity, we proposed the extended-NBS (e-NBS) to search for altered connected components in groups of directed networks [47]. The method overview is shown in **Figure 3**. We search for strongly connected components (SCC) and weakly connected components (WCC) with and without direction information. A classical depth-first search algorithm was adopted when searching for SCCs and WCCs. The edge-wise p-value was utilized to filter for candidate connections and construct a difference network. Since there is no consensus on how to choose the pre-defined p-value threshold, we changed it within a certain range to test method performance. Specifically, an edge is kept if the p-value is less than the pre-define

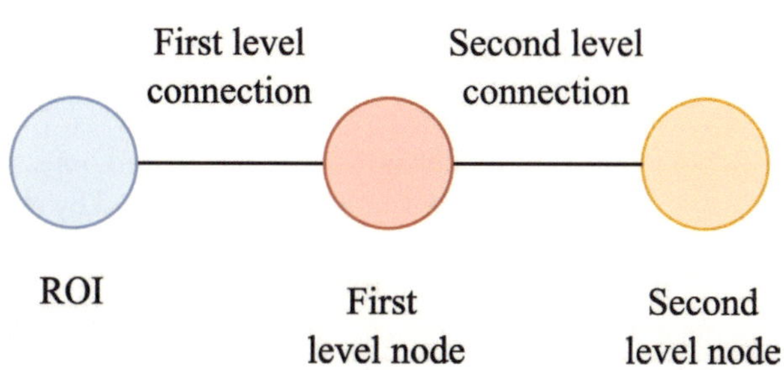

Figure 4.
Two-step connected component. The first level node is directly connected to the ROI in the binarized difference network, while the second level node is connected with the first-level node.

p-value threshold. For edge-wise comparison, we also tried to use two-sample t-test and the non-parametric Mann-Whitney test. The e-NBS method, together with the CCM-based directed connection estimation method, was verified using a dataset of spinal cord injury patients and healthy controls.

Moreover, we note that given the framework of e-NBS, one can define connected components that suit research needs. For example, in a study of motor function alteration following spinal cord injury, researchers are interested in connections related to sensorimotor areas and visual regions. The connected component can be defined as significant different connections related to these regions of interest. Furthermore, we can define two-step connected components that comprise connections directly related to the ROIs in the binarized difference network, and connections related to regions (first level nodes) that connect with ROIs (**Figure 4**). Either way, the permutation test in e-NBS makes it possible to draw conclusion on the significance of the defined component.

4. Clinical applications

The resting-state fMRI has been applied to clinical research and applications, mainly investigating pathophysiological mechanisms and searching for sensitive biomarkers for early diagnosis [6, 7]. The prognosis predictability of rs-fMRI is intriguing as well [53–55]. In glioma research, resting-state fMRI has also shown potential in diagnosis and treatment planning. Here we introduce three examples of applications and related works.

4.1 Neurorehabilitation

It has been shown that changes in both brain function and structure occur following central nervous lesions, such as spinal cord injury [56] and cerebral stroke [57]. According to the theory of neuroplasticity, the brain function continues to change during rehabilitation, and it is the theoretical and physiological basis for individualized neurorehabilitation as well as assistive rehabilitation technologies, such as transcranial direct current stimulation (tDCS) [58–60] and brain-computer interfaces (BCI) [61, 62]. We performed a study on spinal cord injury patients and investigated the alteration of grey matter volume extracted from structural MRI and functional connectivity related to the sensorimotor area, combining clinical assessments [63]. We found that that the alteration of anatomical structure features and the brain network connectivity in the sensorimotor area were non-concomitant following spinal cord injury, and the functional connectivity within the sensorimotor area had a significant correlation with clinical sensory scores, indicating the potential of FC as a prediction biomarker.

Another issue related to neurorehabilitation is the automated objective evaluation of rehabilitation progress. Traditionally, patient recovery is assessed by clinical measurements, which can only reflect behavioral improvements and might include subjective bias. We proposed a distance-based rehabilitation evaluation method that takes resting-state fMRI data of patients and healthy controls as input (**Figure 5**) [64]. We hypothesize that the sample point distribution of patients and healthy controls in the feature space is dichotomous. A support vector machine (SVM) classifier was first trained using significantly different functional connectivity of healthy controls and the first scanning

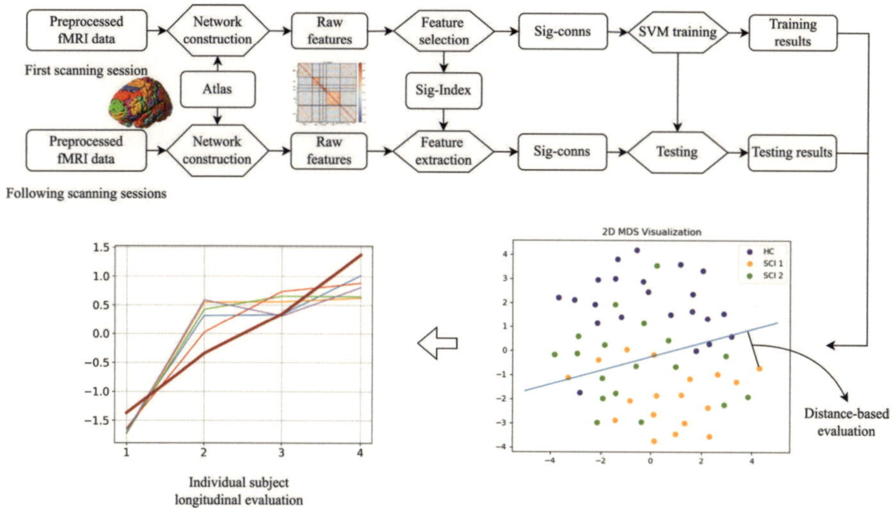

Figure 5.
Method overview of the distance-based rehabilitation evaluation framework.

session of patients. The distance of the patient sample points to the separating hyperplane was calculated and used to evaluate patient recovery. If the patient recovered, the sample point of the patient would move toward healthy controls and the distance would decrease. The method was verified using both group level and individual longitudinal data, and the distance evaluation was consistent with clinical measurements.

On the other hand, a stroke could lead to certain movement disabilities. Motor-related brain function alteration after stroke and during recovery is of great interest. Brain-Computer Interface (BCI) systems are helpful in motor recovery, possibly by stimulating neuroplasticity following brain activity [65]. The brain network reorganization of stroke patients after BCI training is of great significance. We conducted an experiment to investigate the functional changes after BCI training and their relations to clinical scores [66]. Functional connectivity was calculated using data collected before and after training and we searched for significant increased FC in groups with and without BCI training. The correlation between FC and clinical scores was also calculated. We found increased FC between certain cerebral and subcortical regions and the inter-hemisphere FC was positively correlated with motor scores.

4.2 Multiple system atrophy

Multiple system atrophy (MSA) is a neurodegenerative disease typically characterized by parkinsonism, cerebellar ataxia syndrome, and autonomic nervous dysfunction [67]. It is further divided into two subtypes, MSA with predominant parkinsonism (MSA-P) and MSA with predominant cerebellar ataxia (MSA-C) [67]. Previous studies mainly investigated the structural abnormalities related to MSA patients and compared subtypes of MSA with Parkinson's Disease (PD) as well as healthy controls [68–71]. The functional alteration induced by MSA is also studied by calculating regional homogeneity (ReHo) [72], the amplitude of low-frequency fluctuations (ALFF) [73], as well as functional and effective connectivity [74, 75].

The dynamic functional features of MSA-C patients not thoroughly investigated before. We conducted an experiment on MSA-C patients and proposed a method to

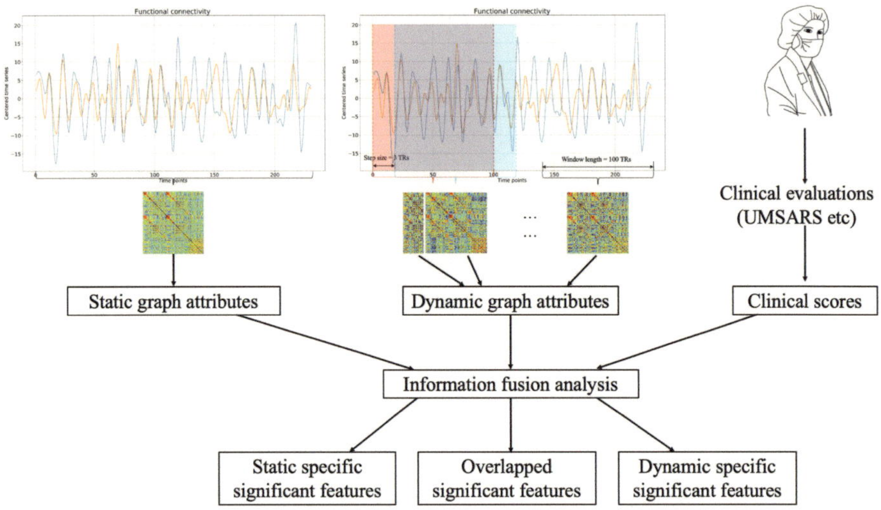

Figure 6.
The coalition analysis of rs-fMRI data combining static, dynamic functional connectivity as well as clinical information.

combine static and dynamic functional connectivity features, as well as clinical scores (**Figure 6**) [51]. The static and dynamic brain networks were constructed using methods described in section 2.3 and static and dynamic graph theory attributes were calculated. Statistical comparisons and correlation analysis were carried out and significant different and correlated features were found. The significant regions mainly covered the cerebellum and certain cerebral areas, which is consistent with prior knowledge. The dynamic features showed the highest area under the curve (AUC) value during receiver-operating characteristic (ROC) analysis, indicating the potential of dynamic features in disease diagnosis.

Apart from structural and functional analysis, multimodal research on MSA is getting more attention. We also tried to combine structural, diffusion, tractography, and functional features extracted from T1, DTI, and fMRI to search for sensitive biomarkers for MSA-C patient diagnosis (**Figure 7**) [76]. The T1 data were processed to produce grey matter and white matter probability maps. We performed tractography on DTI data and counted the number of tracts crossing each brain region. The fraction anisotropy (FA) and mean diffusivity (MD) maps were also obtained. For rs-fMRI, we calculated functional connectivity and constructed brain networks. The extended network-based statistics for the undirected network were adopted to search for significant different connected components between the two groups. By using the AAL atlas, feature maps extracted from different modalities were converted to feature vectors and networks. After that, significant analysis was performed with false discovery rate correction and we identified significant different features, mainly distributed in cerebellar and certain cerebral regions. The correlation of these features with clinical scores was also tested. We also searched for sensitive biomarkers in disease diagnosis by applying a nested leave-one-out cross-validation framework and evaluated classification performance using the significant features of each region with a support vector machine (SVM) classifier, as shown in **Figure 7**. The identified biomarkers were mainly cerebellar regions. Different modalities contain complementary information. Merging multimodal

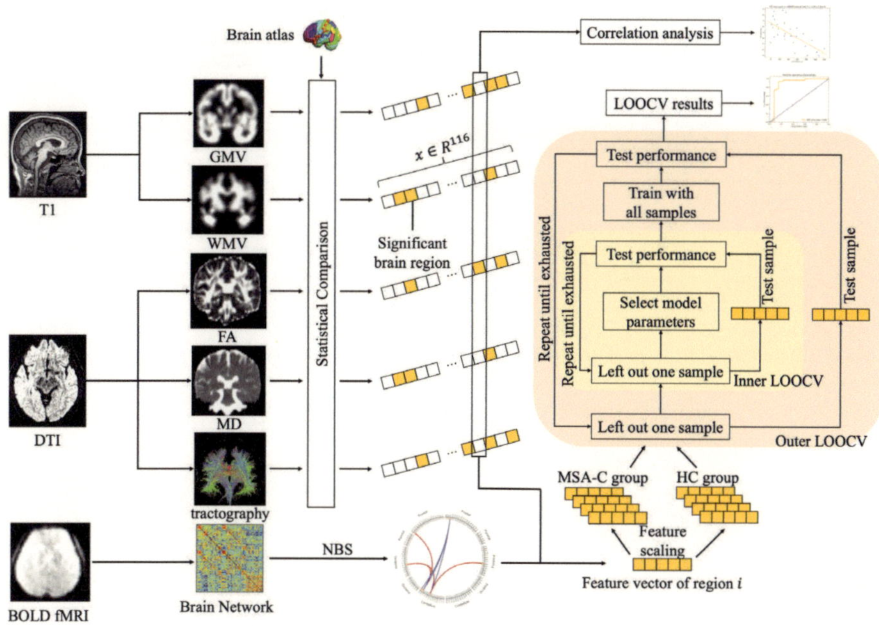

Figure 7.
The multimodal MRI feature fusion framework and the nested leave-one-out cross-validation procedure. GMV: grey matter volume; WMV: white matter volume; FA: fractional anisotropy; MD: mean diffusivity; NBS: network-based statistics; LOOCV: leave-one-out cross-validation.

data and clinical variables together can further reveal the neurological alteration related to the disease as well as increase the accuracy, robustness, and generalization of the disease diagnosis algorithm.

4.3 Glioma

Glioma stems from the canceration of neurogliocyte and is the most common tumor in the human brain [77]. It has an intensive impact on the structure of the brain and further on the corresponding physiological functions. Different locations of the glioma will result in different functional alterations and prognosis outcomes. For a high-level glioma, it is highly likely to relapse even after being excised in a surgery [78]. As a result, it is necessary to analyze the brain function changes according to the location and volume of glioma for both diagnosis and treatment. We proposed a framework of multilevel functional network analysis to find the functional network characteristics of glioma patients [79]. The multilevel network consists of a hemisphere functional network, glioma voxel local network, and glioma region local network, as illustrated in **Figure 8**. The hemisphere functional network was constructed based on regions from a single hemisphere in the AAL atlas excluding cerebellar parcellation (**Figure 9**). The glioma voxel local network is constructed at the voxel level in the region of glioma that is extracted by a tumor segmentation method. And glioma region local network is also constructed at the voxel level, but within each atlas region containing the glioma. A ratio, defined as the number of voxels in an AAL area that belongs to the segmented glioma region over the total voxel number of the area, is used as the threshold for selecting areas containing the glioma in the AAL atlas.

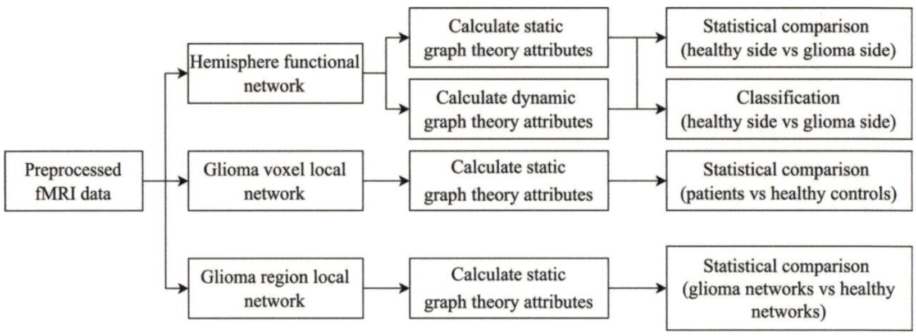

Figure 8.
A framework of multilevel functional network analysis for finding the functional network characteristics of glioma patients.

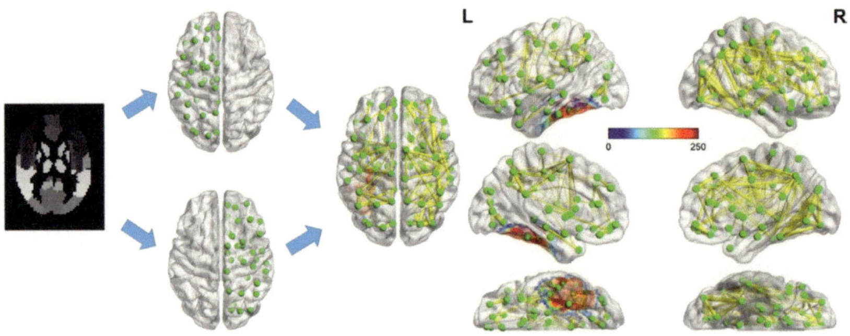

Figure 9.
The process of the construction of the hemisphere functional networks is based on the AAL atlas of a glioma patient. The green dots stand for the nodes of the functional network. The yellow line segments represent the weighted edges whose thickness reflects the weight. The colored area shows the tumor region and different colors reflect the possibility of whether a voxel belongs to the tumor.

Network features, including connectivity strength, characteristic path length, average nodal betweenness centrality, and average nodal clustering coefficient, were calculated for all networks. The network connectivity strength was defined as the average z-scores of all edges. Network characteristic path length equals the average of shortest paths between each pair of nodes in the network. Nodal attributes, including betweenness centrality and clustering coefficients, are calculated at each node within the network and averaged as network features. For hemisphere functional networks, both static and dynamic functional connectivity were investigated. Since the period of the BOLD signal induced by the hemodynamic response of neuronal activity is about 20s [80], during the reconstruction of dynamic networks, a sliding window with a length of 50s and a step size of 10s was selected. Each glioma patient received functional scanning lasting for 460s. As a result, the sliding time window extracted 46 sub-signals with a length of 50s and constructed dynamic brain networks with 46-time slices.

In this study, 38 patients with tumors in one side of the brain were enrolled. We constructed 38 positive and 38 negative hemisphere functional networks. Among these patients, 15 subjects had glioma area segmentation. Moreover, 15 healthy subjects were collected as the control group. The local network analysis was performed on 15 patients with segmentation and 15 healthy controls. We used the two-sample t-test to evaluate the significant difference of each feature between hemisphere functional networks constructed on the healthy side and the glioma side. The glioma voxel local networks and glioma region local networks were constructed at the same location of glioma segmentation in data collected from healthy controls as well. Statistical comparison was performed to compare network features of glioma voxel local networks and glioma region local networks from patients and healthy controls. There were 41 glioma region local networks constructed from 15 patients, and for comparison, 41 local networks were estimated from healthy controls.

We also investigated the classification performance using hemisphere functional networks. Given that the sample size is small (38 networks with glioma and 38 networks with healthy tissue), linear support vector machine (SVM) was chosen as the classifier. Static and dynamic network features were extracted and aligned into a feature vector of dimensions 4 and 184 (46×4) as the input to the classifier, and the leave-one-out cross-validation method is employed to evaluate the performance. The results showed that both dynamic and static features can distinguish the normal and abnormal networks. In addition, dynamic features obtained 100% accuracy in our dataset, while static features showed 71.5% accuracy.

Results revealed by the multilevel functional network analysis method showed that the existence of glioma changed certain features of the normal functional networks. Our work finds that glioma weakened the connection strength of the global and local functional networks. Moreover, it decreased the clustering degree of the nodes in both local functional networks, indicating that glioma may destruct the non-randomness and the small-world property of brain networks.

Previous studies have already investigated how glioma alters functional connectivity [80–83]. We find that glioma attenuates the connectivity of functional networks, which is in accordance with previous studies. Moreover, we also involved network features other than connectivity. Our study emphasized the characteristic features, such as betweenness centrality, clustering coefficient, and characteristic path length, which were not covered by previous research.

5. Future directions

Despite progress in recent years, there are lots of work to be done in developing new methods for constructing and analyzing brain networks, as well as performing group and individualized analysis. In this section, we propose some possible directions in the field of brain network research.

Network science has been used to analyze brain networks and advanced methods need to be developed to characterize the topological features of brain networks. The algebraic topological data analysis (TDA) method provides a new way to analyze the interactions between a set of nodes instead of bilateral connections. TDA could act as a complement to graph theoretical analysis in describing the topology characteristic of brain networks. More advanced network theory concepts, such as algebraic topology, have also been introduced to the analysis of brain networks [5]. Moreover, artificial neural networks and deep learning methods have been shown to be powerful in analyzing graph data. On the one hand, before network construction, models, such as Recurrent Neural Network (RNN) and Transformer, that were originally proposed to process sequential data, such as natural language and voice, can be applied to analyze the BOLD time series, both with and without preprocessing. Since the network perspective mainly models the inter-relationships between signals of spatially distinct regions, applying deep learning models directly to the time series could possibly extract information complementary to statistical dependency, as described by functional connectivity. On the other hand, after constructing brain networks using functional connectivity, directed connectivity, or DTI fiber tracking, Graph Neural Network (GNN) or Graph Convolutional Network (GCN) could be utilized to merge these multimodal networks and combine both edge-wise features (connections) and nodal features, such as graph theory attributes. GNN was proposed to directly analyze graphs that can model relationships between nodes and perform inference on node, edge, or graph level. Applying GNN to brain networks, especially multilevel static and dynamic brain networks, could possibly extract useful features and enable multimodal information fusion.

On the application side, multiple group comparison methods have been developed. However, for clinical application, individualized diagnosis and treatment are crucial. How to transform conclusions derived from group research into individual situations is a challenging question. We define "healthy templates" as a set of methods to delineate characteristics of a healthy population. The healthy templates describe the distribution of features of healthy people and need to be built for each feature extracted from different modalities. In its most basic form, the healthy template can be a value range given a specific feature. Subjects whose feature value falls within this value range would be considered to be normal, similar to the interpretation of a blood test result. Open-source datasets are valuable resources in the construction of healthy templates. However, the site effect of MRI data is a crucial issue and multi-site data harmonization techniques need to be adopted when combining data from different scanning locations. Several methods have been proposed for harmonization but their utility remains to be tested [84, 85]. With low variance healthy templates, individualized precise treatment planning and prognosis prediction would become possible.

6. Conclusion

The human brain is modeled as a functionally inter-connected network. Resting-state functional magnetic resonance imaging enables observing brain spontaneous activity *in vivo*. In this chapter, we reviewed the process of rs-fMRI data as well as group analysis methods. Different node definitions and edge estimation were discussed during the network construction stage. Nodes can be defined at the voxel level or with the help of a brain atlas. Lesions, such as glioma segmentation result, can also guide node definition. Edges are estimated in static, dynamic as well as directed scenarios. We presented two major methods to compare groups of brain networks data, significance analysis, and network-based statistics. Combined with the brain atlas, whole-brain networks are characterized by graph theory attributes developed in network science. Network-based statistics enables the direct comparison of groups of brain networks. We also discussed the clinical application of rs-fMRI data analysis in neurorehabilitation, multiple system atrophy, and glioma patients. At last, future research directions are discussed, with an emphasis on network science, novel deep learning models, and individualized clinical applications.

Acknowledgements

This work was supported by a grant from the Tsinghua University Initiative Scientific Research Program (No. 20131089382) and the National Natural Science Foundation of China (No. 61171002, 60372023). We would like to thank Zexuan Hao

DOI: http://dx.doi.org/10.5772/intechopen.104827

and Ziliang Zhang from the Department of Electronic Engineering, Beijing National Research Center for Information Science and Technology (BNRist), Tsinghua University for useful advice during the experiment and manuscript development.

Conflict of interest

The authors declare no conflict of interest.

Author details

Yunxiang Ge and Weibei Dou*
Department of Electronic Engineering, Beijing National Research Center for Information Science and Technology (BNRist), Tsinghua University, Beijing, China

*Address all correspondence to: douwb@tsinghua.edu.cn

References

[1] Biswal B, Zerrin Yetkin F, Haughton VM, Hyde JS. Functional connectivity in the motor cortex of resting human brain using echo-planar mri. Magnetic Resonance in Medicine. 1995;**34**(4):537-541

[2] Calhoun VD, Liu J, Adalı T. A review of group ICA for fMRI data and ICA for joint inference of imaging, genetic, and ERP data. NeuroImage. 2009;**45**(1):S163-S172

[3] van den Heuvel MP, Hulshoff Pol HE. Exploring the brain network: A review on resting-state fMRI functional connectivity. European Neuropsychopharmacology. 2010;**20**(8):519-534

[4] Rosazza C, Minati L. Resting-state brain networks: Literature review and clinical applications. Neurological Sciences. 2011;**32**(5):773-785

[5] Bassett DS, Sporns O. Network neuroscience. Nature Neuroscience. 2017;**20**(3):353-364

[6] Fox M, Greicius M. Clinical applications of resting state functional connectivity. Frontiers in Systems Neuroscience. 2010;**4**(19)

[7] Lee MH, Smyser CD, Shimony JS. Resting-State fMRI: A Review of Methods and Clinical Applications. American Journal of Neuroradiology. 2013;**34**(10):1866-1872

[8] Yan C, Zang Y. DPARSF: A MATLAB toolbox for" pipeline" data analysis of resting-state fMRI. Frontiers in Systems Neuroscience. 2010;**4**(13)

[9] Murphy K, Fox MD. Towards a consensus regarding global signal regression for resting state functional connectivity MRI. NeuroImage. 2017;**154**:169-173

[10] Grajauskas LA, Frizzell T, Song X, Darcy R. White matter fMRI activation cannot be treated as a nuisance regressor: Overcoming a historical blind spot. Frontiers in Neuroscience. 2019;**13**(1024)

[11] Shirer WR, Jiang H, Price CM, Ng B, Greicius MD. Optimization of rs-fMRI Pre-processing for Enhanced Signal-Noise Separation, Test-Retest Reliability, and Group Discrimination. NeuroImage. 2015;**117**:67-79

[12] Woletz M, Hoffmann A, Tik M, Sladky R, Lanzenberger R, Robinson S, et al. Beware detrending: Optimal preprocessing pipeline for low-frequency fluctuation analysis. Human Brain Mapping. 2019;**40**(5):1571-1582

[13] Yourganov G, Fridriksson J, Stark B, Rorden C. Removal of artifacts from resting-state fMRI data in stroke. NeuroImage: Clinical. 2018;**17**:297-305

[14] Ge Y, Pan Y, Dou W. Analysis of BOLD fMRI signal preprocessing pipeline on different datasets while reducing false positive rates. BIBE 2018; International Conference on Biological Information and Biomedical Engineering; 2018

[15] Du H, Xia M, Zhao K, Liao X, Yang H, Wang Y, et al. PAGANI Toolkit: Parallel graph-theoretical analysis package for brain network big data. Human Brain Mapping. 2018;**39**(5):1869-1885

[16] Wang X. Brain Function Analysis Method for Glioma Patient by Using Information Combination of MRI and

fMRI Signal. Beijing, China: Tsinghua University; 2014

[17] Brodmann K. Vergleichende Lokalisationslehre der Grosshirnrinde in ihren Prinzipien dargestellt auf Grund des Zellenbaues. Barth; 1909

[18] Tzourio-Mazoyer N, Landeau B, Papathanassiou D, Crivello F, Etard O, Delcroix N, et al. Automated anatomical labeling of activations in SPM using a macroscopic anatomical parcellation of the MNI MRI single-subject brain. NeuroImage. 2002;**15**(1):273-289

[19] Rolls ET, Joliot M, Tzourio-Mazoyer N. Implementation of a new parcellation of the orbitofrontal cortex in the automated anatomical labeling atlas. NeuroImage. 2015;**122**:1-5

[20] Rolls ET, Huang C-C, Lin C-P, Feng J, Joliot M. Automated anatomical labelling atlas 3. NeuroImage. 2020;**206**:116189

[21] Fan L, Li H, Zhuo J, Zhang Y, Wang J, Chen L, et al. The human brainnetome atlas: A new brain atlas based on connectional architecture. Cerebral Cortex. 2016;**26**(8):3508-3526

[22] Joliot M, Jobard G, Naveau M, Delcroix N, Petit L, Zago L, et al. AICHA: An atlas of intrinsic connectivity of homotopic areas. Journal of Neuroscience Methods. 2015;**254**:46-59

[23] Luo H, Dou W, Pan Y, Wang Y, Mu Y, Li Y, et al. Joint analysis of multi-level functional brain networks. 2016 9th International Congress on Image and Signal Processing. BioMedical Engineering and Informatics (CISP-BMEI). Oct. 2016;**2016**:15-17

[24] Gong Y, Wu H, Li J, Wang N, Liu H, Tang X. Multi-granularity whole-brain segmentation based functional network analysis using resting-state fMRI. Frontiers in Neuroscience. 2018;**12**(942)

[25] Doucet GE, Lee WH, Frangou S. Evaluation of the spatial variability in the major resting-state networks across human brain functional atlases. Human Brain Mapping. 2019;**40**(15):4577-4587

[26] Wang J, Wang L, Zang Y, Yang H, Tang H, Gong Q, et al. Parcellation-dependent small-world brain functional networks: A resting-state fMRI study. Human Brain Mapping. 2009;**30**(5):1511-1523

[27] Wu Z, Xu D, Potter T, Zhang Y, TAsDNI. Effects of brain parcellation on the characterization of topological deterioration in Alzheimer's disease. Frontiers in Aging Neuroscience. 2019;**11**(113)

[28] Allen EA, Damaraju E, Plis SM, Erhardt EB, Eichele T, Calhoun VD. Tracking whole-brain connectivity dynamics in the resting state. Cerebral Cortex. 2014;**24**(3):663-676

[29] Sakoğlu Ü, Pearlson GD, Kiehl KA, Wang YM, Michael AM, Calhoun VD. A method for evaluating dynamic functional network connectivity and task-modulation: Application to schizophrenia. Magnetic Resonance Materials in Physics, Biology and Medicine. 2010;**23**(5):351-366

[30] Calhoun Vince D, Miller R, Pearlson G, Adalı T. The cronnectome: Time-varying connectivity networks as the next frontier in fMRI data discovery. Neuron. 2014;**84**(2):262-274

[31] Wang Z, Zhang X, Dou W, Zhang M, Chen H, Lu M, et al. Best window width determination and glioma analysis application of dynamic brain network measure on resting-state functional magnetic resonance imaging. Journal of

Medical Imaging and Health Informatics. 2016;**6**(7):1735-1740

[32] Leonardi N, Van De Ville D. On spurious and real fluctuations of dynamic functional connectivity during rest. NeuroImage. 2015;**104**:430-436

[33] Mokhtari F, Akhlaghi MI, Simpson SL, Wu G, Laurienti PJ. Sliding window correlation analysis: Modulating window shape for dynamic brain connectivity in resting state. NeuroImage. 2019;**189**:655-666

[34] Zhang W, Wang Z, Dou W, Wang X, Lu M, Zhang M, et al. Dynamic features extraction method of resting-state BOLD-fMRI signal and its application to brain data classification between normal and glioma. 2014 12th International Conference on Signal Processing (ICSP); 2014

[35] Rashid B, Damaraju E, Pearlson GD, Calhoun VD. Dynamic connectivity states estimated from resting fMRI Identify differences among Schizophrenia, bipolar disorder, and healthy control subjects. Frontiers in Human Neuroscience. 2014;**8**(897)

[36] Yuan H, Phillips R, Wong CK, Zotev V, Misaki M, Wurfel B, et al. Tracking resting state connectivity dynamics in veterans with PTSD. NeuroImage: Clinical. 2018;**19**:260-270

[37] Zhu H, Huang J, Deng L, He N, Cheng L, Shu P, et al. Abnormal dynamic functional connectivity associated with subcortical networks in Parkinson's disease: A temporal variability perspective. Frontiers in Neuroscience. 2019;**13**(80)

[38] Aggarwal P, Gupta A. Multivariate graph learning for detecting aberrant connectivity of dynamic brain networks in autism. Medical Image Analysis. 2019;**56**:11-25

[39] Xia Y, Chen Q, Shi L, Li M, Gong W, Chen H, et al. Tracking the dynamic functional connectivity structure of the human brain across the adult lifespan. Human Brain Mapping. 2019;**40**(3):717-728

[40] Liu J, Liao X, Xia M, He Y. Chronnectome fingerprinting: Identifying individuals and predicting higher cognitive functions using dynamic brain connectivity patterns. Human Brain Mapping. 2018;**39**(2):902-915

[41] Mclntosh A, Gonzalez-Lima F. Structural equation modeling and its application to network analysis in functional brain imaging. Human Brain Mapping. 1994;**2**(1-2):2-22

[42] Friston KJ, Harrison L, Penny W. Dynamic causal modelling. NeuroImage. 2003;**19**(4):1273-1302

[43] Deshpande G, LaConte S, James GA, Peltier S, Hu X. Multivariate Granger causality analysis of fMRI data. Human Brain Mapping. 2009;**30**(4):1361-1373

[44] Granger CWJ. Investigating causal relations by econometric models and cross-spectral methods. Econometrica. 1969;**37**(3):424-438

[45] Razi A, Seghier ML, Zhou Y, McColgan P, Zeidman P, Park H-J, et al. Large-scale DCMs for resting-state fMRI. Network Neuroscience. 2017;**1**(3):222-241

[46] Friston KJ, Kahan J, Biswal B, Razi A. A DCM for resting state fMRI. NeuroImage. 2014;**94**:396-407

[47] Ge Y, Yang Z, Feng Y, Pan Y, Dou W. Extended network-based statistics for measuring altered directed connectivity components in the human brain. 2021 IEEE International Conference on Bioinformatics and Biomedicine (BIBM); 2021

[48] Sugihara G, May R, Ye H, Hsieh C-H, Deyle E, Fogarty M, et al. Detecting causality in complex ecosystems. Science. 2012;**338**(6106):496-500

[49] Duggento A, Guerrisi M, Toschi N. Echo state network models for nonlinear Granger causality. Philosophical Transactions of the Royal Society A: Mathematical, Physical and Engineering Sciences. 2021;**379**(2212):20200256

[50] Korhonen O, Zanin M, Papo D. Principles and open questions in functional brain network reconstruction. Human Brain Mapping. 2021;**42**(11):3680-3711

[51] Zheng W, Ge Y, Ren S, Ran W, Zhang X, Tian W, et al. Abnormal static and dynamic functional connectivity of resting-state fMRI in multiple system atrophy. Aging (Albany NY). 2020;**12**(16):16341-16356

[52] Zalesky A, Fornito A, Bullmore ET. Network-based statistic: Identifying differences in brain networks. NeuroImage. 2010;**53**(4):1197-1207

[53] Arbabshirani MR, Plis S, Sui J, Calhoun VD. Single subject prediction of brain disorders in neuroimaging: Promises and pitfalls. NeuroImage. 2017;**145**:137-165

[54] Kim B, Winstein C. Can neurological biomarkers of brain impairment be used to predict poststroke motor recovery? A systematic review. Neurorehabilitation and Neural Repair. 2017;**31**(1):3-24

[55] Stinear CM, Ward NS. How useful is imaging in predicting outcomes in stroke rehabilitation? International Journal of Stroke. 2013;**8**(1):33-37

[56] Solstrand Dahlberg L, Becerra L, Borsook D, Linnman C. Brain changes after spinal cord injury, a quantitative meta-analysis and review. Neuroscience & Biobehavioral Reviews. 2018;**90**: 272-293

[57] Thiel A, Vahdat S. Structural and resting-state brain connectivity of motor networks after stroke. Stroke. 2015;**46**(1):296-301

[58] Chen JL, Schlaug G. Increased resting state connectivity between ipsilesional motor cortex and contralesional premotor cortex after transcranial direct current stimulation with physical therapy. Scientific Reports. 2016;**6**(1):23271

[59] Allman C, Amadi U, Winkler AM, Wilkins L, Filippini N, Kischka U, et al. Ipsilesional anodal tDCS enhances the functional benefits of rehabilitation in patients after stroke. Science Translational Medicine. 2016;**8**(330):330re1-re1

[60] Caeyenberghs K, Clemente A, Imms P, Egan G, Hocking DR, Leemans A, et al. Evidence for training-dependent structural neuroplasticity in brain-injured patients: A critical review. Neurorehabilitation and Neural Repair. 2018;**32**(2):99-114

[61] Athanasiou A, Klados MA, Pandria N, Foroglou N, Kavazidi KR, Polyzoidis K, et al. A systematic review of investigations into functional brain connectivity following spinal cord injury. Frontiers in Human Neuroscience. 2017;**11**(517)

[62] Hu M, Ji F, Lu Z, Huang W, Khosrowabadi R, Zhao L, et al. Differential amplitude of low-frequency fluctuations in brain networks after BCI training with and without tDCS in stroke. 2018 40th Annual International Conference of the IEEE Engineering in Medicine and Biology Society (EMBC); 2018

[63] Pan Y, Dou W-B, Wang Y-H, Luo H-W, Ge Y-X, Yan S-Y, et al. Non-concomitant cortical structural and functional alterations in sensorimotor areas following incomplete spinal cord injury. Neural Regeneration Research. 2017;**12**(12):2059-2066

[64] Ge Y, Pan Y, Wu Q, Dou W. A distance-based neurorehabilitation evaluation method using linear SVM and resting-state fMRI. Frontiers in Neurology. 2019;**10**(1105)

[65] Song J, Nair VA, Young BM, Walton LM, Nigogosyan Z, Remsik A, et al. DTI measures track and predict motor function outcomes in stroke rehabilitation utilizing BCI technology. Frontiers in Human Neuroscience. 2015;**9**(195)

[66] Wu Q, Yue Z, Ge Y, Ma D, Yin H, Zhao H, et al. Brain functional networks study of subacute stroke patients with upper limb dysfunction after comprehensive rehabilitation including BCI training. Frontiers in Neurology. 2020;**10**(1419)

[67] Gilman S, Wenning GK, Low PA, Brooks DJ, Mathias CJ, Trojanowski JQ, et al. Second consensus statement on the diagnosis of multiple system atrophy. Neurology. 2008;**71**(9):670-676

[68] Huppertz H-J, Möller L, Südmeyer M, Hilker R, Hattingen E, Egger K, et al. Differentiation of neurodegenerative parkinsonian syndromes by volumetric magnetic resonance imaging analysis and support vector machine classification. Movement Disorders. 2016;**31**(10):1506-1517

[69] Planetta PJ, Kurani AS, Shukla P, Prodoehl J, Corcos DM, Comella CL, et al. Distinct functional and macrostructural brain changes in Parkinson's disease and multiple system atrophy. Human Brain Mapping. 2015;**36**(3):1165-1179

[70] Pellecchia MT, Barone P, Mollica C, Salvatore E, Ianniciello M, Longo K, et al. Diffusion-weighted imaging in multiple system atrophy: A comparison between clinical subtypes. Movement Disorders. 2009;**24**(5):689-696

[71] Chen B, Fan G, Sun W, Shang X, Shi S, Wang S, et al. Usefulness of diffusion-tensor MRI in the diagnosis of Parkinson variant of multiple system atrophy and Parkinson's disease: A valuable tool to differentiate between them? Clinical Radiology. 2017;**72**(7):610-615

[72] You H, Wang J, Wang H, Zang Y-F, Zheng F-L, Meng C-L, et al. Altered regional homogeneity in motor cortices in patients with multiple system atrophy. Neuroscience Letters. 2011;**502**(1):18-23

[73] Wang N, Edmiston EK, Luo X, Yang H, Chang M, Wang F, et al. Comparing abnormalities of amplitude of low-frequency fluctuations in multiple system atrophy and idiopathic Parkinson's disease measured with resting-state fMRI. Psychiatry Research: Neuroimaging. 2017;**269**:73-81

[74] Yao Q, Zhu D, Li F, Xiao C, Lin X, Huang Q, et al. Altered functional and causal connectivity of cerebello-cortical circuits between multiple system atrophy (Parkinsonian Type) and Parkinson's disease. Frontiers in Aging Neuroscience. 2017;**9**(266)

[75] Ren S, Zhang H, Zheng W, Liu M, Gao F, Wang Z, et al. Altered functional connectivity of cerebello-cortical circuit in multiple system atrophy (Cerebellar-Type). Frontiers in Neuroscience. 2019;**12**(996)

[76] Ge Y, Zheng W, Li Y, Dou W, Ren S, Chen Z, et al. Altered Brain Volume,

Microstructure Metrics and Functional Connectivity Features in Multiple System Atrophy. Frontiers in Aging Neuroscience. 2022;**14**:799251. DOI: 10.3389/fnagi.2022.799251

[77] Mamelak AN, Jacoby DB. Targeted delivery of antitumoral therapy to glioma and other malignancies with synthetic chlorotoxin (TM-601). Expert Opinion on Drug Delivery. 2007;**4**(2):175-186

[78] Wong ET, Brem S. Taming glioblastoma: Targeting angiogenesis. Journal of Clinical Oncology. 2007;**25**(30):4705-4706

[79] Li Y. Multi-level Functional Network Characteristic Analysis for Glioma Patients Based on fMRI Information. Beijing, China: Tsinghua University; 2015

[80] Constable RT. Challenges in fMRI and its limitations. In: Faro SH, Mohamed FB, editors. Functional MRI: Basic Principles and Clinical Applications. New York, NY: Springer New York; 2006. pp. 75-98

[81] Manglore S, Bharath R, Panda R, George L, Thamodharan A, Gupta A. Utility of resting fMRI and connectivity in patients with brain tumor. Neurology India. 2013;**61**(2):144-151

[82] Esposito R, Mattei PA, Briganti C, Romani GL, Tartaro A, Caulo M. Modifications of default-mode network connectivity in patients with cerebral glioma. PLOS ONE. 2012;**7**(7):e40231

[83] Harris RJ, Bookheimer SY, Cloughesy TF, Kim HJ, Pope WB, Lai A, et al. Altered functional connectivity of the default mode network in diffuse gliomas measured with pseudo-resting state fMRI. Journal of Neuro-Oncology. 2014;**116**(2):373-379

[84] Fortin J-P, Parker D, Tunç B, Watanabe T, Elliott MA, Ruparel K, et al. Harmonization of multi-site diffusion tensor imaging data. NeuroImage. 2017;**161**:149-170

[85] Fortin J-P, Cullen N, Sheline YI, Taylor WD, Aselcioglu I, Cook PA, et al. Harmonization of cortical thickness measurements across scanners and sites. NeuroImage. 2018;**167**:104-120

Section 2

Neuropathophysiology

Chapter 7

Neuroimaging in Common Neurological Diseases Treated by Anticoagulants

Pipat Chiewvit

Abstract

Stroke imaging/Cerebral Venous sinus thrombosis/Arterial dissecting disease in Head and Neck regions/Neurocomplication of anticoagulation therapy. Nowsday, anticoagulant drugs are common drugs used in daily practice for patients in neurology clinic. Anticoagulant treatment used for treated symptomatic patients as well as for prophylaxis therapy in asymptomatic patients. The purpose of this chapter based on the review of essential neuroimaging in the most common neurological conditions that benefit from treatment with anticoagulant drugs such as ischemic stroke, cerebral venous sinus thrombosis, and arterial dissecting disease of head and neck arteries and will be enclosed with neuroimaging in case of neurocomplication by anticoagulant therapy.

Keywords: stroke, intracerebral hemorrhage, cerebral infarction, CT, MRI, CTA, CTV, CTP, MRA, MRV, angiography, cerebral venous thrombosis, arterial dissection, neurocomplication, anticoagulation

1. Introduction

Stroke is a neurovascular disease with a high incidence throughout the world; according to the global stroke fact sheet 2019 by World Stroke Organization, stroke is the second leading cause of death and the third leading cause of disability. As up to 13.7 million new cases of stroke each year, globally one in four people over age 25 will have a stroke in his or her lifetime. About 5.5 million people die from stroke annually, and 39% of all who die from stroke are under 70 years old [1]. The burden of stroke is due to widely prevalent risk factors such as hypertension, diabetes, smoking, metabolic syndrome, and behavioral factors. Modification risk factors and administration preventive therapy are the keys to decrease stroke burden. Besides of the systemic disease such as hypertension can cause stroke, a number of cardiac diseases and vascular diseases affected intracranial arteries such as non-valvular atrial fibrillation, prosthetic heart valve, dilated cardiomyopathy, atrial appendage thrombus and atrial myxoma, atherosclerosis disease of aortic arch, neck arteries and intracranial arteries, arterial dissection etc. are the common causes of stroke [2–4]. Those conditions are prone to cause an embolic stroke (cardioembolic or artery to artery embolism). Antithrombotic agents such as antiplatelet drugs and anticoagulants are the drugs that are efficient in the prevention of embolic stroke. Thus, a majority of patients who have high risk factors in the group of stroke experience and inexperience stroke might have a regular anticoagulant administration. Regular monitoring of prothrombin time and International normalized ratio (INR) to predict the bleeding complications. Because ischemic and hemorrhagic strokes have different causes and effects on the body, both require different treatments. The accurate diagnosis types of strokes as ischemic stroke or hemorrhagic stroke is important. Neuroimaging is the investigate procedure to make a correct diagnosis type of stroke and extension of stroke lesion that take part in triage stroke 2patients for further proper management.

Guideline recommendation for stroke prevention (ischemic stroke/TIA) [5]:

- To evaluate the cause of ischemic stroke.
- Getting rid of all risk factors is important for secondary stroke prevention such as well-controlled DM/hypertension, pause of smoking, and decreased lipid level.
- Changing patients to a healthy diet (low salt, low cholesterol rich diet) and promoting physical activity such as exercise.
- The administration of antithrombolytic therapy in the patients who have no contraindication to antiplatelet or anticoagulant treatment such as combination of antiplatelets and anticoagulation or dual antiplatelet therapy is not recommended in long term treatment. For the short term treatment, dual antiplatelet therapy is recommended in specific patients such as early arriving minor stroke, high risk transient ischemic stroke or severe sympatimatic intracranial stenosis etc.
- Atrial fibrillation remains a common and high-risk condition for second ischemic stroke. Anticoagulation is usually recommended if the patient has no contraindications. Heart rhythm monitoring for occult atrial fibrillation is usually recommended if no other cause of stroke is discovered.
- Extracranial carotid artery disease is an important and treatable cause of stroke. Patients with severe stenosis ipsilateral to a nondisabling stroke or transient ischemic attack who are candidates for intervention should have the stenosis fixed, likely relatively early after their ischemic stroke. The choice between carotid endarterectomy and carotid artery stenting should be driven by specific patient comorbidities and features of their vascular anatomy.
- Patients with severe intracranial stenosis in the vascular territory of ischemic stroke or transient ischemic attack should not receive angioplasty and stenting as first-line therapy for preventing recurrence. Aggressive medical management of risk factors and short-term dual antiplatelet therapy is preferred.
- Patients with embolic stroke of the uncertain source should not be treated empirically with anticoagulants or ticagrelor because it was found to be of no benefit.

Stroke has three major subtypes:

1. Ischemic stroke

2. Hemorrhagic stroke

3. Stroke mimic conditions

Neuroimaging modalities as noninvasive techniques for the diagnosis of stroke are mainly based on computed tomography (CT) or magnetic resonance imaging (MRI) studies such as noncontrast CT study, CT angiography (single-phase or multiphase CTA), CT perfusion (CTP), CT venography (CTV) and dual-energy CT (DECT) or MRI, MR angiography, MR perfusion, etc. invasive study such as digital subtraction catheterized angiography (DSA) is usually preserved for further treatment by endovascular options such as mechanical thrombectomy, coiling aneurysm, etc. Imaging plays a key role in current guidelines for thrombolysis. The knowledge of typical early ischemic signs in nonenhanced computed tomography (CT) is necessary.

Computed Tomography (CT): According to "Time is brain" the patient in stroke condition in the emergency department must be transferred fast and must have a correct diagnosis. Noncontrast CT scan (NCCT) must be performed as soon as possible after an activated stroke. NCCT is preferred in most institutes/centers and used as the gateway in triage patients who have a clinical stroke, which may be caused by an ischemic process or hemorrhagic process or even mimic stroke conditions (e.g., vascular malformation or tumor). If there is a need for further investigation, contrast-enhanced study can be performed such as CTA, CTV, CTP, multiphase CTA, or even regular contrast-enhanced CT brain, etc. Nowadays, the complete CT stroke protocol for triage patients for mechanical thrombectomy is composed of NCCT brain, CTA (multiphases), and CT perfusion, which can be performed in a single examination, which can be completed in 15 minutes.

NCCT: Noncontrast CT interpretation based on knowledge of brain vascular anatomy and basic density on CT scan together with clinical information is needed for correct diagnosis. After noncontrast CT scan can rule out intracranial hemorrhage or stroke mimic conditions, the patient can proceed to have an intravenous thrombolysis drug with a recombinant tissue-type plasminogen activator (r-tPA) for treatment. Exceptional case that have contraindication for IV treatment (**Table 1**) [6]. CTA and CTP will be further studies for decision-making in management by interventional therapy.

ACUTE stage (onset to 48hrs.): Ischemic stroke pattern composed of two main components as vascular structures and parenchyma changes. The vascular structures CT finding is "hyperdense artery sign, hyperdense vessel sign" which means acute intraluminal clot obstructs in the intracranial arteries such as middle cerebral artery (MCA), internal carotid artery (ICA), basilar artery appears as high density of blood clot by CT along the course of that artery in linear or dot patterns (**Figures 1** and **2**). However, these findings have high specificity up to 90% but low sensitivity of approximately 30% [7–9]. Whenever encountered with hyperdense artery sign on NCCT brain, always correlation to clinical information as well as compare to the contralateral side and venous structures, such as transverse sigmoid sinus, superior sagittal sinus to reduce false-positive conditions That cause high density in the arteries, veins such as hemoconcentration condition. The early CT findings in parenchymatous change from ischemic stroke are mainly hypoattenuation area from cytotoxic edema that corresponds to the territory of occluded artery, visible by NCCT within the first few hours. For example, when there is proximal middle cerebral artery occlusion, there may be hypoattenuation area in the following structures: basal ganglia, internal capsule,

Absolute contraindication
• Acute ICH including hemorrhagic infarction
• History of ICH (microbleed is not contraindicated)
• BP > 185 mmHg, BPs > 110 mmHg
• Serious head trauma or stroke < 3 months
• Thrombocytopenia (platelet <100,000/mm^3) and coagulopathy (PT > 15, INR > 1.7)
• LMWH within 24 hr (38% risk of sICH, 29% risk of death, 33% favorable outcomes, od 0.84 for sICH, od5.3 for death)
• DTIs (TT is sensitive to presence of DITs)
• Factor Xa inhibitor
• Severe hypoglycemia(<50 mg/dl) and hyperglycemia(>400 mg/dl) may be permitted for IVT
• Early radiographic ischemic changes(>1/3 MCA)
Relative contraindication
• Advanced age (>80 yrs.)
• Mild or improving stroke (NIHSS<5)
• Severe stroke (IVT should be cautiously administered in NIHSS>25 at 3–4.5 hrs.)
• Major surgery within 2 weeks (some 3 months)
• Arterial puncture of noncompressible site
• Recent GI bleeding or GU bleeding
• Seizure at onset
• Recent MI within 3 months
• Intracranial structure abnormality (axial tumor, AVM, aneurysm)
• Dementia

Table 1.
Absolute and relative contraindications to IV rt-PA for acute ischemic stroke [6].

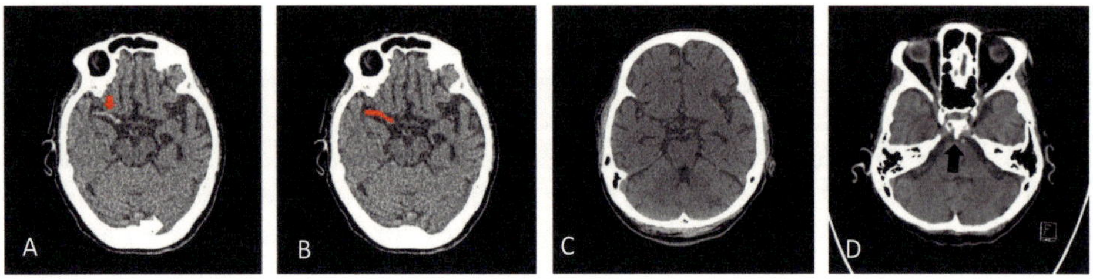

Figure 1.
A, B, C, D NCCT brain demonstrates hyperdense artery sign in horizontal segment (M1 segment) of right middle cerebral artery (red arrow, red highlighted in A,B) and Sylvian segment (m2 segment) of right middle cerebral artery (dot sign, white arrow, C). Hyperdensity of blood clot in patient with clinical basilar artery thrombosis (dot sign, black arrow, D).

insular cortex, frontal operculum, temporal operculum, temporal lobe convexity, frontoparietal convexity. Basically, nature of basal ganglia structures (caudate and lentiform nucleus) are gray matter structures which normally slightly high attenuation to nearby white matter as internal capsule, external capsule. Therefore, acute ischemic

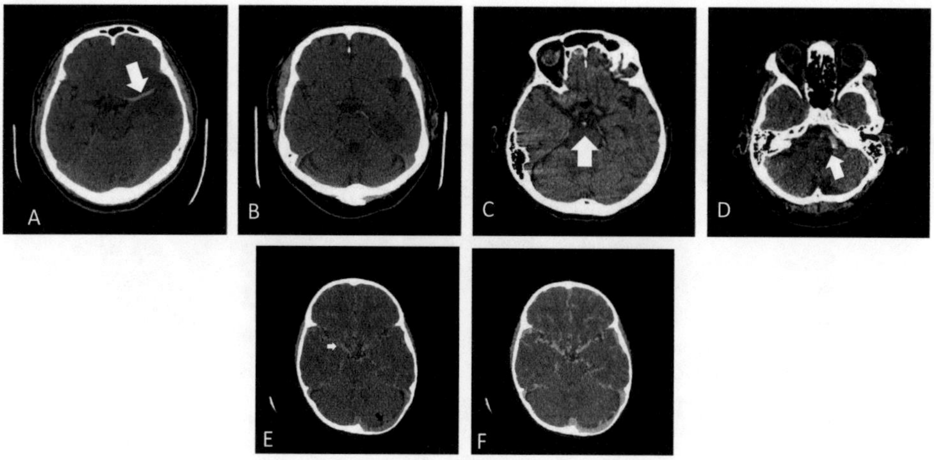

Figure 2.
A, B, C, D NCCT brain in a 59-year-old patient underlying AF with progressive right hemiplegia demonstrates hyperdense artery sign in horizontal segment (M1 segment) of left middle cerebral artery (A). NCCT brain in a 57-year-old male presenting with right homonymous hemianopia; hyperdense left PCA artery sign is demonstrated (B). NCCT brain in a 79-year-old female with alteration of consciousness; hyperdensity basilar artery sign demonstrates thrombosis (dot sign, C). NCCT brain in a 57-year-old man underlying hypertension and dyslipidemia; hyperdense artery sign is demonstrated at left vertebral artery (D). E,F NCCT brain demonstrates false positive hyperdense artery in horizontal segment (M1 segment) of right middle cerebral artery (white arrow, E) as well as contralateral side left MCA and venous structures such as left transverse sinus (black arrow, E). Normal opacification by contrast medium all arteries and venous structures are demonstrated on contrast enhanced CT scan (F). This patient is dehydrated from high fever and headache with hemoconcentration condition.

stroke change in basal ganglia will be hypoattenuation that called "obscuration lentiform nucleus sign" (**Figure 3**). In the same manner, the insular cortex that locates lateral to basal ganglia and external capsule will be hypodensity and loss of normal cortical lobulated pattern are called as "loss of cortical ribbon sign" (**Figure 3**). If cortical and subcortical gray-white matter is involved, NCCT will demonstrate loss of gray-white matter differentiation and effacement of cortical gyri (**Figure 4**). Application of this parenchymatous changes pattern in CT scan of MCA can use in other territories such as ACA, PCA, SCA, AICA and PICA (**Figure 5**). According to the study by Philip AB, Andrew MD, Jinjin Z, Alastair MB et.al (2000), the Alberta stroke program early CT score (ASPECTS) is widely used due to its simplicity and reliability in predicting functional outcomes and symptomatic intracerebral hemorrhage after intravenous treatment by alteplase [10]. It is a quantitative CT grading system precisely evaluating MCA territory by allotting 10 points in 10 different locations that cover entirely the MCA territory. The normal total score is 10. Each point will be subtracted in any one area of early ischemic change (visible hypodensity abnormality); therefore, the total score is 0 referring to ischemic change in the entire MCA territory. The study suggests that if patient CT ASPECTS is 7 or less, the risk of symptomatic ICH with alteplase is 14 times greater than ASPECTS greater than 7. In patients with scores above 7, the rate of symptomatic intracerebral hemorrhage is 1%. When hemorrhagic transformation develops in the area of cerebral infarction, hyperdensity of acute hemorrhage will be encountered (**Figure 6**).

Subacute stage (2 days to 2 weeks): After maximum swelling of hypoxic cell, damage of BBB will increase leakage of intracellular fluid to extracellular fluid space causing vasogenic edema. The infarcted area in subacute stage will be easily

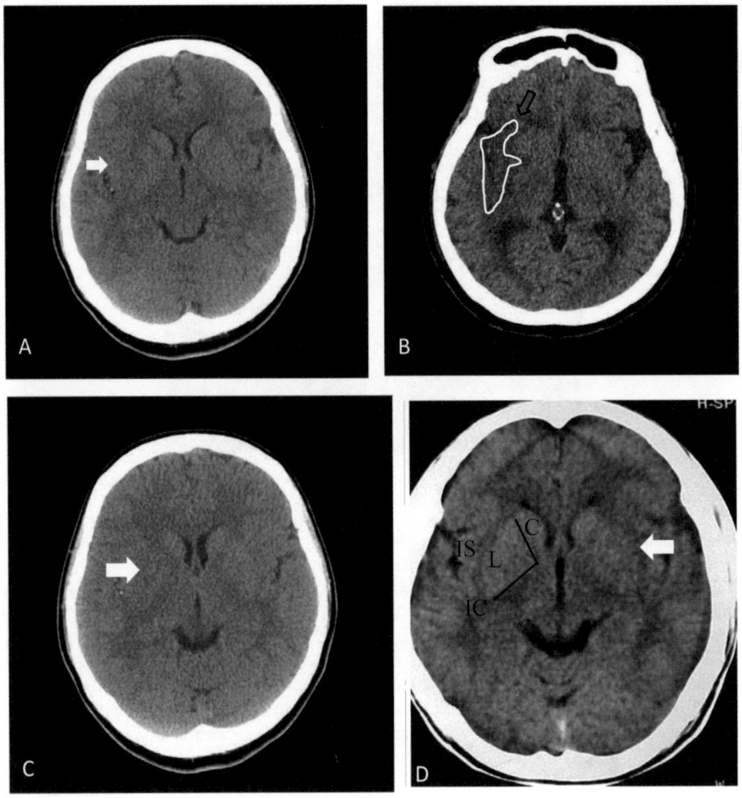

Figure 3.
A, B, C, D NCCT brain at level of basal ganglia demonstrates "loss of insular ribbon sign " in right insular cortex (white arrow, A and white outline area. B) and obscuration of lentiform nucleus sign" in right lentiform nucleus (white arrow, C) and in left lentiform nucleus (white arrow, D).

detected by NCCT scan due to high-contrast resolution between swelling tissue and adjacent normal tissue. In the early subacute stage of the large territory infarction, increased pressure effect to midline structure, ventricular system may collapse or there may be displacement; therefore,subfalcine herniation or uncal or descending transtentorial herniation may occur. In addition, the high incidence of hemorrhagic transformation in the territory infarction (10–43%) [11] will increase this risk for the patient who has antithrombolytic therapy. If such a circumstance occurs, there is a new high attenuation lesion of the acute hemorrhage within the area of infarction on the follow-up CT scan. In the late subacute stage, the edema will subside with the decreased degree of brain swelling. If the patient is performed contrast enhanced CT scan, gyral enhancement is demonstrated in the subacute stage of cerebral infarction (**Figures 6B** and 7).

Chronic, old stage (2 weeks to 2 months): Further resolution in the degree of brain swelling, as well as degeneration of the dead infarcted brain, is observed. The density of old infarction will be decreased to near CSF density level in some cases. Signs of brain volume loss in the old infarction area may occur as ex vacuo dilatation of ipsilateral ventricle expand toward the old infarction area, retrogradely Wallerian degeneration along ipsilateral corticospinal tract in the lower level to the infarction, etc. (**Figure 8**).

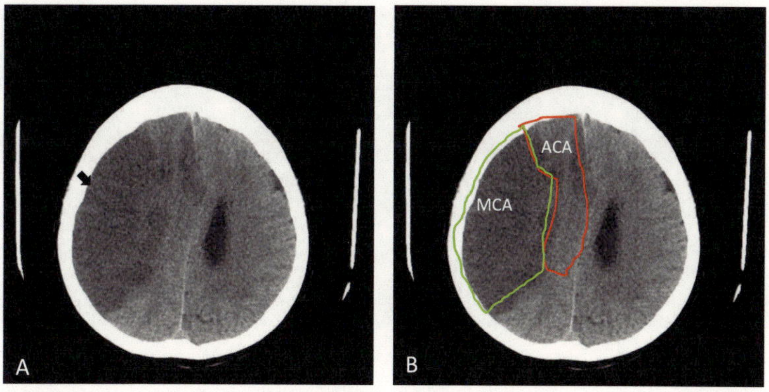

Figure 4.
A, B NCCT brain at level of supraganglionic demonstrates hypodensity areas with "loss of gray-white matter differentiation " and " loss of cerebral sulci" in right frontoparietal lobes (black arrow, A and color outlines. B) The right convexity lesion is according to right MCA territory whereas right frontal parasagittal lesion is according to right ACA territory.

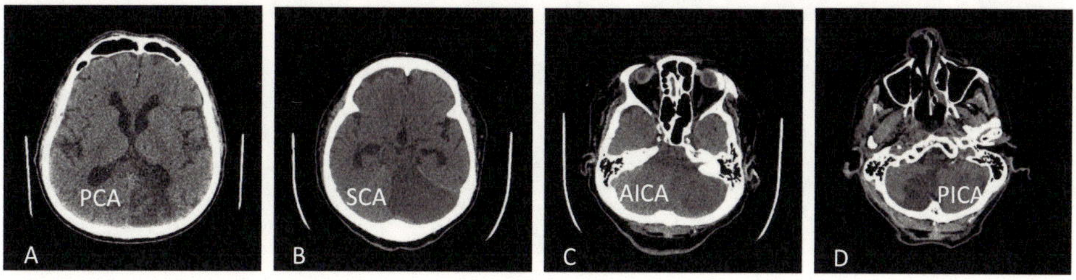

Figure 5.
A, B ,C, D NCCT brain at different levels of posterior circulation territories demonstrates hypodensity areas of acute infarction in left occipital lobe according to left distal PCA territory (A), left superior cerebellum according to left superior cerebellar artery (B), left anterior inferior cerebellum according to AICA territory (C) and right posteroinferior cerebellum according to right PICA territory (D).

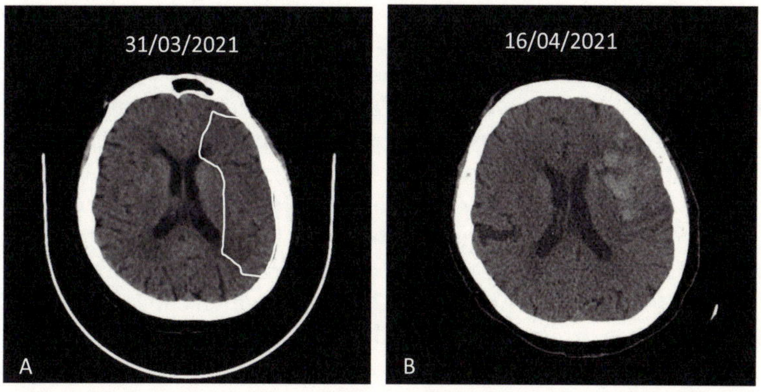

Figure 6.
A, B NCCT brain in acute stroke of left MCA territory demonstrates ill-defined hypodensity areas of acute infarction in left frontoparietal convexity according to left MCA territory (A), and two weeks follow-up study reveals hyperdensity on top of the infarction area compatible with hemorrhagic transformation in subacute stage (B).

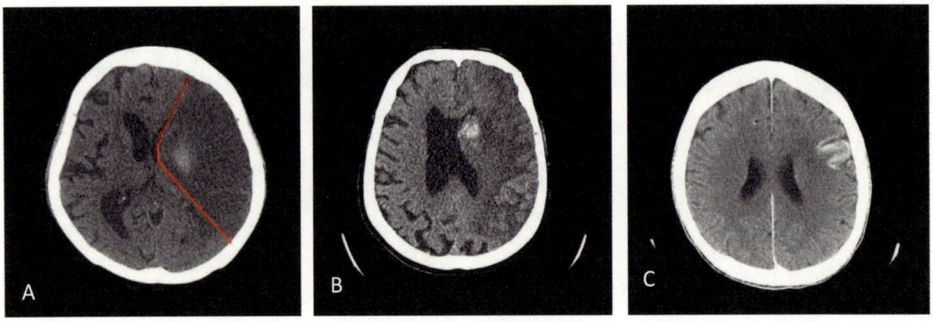

Figure 7.
A, B, C: NCCT brain in subacute stage of left MCA territory infarction (A) increased degree of brain swelling, midline shifted and hemorrhagic transformation developed. Late subacute stage, diminished degree of brain swelling and midline shifted (B). Gyral enhancement along the cortex is demonstrated (C).

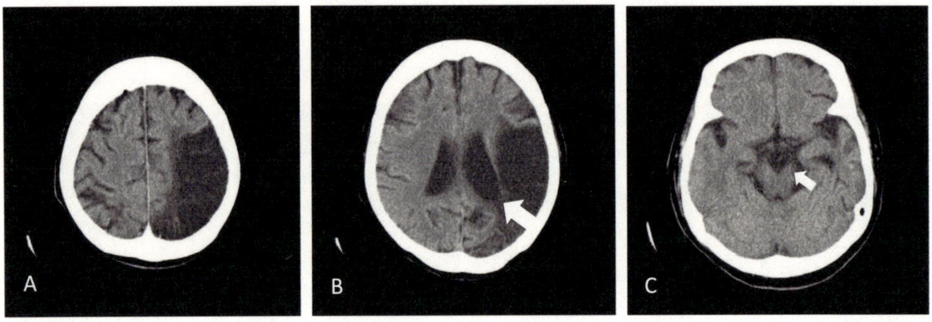

Figure 8.
A, B, C: NCCT brain in old stage of left MCA territory infarction (A) decreased degree of brain swelling to the near, equal to CSF density level. Ex vacuo dilatation of left lateral ventricle (B). Retrogradely Wallerian degeneration is demonstrated in left cerebral peduncle (C).

CT ANGIOGRAPHY (CTA): It is the noninvasive technique to visualize intracranial arteries by means of noncatheterization technique. In acute ischemic stroke, CTA is increasingly used for imaging the vessel and a combination of NCCT and CTA brain provides sufficient information to determine eligibility for thrombectomy in the first 6 hours. CTA can be performed in single-phase or multiphases during one injection of contrast medium. The outstanding advantage of multiphase (three phases) CTA over a single phase is in the detection of collateral circulation to the area of infarction that assists in decision making for further treatment whether the patient has poor collateral circulation or not due to the high risk of symptomatic ICH after IV Alteplase in acute ischemic stroke with poor collateral circulation [10]. By using multiphase CTA brain, the study can be performed in three contiguous phases of CTA as 1st phase, 2nd phase, and 3rd phase, which can be classified the stroke patients in three groups of collateral circulation: Good, Intermediate, Poor collateral circulation (**Figure 9**). In some center, CTA protocol in single phase and 1st phase of multiphase CTA, the coverage of artery extends to aortic arch level to include the origin of three main arteries as the right brachiocephalic trunk, left common carotid artery (CCA), and left subclavian arteries that main trunks from those arteries as bilateral internal carotid artery (ICA) and bilateral vertebral arteries provide bloodstream to intracranial level. CTA brain findings in a patient with acute

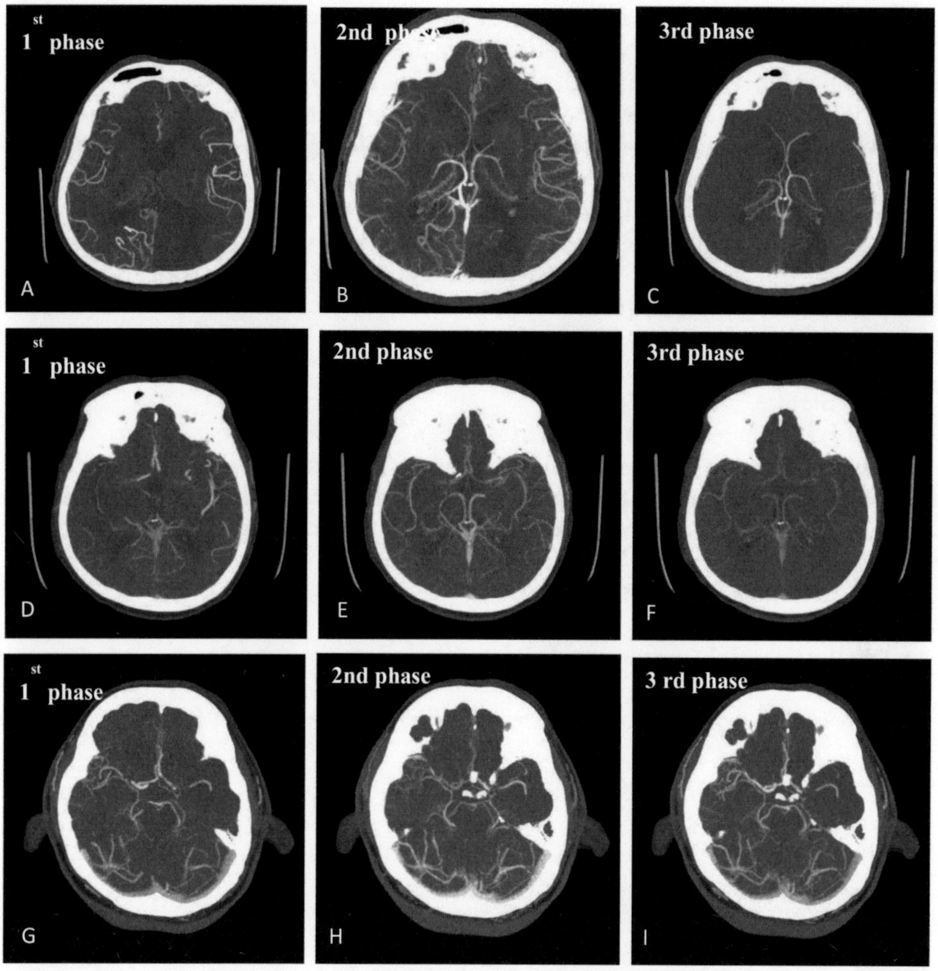

Figure 9.
A-I Multiphases CTA brain in three patients, 1st row (A, B, C) 1st case left proximal MCA occlusion with good collateral score, 2nd row (D, E, F) 2nd case right proximal MCA occlusion with fair collateral score and 3rd row (G,H, I) 3rd case left proximal MCA occlusion with poor collateral score.

ischemic stroke by arterial occlusion are abrupt disappearance of contrast medium opacification at the thrombus site as "artery cut off sign". Distal to the thrombus, the distal run-off arteries can be variable case by case reconstitute by collateral circulation. In addition, CTA provides information on the arteries' status such as intracranial arteriosclerosis disease, vasculopathy, dissecting artery, intracranial aneurysm, cerebral AVM, dural AVM, assessment degree of arterial stenosis, and screening whole axis blood supplied intracranial system from aortic arch to vertex.

CT Perfusion (CTP): It is performed by monitoring only the first pass of an iodinated contrast agent bolus through the cerebral circulation. This principle is used to generate time-attenuation curves for an arterial ROI, a venous ROI, and each pixel. use a semiautomatic option, which consists of first manually tracing a large ROI around the vessel and then letting the software automatically select an accurate ROI. The arterial ROI is optimally selected in one unaffected vessel that is perpendicular to the acquisition plane, either one of the anterior cerebral arteries (ACAs) or the contralateral MCA. In emergency settings, we prefer to select an ACA as the default arterial input function for simplicity because it has been shown to be adequate. The venous ROI is placed over the superior sagittal sinus or torcular Herophili. The software will generate color images of cerebral blood flow (CBF), cerebral blood volume (CBV), and mean transit time (MTT). In the emergency settings, the quick interpretation of CTP is first seen MTT images for the area of increased MTT, if the area of image match to the clinical setting, then further analysis to CBF, CBV in the same area match to MTT image. If CBF and CBV are matched decreased, this is all area of the infarct core (no penumbra) but CBF decreased with normal CBV (or increased) will suggest ischemic penumbra (**Figure 10**).

Summary Findings: CT Brain in Stroke

Acute Stroke (Day 0 to Day2)

Vascular Changes: - Hyperdense artery sign

P**arenchymatous Changes:** - Faint hypodensity area/Loss of gray-white matter differentiation/effacement of cerebral, cerebellar sulci/narrowing of subarachnoid spaces. CT signs for middle cerebral artery (MCA) territory infarction-loss of insular ribbon sign, obscuration of the lentiform nucleus may present.

Subacute Stage (Day 3 to 2 weeks):

- Usually no visible intraluminal clot density
- Increased sharpness (well-demarcated area) of hypodensity of the infarcted area
- Increased mass effect from cytotoxic edema

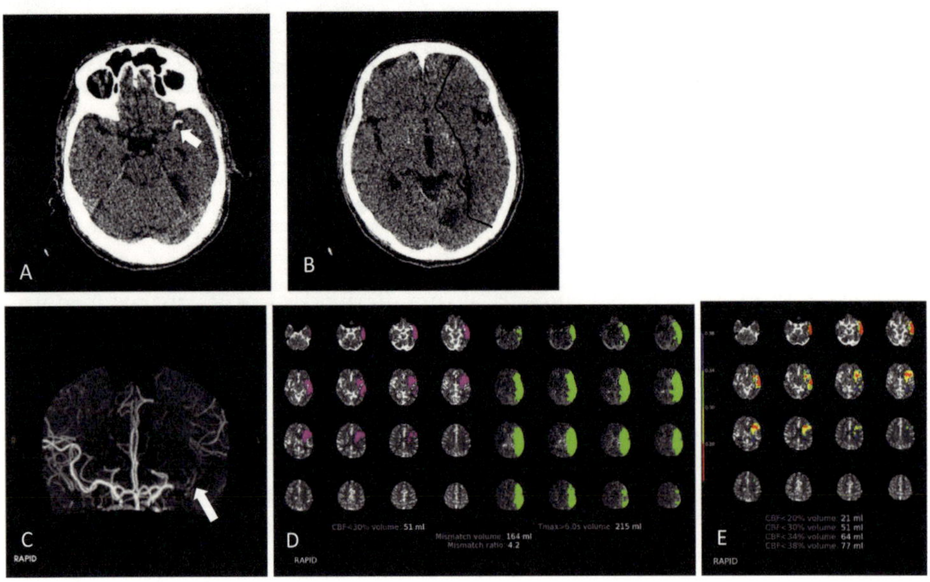

Figure 10.
A-E CT-CTA-CTP: Case Acute ischemic stroke presents with right hemiparesis, NCCT brain demonstrated hyperdense clot sign at left M1 segment (A) and early ill-defined hypodensity areas in left basal ganglia and left frontotemporoparietal lobes (B). CTA brain (C) complete occlusion of left proximal MCA. CTP brain (D,E) mismatch defect (CBF-CBV map) in left cerebral hemisphere.

- Potential to hemorrhagic transformation (looking for hyperdensity lesion among the infarcted area)

Chronic Stage (more than 2 weeks)

- Decreased degree of brain edema.
- Hypodensity of the infarction is equal to CSF density.
- Signs of brain volume loss at old infarcted area such as ex vacuo dilatation of adjacent ventricle, widening of adjacent subarachnoid spaces, such as sulci, folia, and fissure dilatation.
- Retrograde Wallerian degeneration along the corticospinal tract may occur at the level of cerebral peduncle and pons.

MRI (Magnetic Resonance Imaging): In stroke imaging, MRI is one of the noninvasive techniques in the diagnosis of ischemic stroke and other types of strokes such as hemorrhage and stroke mimic conditions [12].

The basic principle of MR machine comprises a magnet embedded within the MRI scanner imaging use of radiofrequency (RF) wave to the brain and activate proton in all different tissues in the area scanned to spin, after that the proton is returning to its original state known as precession with release RF wave to the receiver coil. The different tissue types within the brain return at different rates and it allows us to visualize and differentiate tissues in the brain. MRI is superior to CT in terms of lack of radiation, excellent tissue contrast resolution in spatial and temporal resolution between different normal tissues in the brain, and also well discriminate between normal tissue and the abnormalities. Nevertheless, some disadvantages of the MRI study are it is a long duration scan, it needs the cooperation of the patient, it needs close monitoring, it is high cost, and it is not widely available in all hospitals.

MRI Acute stage: Stroke protocol (Axial DWI/ADC map, T2W, T2W_FLAIR, SWI) (**Figure 11**)

- Diffusion weighted image (DWI) is a very important pulse sequence for diagnosing acute ischemic stroke, DWI can early detect acute ischemic stroke 80–90% of cases. Restrictive diffusion lesion in DWI is the area of cytotoxic edema that will be bright (hypersignal) and typically together with a low ADC value in the ADC map image (hyposignal) (**Figures 11** and **12**). The pattern of DWI will

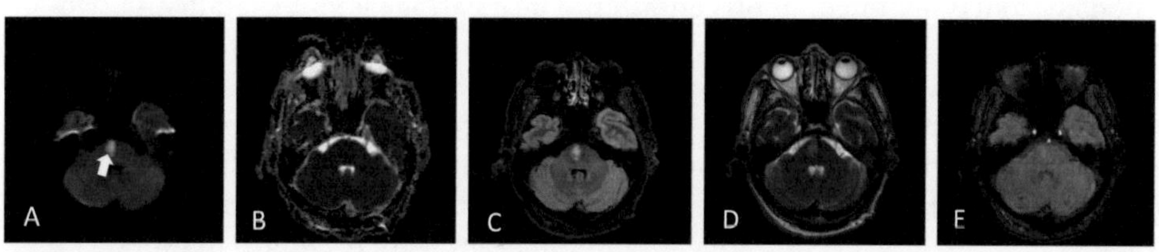

Figure 11.
A-E MRI brain (stroke protocol) composed of Axial DWI/ADC (A,B), T2W_FLAIR (C), T2WSE (D), SWI (E) for basic evaluation of acute infarction, exclusion of hemorrhagic stroke.

differentiate large vessel occlusion, small vessel infarction or cardiac emboli, or border zone infarction.

- T2WSE is for stroke mimic condition.

- T2W_FLAIR looks for the nonmatched area (T2W_FLAIR-DWI) in the acute stage, which probably represents an area of penumbra, evaluates white matter leukoaraiosis, and some time will help in the diagnosis of subarachnoid hemorrhage by MRI study.

- SWI (susceptibility weighted image) uses for the detection whether hemorrhagic transformation in area of infarction and also provides detection of cerebral hemorrhage prior to treatment or follow up after IV thrombosis treatment and also detect cerebral microbleeds (CMBs). The paramagnetic substances in hemorrhage display blooming of hyposignal (dark signal, signal void) lesion which is in contrast to CT scan hemorrhage is hyperdensity (white color).

Subacute-chronic stage

On the subacute stage of infarction, there is no longer visualization of the hypersignal restrictive diffusion but hypersignal in T2W_FLAIR and T2WSE are well-delineated (**Figure 13**) in the early stage of subacute infarction, degree of mass effect from brain edema is still present.

Chronic stage of infarction, the infarcted brain tissue will be phagocytosis by macrophage with fluid replaced as old CSF lesion in the brain, therefore, chronic or old infarction will have a signal intensity similar to CSF, hyposignal T2W_FLAIR and hypersignal in T2WSE (**Figure 13**).

MR Angiography (Figure 14): Similar to CT scan that uses CTA for evaluation of vessel lumen, MR imaging also has 3D time-of-flight (3D-TOF) use of the movement of blood, and the flowing blood that enters the volume imaged will produce a signal for MRA imaging in the assessment of the intracranial and extracranial arteries. The advantage of MRA is that it is a nonradiation and noncontrast medium administration. The interpretation of MRA is almost similar to CTA for diagnosis of stenosis, occlusion, aneurysm, and AVM. The limitations in interpretation of MRA are inability to visualize calcium deposit at blood vessel, slow flow phenomenon, complex flow phenomenon etc. Those circumstances may cause overestimate of occlusion in the noncontrast MRA study. The hypersignal T1 lesion such as subacute hemorrhage will persist in TOF technique of MRA and may influence in MRA analysis. If one needs a definite evaluation of whether it is a true occlusion, true severe stenosis, or merely complex flow phenomenon, contrast enhanced MRA is helpful.

Cerebral Venous Sinus Thrombosis (CVST) [13–17]: Stroke on the cerebral venous system is not uncommon condition. The mean age is young adult and two-thirds of patients are women. It is caused by complete or partial occlusion of either cerebral venous system such as dural venous sinus (superior sagittal sinus, inferior sagittal sinus, transverse and sigmoid sinuses), cortical veins, deep venous system (such as thalamostriate vein, internal cerebral vein or basal vein of Rosenthal, vein of Galen and straight sinus) or in combinations. The mainstay challenge in diagnosis of CVST is due to nonspecific clinical symptoms, and widely clinical manifestations often mimicking other acute neurological conditions. Headache is the most common symptom of CVST in about 90% of cases and may be localized or diffuse

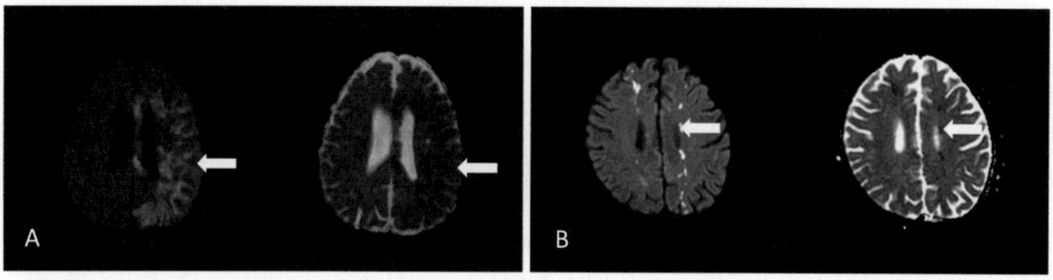

Figure 12.
A–B MRI brain (ACUTE INFARCTION) DWI and ADC map demonstrates restrictive diffusion lesion (hypersignal in DWI and hyposignal in ADC map) in left body of corpus callosum and left frontoparietal convexity compatible with acute cerebral infarction (left ACA and MCA territory, **Figure 11A***) and borderzone infarction array punctate restrictive diffusion lesions (hypersignal DWI/ hyposignal ADC) in central of both frontoparietal lobes.*

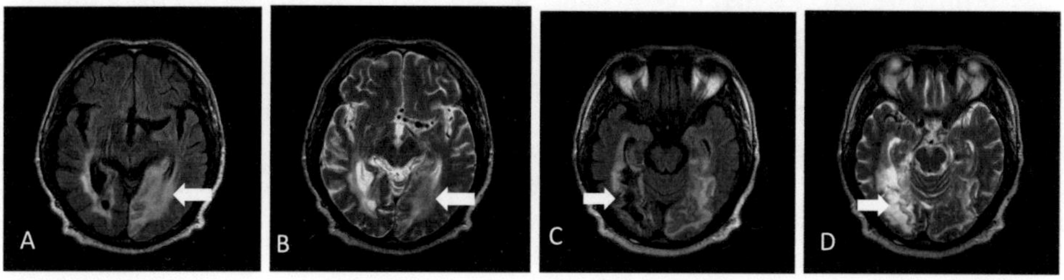

Figure 13.
A, B, C, D: MRI brain (subacute and chronic infarction) Axial T2W_FLAIR (12 A) and Axial T2WSE (12B) reveals hypersignal lesion in left occipital lobe on T2W_FLAIR and T2WSE with pressure effect to compress left occipital horn suggests of subacute cerebral infarction Axial T2W_FLAIR (12 C) and Axial T2WSE (12D) reveals old infarction in right occipital lobe which hyposignal in T2_FLAIR and hypersignal in T2WSE.

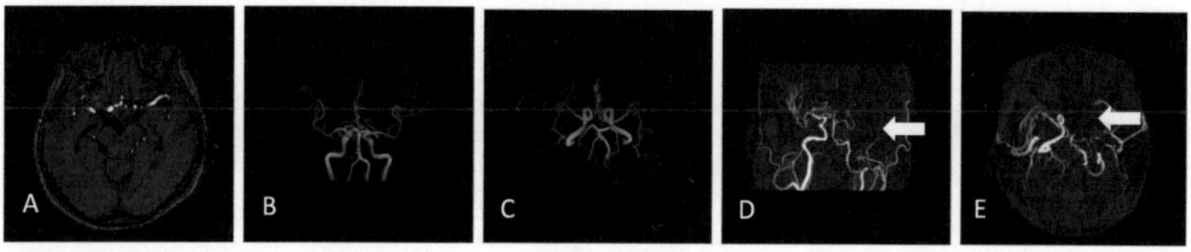

Figure 14.
A–E: MRA brain uses 3D TOF technique, Normal patency arterial lumen have high signal of MRA along course of arteries in source image (A), MIP images in coronal and axial planars (B,C). Disappearance of MRA signal in left ICA, left MCA and bilateral ACAs (D, E) could be due to severe arterial stenosis or occlusion of those arteries.

or migraine-like and aura. Even though, further investigation by neuroimaging for detection CVST is suggested whenever, new onset of headache, persistent, worse on Valsalva, not response to regular medical treatment in the patient with have CVST risk factors. Stroke-like focal neurological symptoms are about 40% of cases with motor symptoms followed by visual impairment and aphasia and some cases with seizures. Nowadays, an increasing concern of this condition leads to increased detection of CVST patients. The patient is suspected of CVST and needs urgent neuroimaging to confirm diagnosis, such as CT and MRI to visualize blood clots or thrombosis in the cranial venous system.

CT brain (NCCT, CECT): It is the investigation modality of choice for the patients who are in the emergency department. NCCT can confirm or exclusion diagnosis by detecting acute thrombosis or blood clot (direct sign) as hyperdensity lesion in the course of dural venous sinuses, cortical vein, or deep venous structures and also evaluation the brain parenchyma for brain edema, intracranial hemorrhage (indirect sign). The density of blood clot is variable in density depend on the stage of thrombus, the acute thrombosis, blood clot is in hyperdensity (average 50–70 HU) in contrast to the brain (30–40 HU). If acute thrombus is along superior sagittal sinus (SSS) on axial CT scan suggests "hyperdense delta sign" (**Figure 15A**) and whenever along cortical veins on axial CT scan, we can call "cord sign" (**Figure 15B**) Even though, NCCT is more sensitive in the detection of deep cerebral venous thrombosis and cortical vein thrombosis than dural venous sinus thrombosis.

Similar to the evaluation of hyperdense clot sign in acute ischemic stroke, before interpretation of the hyperdensity lesion such as acute thrombus, a complete evaluation of intracranial vascular system density is needed to get rid of the false sign of hyperconcentration blood.

The brain parenchyma changes (**Figures 16A** and **B**) such as brain swelling, vasogenic brain edema related to the site of venous occlusion. Intracranial hemorrhage can occur in the brain parenchyma as hemorrhagic venous infarction, or in the extra-axial locations such as subarachnoid hemorrhage, subdural hemorrhage usually relates to the site of venous occlusion.

Contrast enhanced CT brain (CECT): In the case of equivocal in CVT diagnosis and no contraindication for contrast medium, the contrast enhanced CT by intravenously contrast medium administration is suggested either by injector or manually. CT density of intraluminal thrombus is relative lower than density of contrast medium, therefore, appear as filling defect nonopacification by contrast medium in course of thrombosed venous sinuses, veins. If the circumstance occurs in superior sagittal sinus thrombosis on axial CT scan, it was called "empty delta sign" (**Figure 15E**).

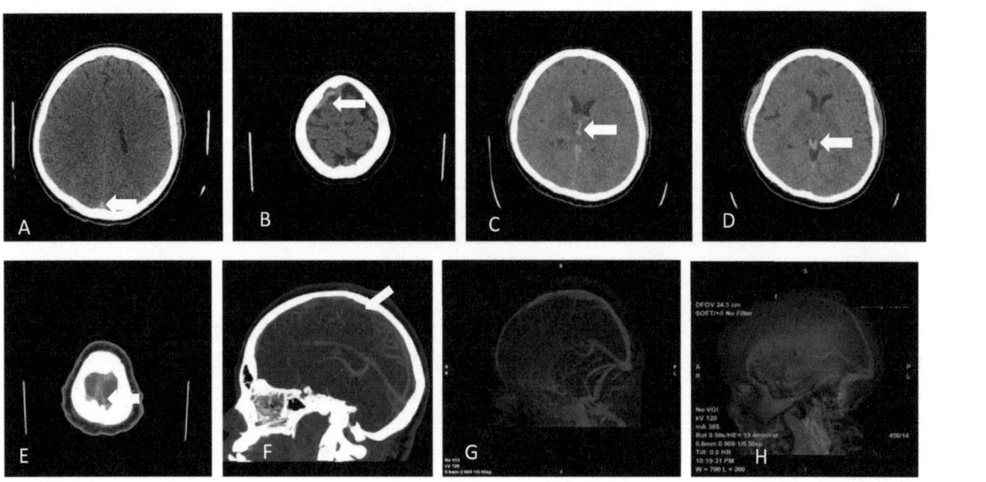

Figure 15.
A-H: NCCT, CTV in cerebral venous thrombosis NCCT brain (A, B) "hyperdense delta sign" in posterior of superior sagittal sinus (SSS) (A) and "hyperdense cord sign" in right cortical vein (B), bilateral internal cerebral veins (C,D) CT Venography reveals filling defect in SSS (E, F) Volume rendering technique (G, H) normal CTV (G) and nonopacification of contrast in SSS in case of SSS thrombosis (H).

CT venography (CTV): Slightly different to CECT, CT venography used bolus contrast medium injection of 75–100 ml. by injector with an acquisition delay and automated CT study in the venous phase. CTV will provide an accurate detail in the evaluation of cerebral venous system. Before evaluation and interpretation, thin axial section contrast enhanced CT venography will be used to perform post processing rendering techniques such as maximum intensity projection (MIP) (**Figure 15F**) or surface, shaded displays (SSD) and volume rendering (VR) (**Figure 15H**) which all acquired from post-processing work station system send into Picture Archiving and Communication System (PACS). Partial or nonopacification of contrast medium in case of incomplete or complete occlusion of sinuses, and veins are demonstrated

Magnetic Resonance Imaging (MRI) brain: It is the noninvasive neuroimaging technique without exposure to the radiation. The patient who has contraindication to CT scan or equivocal diagnosis by CT scan are indicated for MRI study for diagnosis CVT. Similar to CT, the interpretation must be evaluation two main findings such as intraluminal clot sign along cranial venous system and parenchymatous sign, the change of the brain resulting from venous occlusion.

Intraluminal thrombus: Due to variable stage of the intraluminal clot, MRI on T1W and T2W have variable signal of intraluminal thrombus due to alteration in hemoglobin oxygenation and iron oxidation state in the blood clot such as acute stage (less than 7 days) thrombus signal is isosignal T1W and hyposignal T2W, subacute stage 7 days–15 days) T1W is hypersignal (**Figure 17A** and **B**) and T2W is iso- to hypersignal and chronic stage (greater than 15 days) T1W is isosignal and T2 W is hypersignal. Thus, the diagnosis of acute thrombus by MRI is difficult due to signal that seems like normal flow void signal in T2W [13]. Susceptibility weighted image (SWI) is pulse sequences that assist in evaluation of hemorrhagic component by blooming susceptibility effect of the hemorrhagic component to be slightly larger than regular size and easily to visualization (**Figure 17**).

After administration of contrast medium, T1W contrast enhances study, there is nonopacification segment of thrombosed sinuses, and veins cause filling defect of those venous structures (**Figure 17**) such as an empty delta sign.

Parenchymatous changes: These are indirect signs secondary to thrombus such as brain edema in the brain region related to the site of venous occlusion, hemorrhagic venous infarction, subdural hemorrhage, and subarachnoid hemorrhage.

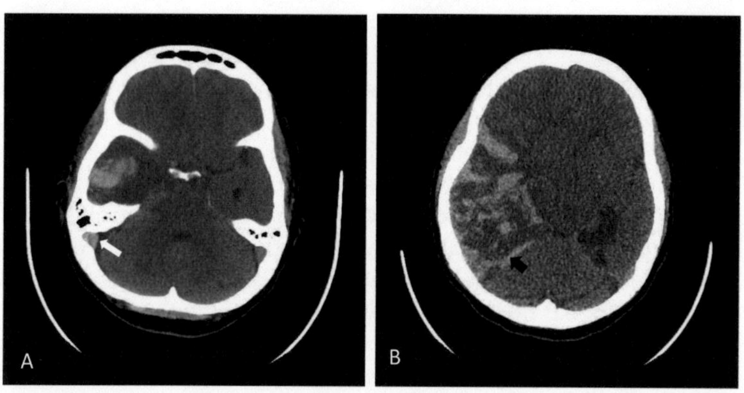

Figure 16.
A–B Axial NCCT brain reveals cerebral venous sinus thrombosis with "hyperdense delta sign" (white arrow) in right sigmoid sinus with hemorrhagic venous infarction in right temporal lobe, small subdural hemorrhage is noted along right tentorium cerebelli (black arrow).

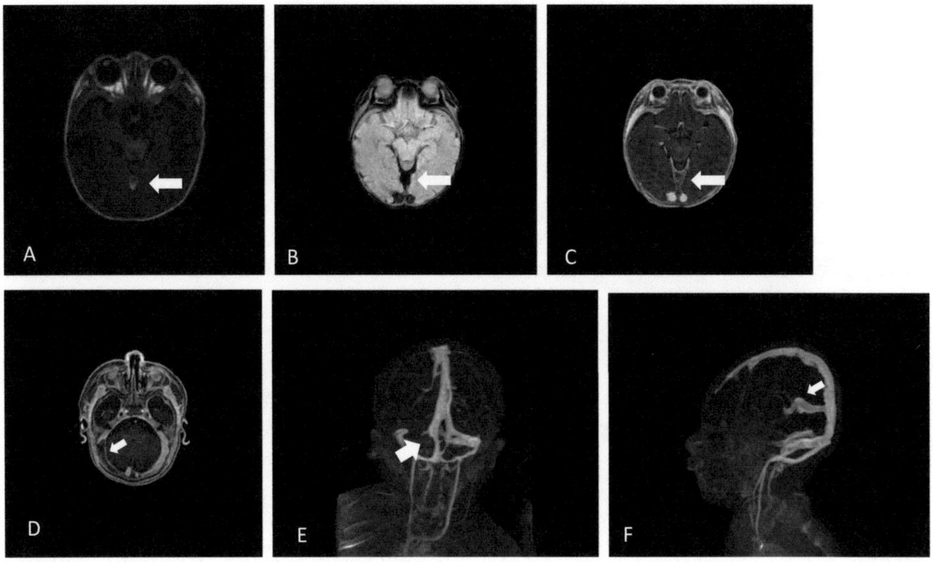

Figure 17.
A–F MRI and MRV brain in patient with deep cerebral vein and dural venous thrombosis. Axial T1W (A) reveals hypersignal T1 of an intraluminal thrombosis in the vein of Galen, Axial SWI (B) demonstrates blooming susceptibility effect of paramagnetic substance(clot) and post contrast enhanced T1W (C) demonstrates filling defect of noncontrast medium opacification. In addition, nonopacification of right transverse sinus is also detected in the same patient (D). 3D CE-MRV (contrast enhanced MRV), MIP images demonstrate filling defect in vein of Galen and noncontrast segment n right transverse sinus. All of findings suggest subacute stage of cerebral venous sinus thrombosis.

The parenchymatous abnormalities are better shown by MRI than CT as the area of hyposignal T1W and hypersignal T2 lesion in the brain parenchyma, usually in vasogenic edema and not followed arterial territory. Diffused weighted image (DWI) is the pulse sequence that allows subclassification of parenchymal abnormalities as either primarily vasogenic edema or primarily cytotoxic edema. The hyposignal on ADC mapping. In contrast with arterial ischemic stroke, many parenchymal abnormalities secondary to venous occlusion are reversible. After administration of contrast medium, on the T1W contrast enhanced images, marked enhancement of the subependymal plexus and the medullary veins that run perpendicular to the wall of the lateral ventricles are demonstrated. Hemorrhagic venous infarction can sometimes be present, cause susceptibility artifacts on T2*WI, SWI, and GRE images as blooming dark signal (**Figure 17**) in the infarcted area.

MRV (MR Venography) (**Figure 17**): It is an MRI technique for demonstration of cerebral venous structures. This technique can be performed either by noncontrast enhanced such as time-of-flight (TOF), phase contrast MRV or contrast enhanced study. Usually, we performed with noncontrast technique and normal cerebral venous sinuses will have normal hypersignal of flow along the course of cerebral venous sinuses includes dural venous sinuses, cortical veins, and deep cerebral venous system. If CVT occurs, there is no flow, no MRV hypersignal in the thrombosed venous structures. However, the limitation of noncontrast enhanced study is the flow artifact. To avoid this limitation, the contrast enhanced MRV may be beneficial. Contrast medium signal opacification entirely of the patency cerebral veins.

CVT can result in death or permanent disability, it generally has a favorable prognosis if diagnosed and treated early. CVT is treatable and curable by medical treatment, and the mainstay is prompt anticoagulation with parenteral heparinization.

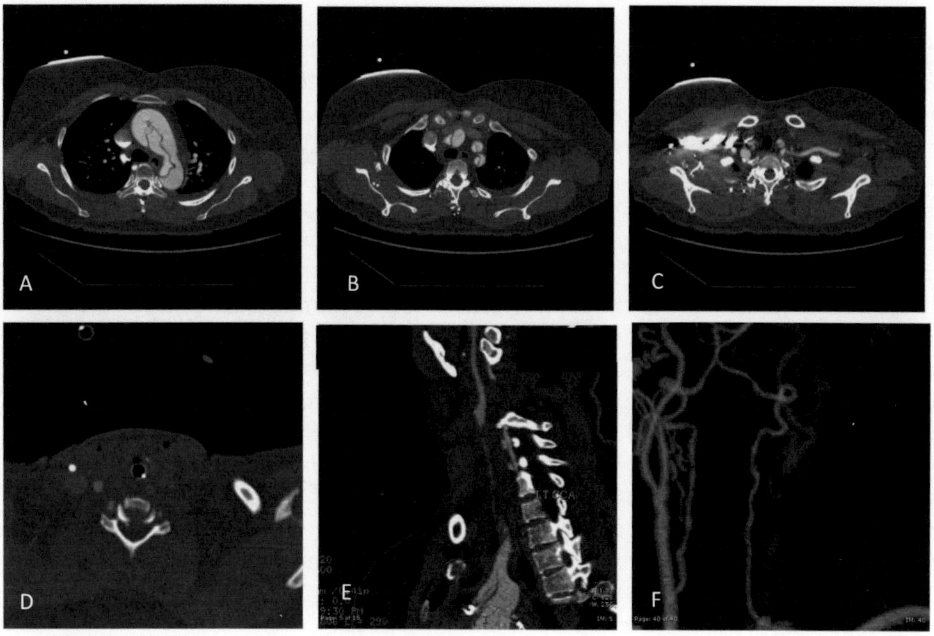

Figure 18.
A-F CT Angiography of a 50-year-old woman, known case of aortic dissection type A, presented with quadriparesis for 5 hours suspected of carotid dissection. Axial CTA (A–D) demonstrates radiolucency lines in the aortic arch (A), dissection extends to right brachiocephalic artery, left CCA and left subclavian artery (B), bilateral CCAs (C) cause near total occlusion LCCA with eccentric residual lumen (D). MPR in coronal plane (E) with typical string sign and 3D Volume rendering image (F) decreased contrast opacity in the remaining LCCA and LICA.

Craniocervical Arterial Dissection: [18–21] It is one of the causes of stroke up to 25% of cases. The patients are in young and middle-aged group. An accurate and prompt diagnosis of this condition is crucial because timely and appropriate therapy can significantly reduce the risk of stroke and long-term sequelae. Because of the great diversity in the clinical features of craniocervical artery dissection, imaging plays a primary role in its diagnosis, nowadays by noninvasive diagnostic imaging techniques such as CT angiography and MRI with MR angiography. To achieve an accurate diagnosis of craniocervical artery dissection, it is important to understand pathologic features (intimal tear, intramural hematoma, and dissecting aneurysm) and the spectrum of imaging findings of CT angiography, magnetic resonance (MR) imaging with MR Angiography, and conventional angiography; and potential pitfalls in image interpretation.

The causes of arterial dissection are traumatic and spontaneous in origin. Head and neck trauma or minor trauma as cervical manipulation can trigger in patients with underlying arteriopathy. Connective tissue diseases, such as fibromuscular dysplasia, Ehlers-Danlos syndrome type IV, Marfan syndrome, autosomal dominant polycystic kidney disease, and osteogenesis imperfect are the underlying cause of arteriopathies. When a primary tear occurs in the intima of the arterial wall, the blood stream can penetrate into the depth of the arterial wall, such as tunica media and extends cranially according to the direction of blood stream, then intramural hematoma develops and compression to the arterial lumen. If dissection toward adventitia, it will form dissecting pseudoaneurysm and thromboembolic phenomenon can occur.

Common locations of dissection is the extracranial segment in the cervical level, spontaneous dissection particularly internal carotid artery (ICA) tends to occur at distal to carotid bulb and extends not beyond ICA entry to the petrous bone. For the extracranial vertebral artery (VA) dissection, the most common locations are at V2 segment (foramen transversarium segment) and V3 segment (extravertebral segment). The reason of high incidence in the extracranial arteries dissection is those neck arteries can mobility than intracranial arteries and also trauma against to adjacent bony structures such as styloid process or cervical spines. The most frequent clinical manifestation of carotid territory ischemia (49%–82.5%), whereas dissections without luminal narrowing cause more local signs and symptoms. A completed stroke usually occurs during the first 7 days after the onset of symptoms but can occur up to 1 month later. Local signs and symptoms include head, facial, or neck pain, Horner syndrome, pulsatile tinnitus, and cranial nerve palsy. Headache is frequently the earliest symptom (47%) of patients. The clinical manifestation from vertebral artery dissection is headache or neck pain accompanied or followed by posterior circulation ischemia (57%–84%). Treatment of craniocervical dissection is by medical therapy, anticoagulant drugs are used to prevent thrombosis and embolism caused by extracranial dissection. Endovascular treatment is for the patient who remains symptomatic due to thromboembolic events or subarachnoid hemorrhage from dissecting pseudoaneurysm intracranial location.

Imaging: Even though Digital Subtraction Angiography (DSA) is the gold standard in the diagnosis craniocervical dissection, but noninvasive imaging technique CT Angiography or MR Angiography is widely accepted and used in clinical practice.

CT Angiography (CTA) (**Figure 18**): CTA study in diagnosis craniocervical artery dissection may study from aortic arch up to cervical and intracranial levels continuously. Normal CTA is complete contrast medium opacification of the arterial lumen from origin to the whole course of arteries. Based on pathophysiology, eccentric narrowing of the arterial lumen with mural wall thickening cause "target sign" in axial CTA scan. Other signs such as short segment arterial stenosis, total occlusion, dissecting aneurysm, filling defect, intimal flap, focal stenosis and dilatation (string and pearl sign), and tapering stenosis (flame sign) may be detected. Please note that intimal flap is a rare finding for craniocervical arterial dissection and is common in carotid dissection. The sensitivity of CTA in the diagnosis of cervical artery dissection is 74%-98% and specificity is 84%-100%.

MRI and MR Angiography (MRA) (**Figure 19**): Similar to the CTA study, the MRA study may study in the whole axis from the aortic arch to the intracranial level. Whenever suspicious of extracranial arterial dissection, MRI with fat-suppressed T1 weighted image is a recommended pulse sequence in detection intramural thrombus. If intramural thrombus is in the subacute stage (3 days up to 2 months duration of blood clot), the hematoma will give hypersignal in T1W; therefore, the mural thrombus might be demonstrated for diagnosis of arterial dissection. Luminal stenosis, eccentric shaped lumen (mural wall thickening with the displacement of arterial lumen off midline and external diameter enlargement from the summation of mural hematoma are all possible findings that assist in the diagnosis of arterial dissection. MRI and MRA are excellent imaging methods in diagnosis of carotid artery dissecting artery with sensitivity 87%-99% compare to DSA whereas sensitivity is about 60% for vertebral artery (VA) dissection. The detection in vertebral artery dissection is quite lower than in carotid artery are due to small size of vertebral artery and flow

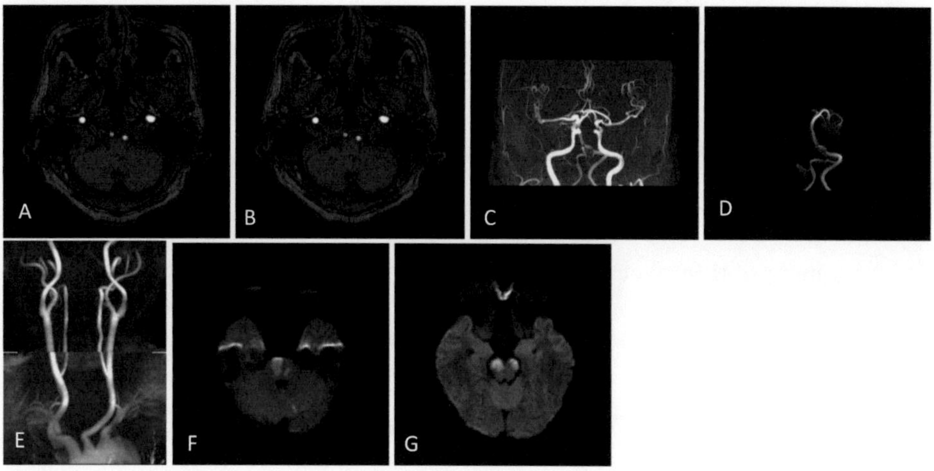

Figure 19.
MRA craniocervical artery of a 69-year-old male, source images MRA intracranial level (A, B) 3D-TOF demonstrates hypointense signal of intimal flap in bilateral V4 segments of vertebral arteries. MRA MIP images (C, D) arterial dissection extends up to vertebrobasilar junction. MRA MIP image (E) at cervical level demonstrates normal MRA of all bilateral CCA, cervical ICAs, cervical VAs. DWI images (F, G) demonstrate restrictive diffusion areas of acute infarction in pons, cerebellum and midbrain from thromboembolism.

related enhancement of paravertebral veins make the diagnosis is difficult. In clinical practice, when equivocal finding in diagnosis VA dissection, CTA is recommended.

Recommendation, when clinical suspicious of carotid artery dissection, we can use either CTA or MRA with similar results, whenever, suspicious of VA dissection, CTA may be superior to MRA.

2. Neurological complication of anticoagulation therapy

The most common intracranial complication during regular treatment by anticoagulant drugs is intracerebral hemorrhage (ICH) [22–25]. Basically, the cause of ICH is classified into two groups as primary cause due to spontaneous rupture of small vessels or amyloid angiopathy as a majority of the patients about 78%-88% and secondary cause associated with tumor, impaired coagulopathy as a minority group. Even though small in number but significant risk factors with ICH such as hemophilia or acute leukemia with thrombocytopenia or patient during treatment by anticoagulant drugs, massive intracranial hemorrhage is often cause of death.

2.1 Cause of intracranial hemorrhage (ICH)

ACQUIRED:

Iatrogenic coagulopathy

- ICH related to aspirin
- ICH related to anticoagulation, heparin, and Coumadin
- ICH related to thrombolytic treatment

Neoplastic coagulopathy

- ICH related to leukemia
- SDH in cancer patient

Rare cause of ICH

- Drug-induced thrombocytopenia
- Uremia
- Alcohol
- Liver transplantation

CONGENITAL:

- Hemophilia

2.2 Physiology of hemostasis

Two important mechanisms against bleeding are blood coagulation and platelet-mediated hemostasis. The coagulation cascade is triggered as soon as blood contacts the injured endothelial lining. The combination of both coagulation cascade and active formation of the platelet plug is effective autoregulation mechanisms in occlusion of a vascular lesion. Coagulation cascade mechanism have its own two main pathways: 1) intrinsic pathway by physical-chemical activation which its role is not well understanding and 2) extrinsic pathway activated by tissue factor released from the damaged cell.

Extrinsic Pathway: When blood vessel wall injuries exposed plasma to tissue factor, then, Factor VII is a plasma protein bind to tissue factor and activated to factor VIIa, this complex will activate factor IX, X to factor IXa, Xa. Factor Xa and its cofactor Va form a phospholipid-bound complex called the prothrombinase complex, which is highly activated on the surface of platelets and cleaves prothrombin (factor II) to thrombin (factor IIa). Thrombin cleaves fibrinogen (factor I) to fibrin (factor Ia), which is covalently cross-linked by factor XIIIa into fibrin strands. Factor VIII binds to vWF (an adhesive protein important for the generation of the initial platelet plug). Thrombin feedback is important to the entire system. Thrombin, one generated is powerful procoagulant further conversion factor V, VIII to factor Va, VIIIa and covert more prothrombin to thrombin. Thrombin will further accelerate the entire cascade in the formation of a large amount of fibrin.

Regulatory mechanism of the coagulation cascade, such as tissue factor pathway inhibitor, antithrombin III, activated protein C and protein S, thrombomodulin, and fibrinolytic system, is to limit the amount of fibrin clot avoiding tissue ischemia and to prevent widespread thrombosis.

2.3 Pathophysiology of bleeding disorders

Coagulopathies leading to intracranial hemorrhage mostly from acquired caused. Iatrogenic coagulopathy from aspirin, anticoagulants, and thrombolytic agent treatment.

Antiplatelet drugs: Aspirin is the most common antiplatelet drug that is used daily in clinical practice such as acute MI and arterial occlusive cardiovascular disease. Its mechanism is to inactivation enzyme cyclooxygenase result in decreased platelet aggregant thromboxane A2. Incidence of intracranial hemorrhage due to aspirin In the 1991 Swedish Aspirin Low-Dose Trial investigators of patients with a history of TIA or minor stroke reported that the prevalence of intracranial hemorrhage was 1.5%. Finally, in 1997, the International Stroke Trial Collaborative Group concluded that administration of 300 mg aspirin daily compared with placebo following acute stroke pre-vented 1.2 ischemic strokes per 100 treated patients but caused in excess of 0.41 ICHs. Aspirin therapy for primary or secondary stroke prevention and primary MI prevention may slightly increase the low baseline risk of ICH but the increased risk is usually outweighed by the benefits of aspirin. Other antiplatelet agents: clopidogrel (Plavix), abciximab (Reopro), etc. act as glycoprotein IIb/IIIa inhibitors. The incidence of ICH in the clopidogrel group was 0.33%, whereas it was 0.47% in the aspirin group. The newer antiplatelet agents seem to be associated with an ICH risk profile similar to that of aspirin.

Anticoagulant drugs: Warfarin, heparin, and enoxaparin are the most common anticoagulants in routine clinical practice.

- Warfarin is an oral anticoagulant that interfere with vitamin K metabolism in the liver and cause impairment synthesis functional coagulation factors II, VII, IX, X and protein C, protein S. Therefore, Warfarin administration can prolong prothrombin time (PT), INR.

- Heparin is administered parenterally and inhibits action of antithrombin III cause prolong of the partial.

- Thromboplastin time (PTT).

- Enoxaparin, low molecular weight heparin, similar action to heparin with a longer half-life time (4.5 hrs.)

Anticoagulant therapy related ICH is about 10%-20% and ICH is the most dreaded and least treatable complication. The mortality rate is about 46%-68%. The distribution of intracranial hemorrhage related to anticoagulant therapy is in the brain parenchyma, intracerebral hemorrhage (ICH) about 70% followed by subdural hemorrhage (SDH). There is no specific predilection in the brain. Risk factors such as hypertension, increasing age, and previous cerebral infarction. The mechanism of spontaneous ICH developing during anticoagulants is not clearly understood. Postulate that patients with underlying hypertension with hemorrhage derived from small vessel vasculopathy, usually have normal hemostatic mechanisms but fail when anticoagulant.

Thrombolytic agent: Fibrinolytic agents have been used in clinical practice, such as acute MI and acute stroke. The examples of drugs in this group are the exogenous substances streptokinase, urokinase, and endogenous tPA. The mechanism in action is to activate the body's fibrinolytic system by converting plasminogen to plasmin. Plasmin binds to fresh fibrin clots, dissolving them and generating fibrinogen degradation products. Intracranial hemorrhage cause by the thrombolytic agent is in lobar hemorrhage 70%–90% and multiple locations in almost one-third of patients. ICH related to rtPA tended to have fluid levels in the hematomas (suggesting continuing or repeated hemorrhages), multiple parenchymal hemorrhages, and blood in multiple compartments (intraventricular, subarachnoid, subdural, and parenchymal). Patients with post-rtPA ICH also tended to suffer a catastrophic clinical course, with dying or ending up in a persistent vegetative state within hours of hemorrhage onset. Thus, in the case of acute ischemic stroke, there is a summary of absolute and relative contraindication to IV rtPA [6].

Absolute and relative contraindications to IV rt-PA for acute ischemic stroke.

Absolute contraindication

- Acute ICH including hemorrhagic infarction.
- History of ICH (microbleed is not contraindicated).
- BP>185mmHg, BPs>110mmHg.
- Serious head trauma or stroke < 3months.
- Thrombocytopenia (platelet <100,000/mm3) and coagulopathy (PT>15, INR>1.7).
- LMWH within 24 hr (38% risk of sICH, 29% risk of death, 33% favorable outcomes, od 0.84 for sICH, od5.3 for death).
- DTIs (TT is sensitive to the presence of DITs).
- Factor Xa inhibitor.
- Severe hypoglycemia(<50mg/dl) and hyperglycemia(>400mg/dl) may be permitted for IVT.
- Early radiographic ischemic changes (>1/3 MCA).

Relative contraindication

- Advanced age (>80yrs.)
- Mild or improving stroke (NIHSS<5)
- Severe stroke (IVT should be cautiously administered in NIHSS>25 at 3–4.5 hrs.)
- Major surgery within 2 weeks (some 3 months)
- Arterial puncture of noncompressible site
- Recent GI bleeding or GU bleeding
- Seizure at onset

- Recent MI within 3 months
- Intracranial structure abnormality (axial tumor, AVM, aneurysm)
- Dementia

CT brain (Figures 20–22): Noncontrast CT brain is the investigation modality of choice for the patient who had a neurological disorder and was suspected of coagulopathy related symptoms. The most common finding is intracerebral hematoma (ICH), followed by SDH. The CT pattern that raises suspicious of coagulopathy related intracranial hemorrhage is multicompartments hemorrhage such as bleed in the parenchymatous and also in subarachnoid space and in subdural space at the time of initial CT study, which leads to suspicion of hemorrhage related to coagulopathy. The density pattern of hemorrhage is varying case by case as homogenous hyperdensity either or heterogeneous hyperdensity, fluid-hemorrhage level. The density of the hyperdensity portion is 60–80 HU.

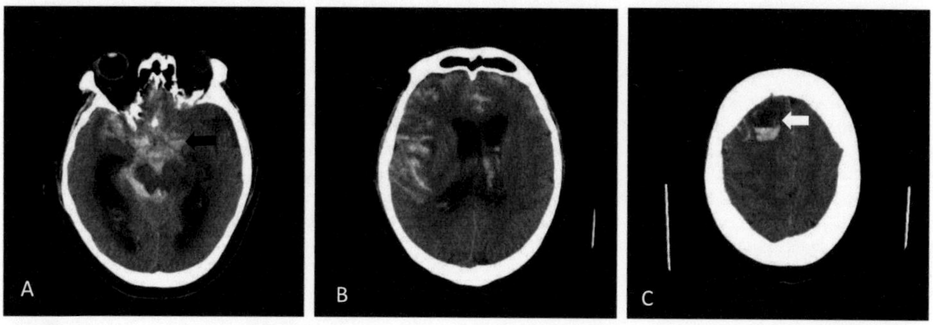

Figure 20.
A,B,C Axial NCCT brain of patient with history of acute aortic dissection type A S/P modified Bentall operation with hemiarch replacement with right carotid artery bypass and resternotomy with clot removal and CABG, on heparin and develops alteration of consciousness. Spontaneous intracranial hemorrhage mainly in subarachnoid space (SAH) (black arrow) with small fluid –level ICH (white arrow) in right frontal lobe.

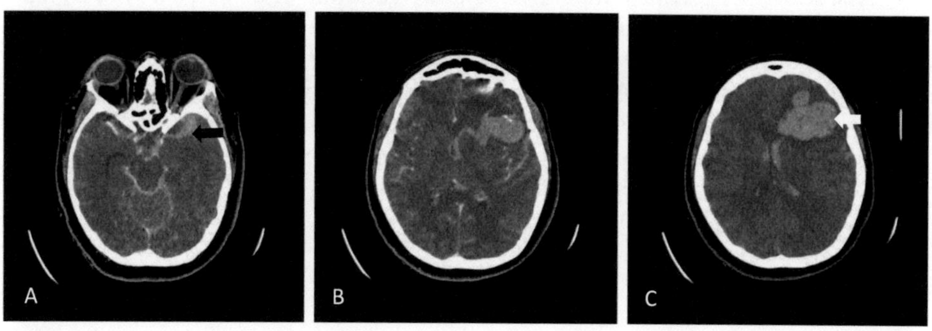

Figure 21.
A,B,C Axial NCCT brain of a patient with congenital heart disease postsurgery (on warfarin), presented with alteration of consciousness. Spontaneous intracranial hemorrhage mainly in subarachnoid space (SAH) (black arrow) and ICH (white arrow) in left frontal lobe.

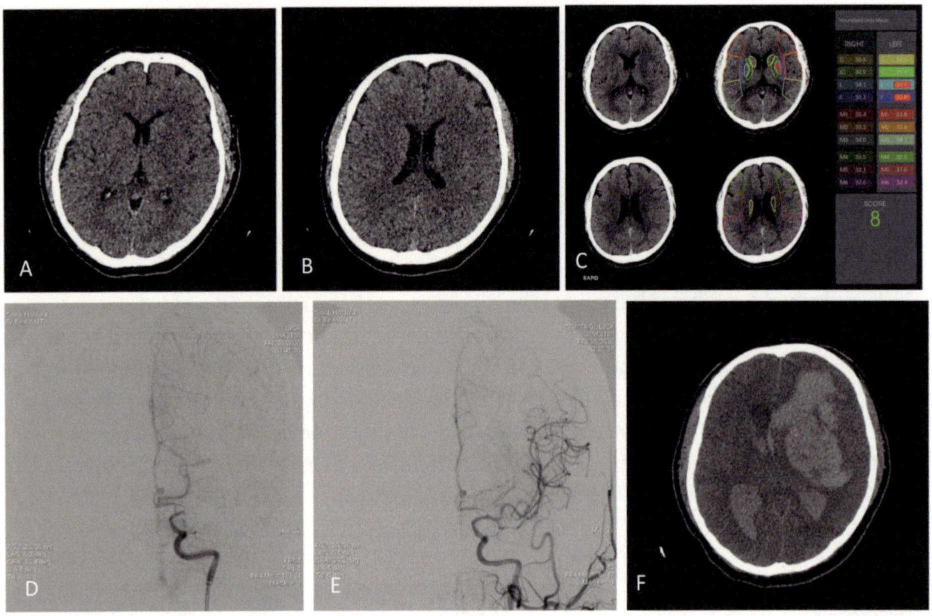

Figure 22.
*A-F A 53-year-old man, with history of DM and HT presented with right hemiparesis and right facial palsy. Acute ischemic stroke was diagnosed at outside hospital. Treatment was giving rTPA at 9.20 p.m. At referred hospital, Axial NCCT at 0.10 a.m. (A, B) reveals ill-defined hypodensity area in left lentiform nucleus, left insular cortex and left high cortical frontal lobe. AI ASPECT score=8 (neuroradiologist reading score=7) Further treatment with mechanical thrombectomy (***Figure 20D, E***) was performed with completely reopening of left MCA and no immediate complication. Axial NCCT brain (07.33 a.m. the day after thrombectomy) reveals ICH in left basal ganglia with IVH.*

3. Summary

Neuroimaging is a noninvasive investigation tool that is essential not only to confirm the diagnosis of emergency neurological diseases such as stroke but also to play a role in triage patients into receiving proper medical treatment, surgical treatment, or intervention treatment. Basic interpretation of CT brain and MRI brain in stroke, CVT, dissection, hemorrhage warrants the understanding of density (CT) or signal intensity (MRI) changes of affected brain tissue in stroke and intraluminal thrombus in each stage of the disease. Further using of advanced neuroimaging investigation such as CTA, multiphase CTA, CTV, and CTP aimed for more specific conditions of the artery disease such as arterial thrombosis, arterial dissection, venous disease such as venous sinus thrombosis, yet, also reduced the need for more invasive investigation such as digital subtraction angiography.

Author details

Pipat Chiewvit
Faculty of Medicine Siriraj Hospital, Department of Radiology, Mahidol University, Bangkok, Thailand

*Address all correspondence to: pipat8999@gmail.com

References

[1] Lioutas VA, Goyal N, Katsanos AH, Krogias C, et al. Clinical outcomes and neuroimaging profiles in non-disabled patients with anticoagulant-related intracerebral hemorrhage. Stroke 2018; **49**(10):2309-2316

[2] Kleindorfer DO, Towfighi A, Chaturvedi S, Cockroft KM, Gutierrez J, Lombardi-hill D et.al. 2021 Guideline for the prevention of stroke in patients with stroke and transient ischemic attack. Stroke 2021;**52**:e3640-ee467

[3] Virani SS, Alonso A, Aparicio HJ, Benjamin EJ, Bittencourt MS, Callaway CW, et al. Heart disease and stroke statistics 2021 update. A report from the American Heart association. Circulation. 2021;**143**:e254-e743

[4] Ferro JM. Cardioembolic stroke: An update. Lancet Neurology. 2003;**2**:177-188

[5] Kleindorfer DO, Cockroft KM, Kernan WN, Meschia JF, Sharrief AZ. 2021 Guideline for the prevention of stroke in patients with stroke and transient ischemic attack. Stroke. 2021;**52**:e364-e467. DOI: 10.1161/STR.0000000000000375

[6] Wann LS, Curtis AB, January CT, Ellenbogen KA, Lowe JE, Estes NA 3rd, et al. 2011 ACCF/AHA/HRS focused update on the management of patients with atrial fibrillation (Updating the 2006 Guideline): A report of the American College of Cardiology Foundation/American Heart Association Task Force on Practice Guidelines. Journal of the American College of Cardiology. 2011;**57**:223-242

[7] Fugate JE, Rabinstein AA. Absolute and Relative contraindications to IV rt-PA for Acute ischemic stroke. The Neuohospitalist. 2015;**5**(3):1101-1121

[8] Jensen-kondering U, Riedel C, Jansen O. Hyperdense artery sign on computed tomography in acute ischemic stroke. World Journal of Radiology. 2010;**2**(9):354-357

[9] ChiengJSL SDR, Chawla A. The hyperdense vessels sign in cerebral computed pearls and pitfall. Singapore Medical Journal. 2020;**61**(5):230-237

[10] Barber PA, Demchuk AM, Zhang J, et-al. Validity and reliability of a quantitative computed tomography score in predicting outcome of hyperacute stroke before thrombolytic therapy. ASPECTS Study Group. Alberta Stroke Programme Early CT Score. Lancet. 2000;**355**(9216):1670-1674

[11] Yaghi S, Willey JZ, Cucchiara B, Goldstein JN, Gonzales NR, Khatri P, et al. Treatment and outcome of hemorrhagic transformation after IV alteplase in acute ischemic stroke: A scientific statement for healthcare professionals from the American heart association/ American stroke association. Stroke. 2017;**48**(12):e343-e361

[12] Marks MP. Cerebral ischemia and infarction. In: Atlas SW, editor. Magnetic Resonance Imaging of the Brain and Spine. 4th ed. Vol. 1. Philadelphia, PA: Williams & Wilkins; 2009. pp. 772-825

[13] Leach JL, Strub WM, Gaskill-Shipley MF. Cerebral venous thrombus signal intensity and susceptibility effects on gradient recalled-echo MR imaging. American Journal of Neuroradiology. 2007;**28**:940-945

[14] Chiewvit P, Piyapittayanan S, Poungvarin N. Cerebral venous thrombosis: Diagnosis dilemma.

Neurochemistry International. 2011;**29**:3(3) e13

[15] Sajjad Z. MRI and MRV in cerebral venous thrombosis. The Journal of the Pakistan Medical Association. 2006;**56**(11):523-526

[16] Canedo-Antelo M, Baleato-Gonzalez S, Mosqueira JA, Casas-Matinez J, Oleaga L, Vilanova JC, et al. Radiological clues to cerebral venous thrombosis. Radiographics. 2019;**39**(6):1611-1628

[17] Chiewvit P, Tritrakarn S, Nilanont Y. Accuracy of noncontrast CT brain in detection of cerebral venous sinus thrombosis. International Journal of Clinical Medicine. 2021;**12**:471-483

[18] Rodallec MH, Marteau V, Gerber S, Desmottes L, Zins M. Craniocervical arterial dissection: Spectrum of imaging findings and differential diagnosis. Radiographics. 2008;**28**:1711-1728

[19] Arnold M, Bousser MG, Fahrni G, Fischer U, Georgiadis D, Gandjour J. Vertebral artery dissection: Presentation findings and predictors of outcome. Stroke. 2006;**37**(10):2499-2503

[20] Mehdi E, Aralasmak A, Toprak H, Yildiz S, Kurtcan S, Kolukisa M, et al. Craniocervical dissections: Radiologic findings, pitfalls, Mimicking disease: A Pictorial review. Current Medical Imaging Reviews. 2018;**14**:207-222

[21] Li Q, Wang J, Chen H, Gong X, Ma N, Goa A, et al. Characterization of craniocervical artery dissection by silmultaneous MR noncontrast angiography and intraplaque hemorrhage at 3T. AJNR. 2015;**15**:1769-1775

[22] Kase CS. Intracerebral hemorrhage: Non-hypertensive causes. Stroke. 1986;**17**:590-595

[23] Mohr JP, Caplan LR, Melski JW, et al. The Harvard Cooperative Stroke Registry: A prospective registry. Neurology. 1978;**28**:754-762

[24] Yarnell P, Earnest M. Primary non-traumatic intracranial hemorrhage. A municipal emergency hospital viewpoint. Stroke. 1976;**7**:608-610

[25] Wijdicks EF, Jack CR Jr. Intracerebral hemorrhage after fibrinolytic therapy for acute myocardial infarction. Stroke. 1993;**24**:554-557

Chapter 8

Sleep Patterns Changes Depending on Headache Subtype and Covariates of Primary Headache Disorders

Füsun Mayda Domaç, Derya Uludüz and Aynur Özge

Abstract

Headache is one of the most common and bothersome problems in neurology practice. The frequency of headache has been substantially increased over the last 30 years due to changes in lifestyle. Controlling the trigger factors and lifestyle changes (e.g. regular sleep, meal time, exercise, etc.) are the first step management strategies in headaches. Sleep and headache have bidirectional effects on each other. While diminished and poor quality of sleep can be a trigger factor for headache (e.g. migraine and tension-type headache (TTH)), some types of headache like hypnic headache and cluster-type headache mainly occur during sleep. Patients with headache may have poor sleep quality, reduced total sleep time, more awakenings, and alterations in architecture of sleep recorded by polysomnography. Progression to chronic forms of headache may also be associated with the duration and quality of sleep. Even though pathophysiology of headache and sleep disorders shares the same brain structures and pathways, sleep disturbances are commonly underestimated and underdiagnosed in headache patients. Clinicians should consider and behold the treatment of accompanying sleep complaints for an effective management of headache.

Keywords: primary headache, sleep, polysomnography

1. Introduction

Headache disordes lead to significant disability worldwide, impairing quality of life, damaging productivity, and substantial burdens of financial cost on both the individual and societies. For this reason, these disorders effectuate a major public health problem in all countries and world regions. In Global Burden Disease (GBD) study 2019, headache disorders have been estimated to account for 46.6 million years lived with disability [YLDs] globally, which has been 5.4% of all YLDs, with 88.2% of them attributed to migraine [1]. The frequency of headache has been substantially increased over the last 30 years due to changes in lifestyle. Controlling the trigger factors and lifestyle changes (e.g. regular sleep, meal time, exercise, etc.) are the first step management strategies in headaches [2]

Sleep disorders and headache have bidirectional effects on each other [3, 4]. The nature of this relationship and whether sleep disturbance or headache has more impact on one another is still poorly understood [5–7]. Headache may be intrinsically related to sleep such as hypnic headache or cluster headache (CH) [8, 9]. Poor quality, and excessive or diminished sleep can be a trigger factor for headache (e.g. migraine and tension-type headache (TTH)), or contrary sleep may have a a relieving effect on a migraine attack [10, 11].

Headache chronicity might be also associated with the duration and quality of sleep [12]. The severity of a sleep disorder may be in an adverse correlation with the intension of the pain [3]. Patients with headache may have poor sleep quality, reduced total sleep time, more awakenings, and alterations in architecture of sleep recorded by polysomnography (PSG) [13]. Prevalance of chronic headache was found to be higher in patients who underwent polysomnographic investigation due to sleep problems [14].

In this section, we aim to discuss the effects and relation of sleep and primary headaches on one another.

2. Common pathophysiology between headache and sleep disorders

Sleep is a recurrent, reversible, periodic, and cyclic active and physiological process that is essential for life [15]. Sleep has essential roles for health such as regulating immune system, releasing hormones (e.g. growth hormone), neurodevelopment, mental health, memory, etc. [16]. Several factors effect the total duration of sleep, time to fall asleep, time to wake-up, and total duration of wakefullness. Sleep and wake cycle is regulated by circadian and homeostatic rhythms. Light is a main factor for circadian rhythmicity, in addition to, time for meals, work or school schedules, social activites, or internal biological clock designate sleep and wake durations [17].

Sleep is composed of four to six recurrent sleep periods each lasting to 80–110 minutes [16]. These periods are constituted of two main sleep stages as rapid eye movement (REM) and non-rapid eye movement (NonREM). NonREM sleep is divided into three stages such as N1, N2, and N3 (slow-wave sleep) [18]. During the early hours of night, NonREM stages dominate the sleep while REM stage dominates the later part of the sleep [16].

Even though pathophysiology of headache and sleep disorders shares the same brain structures and pathways, sleep disturbances are commonly underestimated and underdiagnosed in headache patients [19].

Thalamus is one of the main centers for regulation of sleep and also pain by receiving ascending nociceptive stimulus from trigeminocervical system. Dysregulation in thalamocortical circuits may be predisposing to sleep and headache disorders [20, 21]. Sleep deprivation may induce hyperexcitability and may alter regulation of cortical circuits [22]. During the shift from wakefullnes to sleep, there is an increment in response to tactile, auditory, and propriseptive stimulus. Disturbed sleep seems to escalate pain by decreasing the activity of descending inhibitory pain control system leading to diminish pain treshold [23, 24].

Hypothalamus, containing suprachiasmatic nuclei (SCN) being the main brain structure to maintain sleep-wake cycle, seems to be responsible for the prodromal symptoms of migraine like mood, appetite or sleep changes, fatigue, or yawning [25, 26]. Accompanying autonomic symptoms like nausea, lacrimation, and rinorrhea suspect the role of hypothalamus also during a migraine attack [27].

A reduction in arousal index in REM sleep stage and a reduction in cyclic alternating pattern (CAP) in NonREM sleep stages of migraneous patients may show a dysfunction in the connection of brainstem and hypothalamus both of which are important in the pathophysiology of sleep disorders and migraine [28, 29].

In cluster headache [CH], hypothalamus has a crucial role both in autonomic sypmtoms and periodicity of the headache. During attacks, ipsilateral to the autonomic features hyperactivation in hypothalamus was shown by a positron emission tomography (PET) study [30]. In another study, volume of anterior hypothalamus was found to be increased suggesting a structured alteration in SCN [31].

Neuropeptides orexin A and B that are important for the maintenance of wakefullness are synthesized mainly in lateral and posterior hypothalamus. Orexinergic receptors are located in prefrontal cortex, thalamus, and subcortical areas and also take role in pain modulation, thermoregulation, and autonomic functions except maintaining wakefullness and wake-sleep rhytmicity [25]. Sarchielli et al have found lower levels of orexin in patients with episodic migraine and higher levels in patients with chronic migraine and medication overuse headache suggesting a dysfunction of orexin and a response of hypothalamus to headache [32]. There is an orexin deficiency in narcolepsy, and high prevalance of migraine in patients with narcolepsy may indicate a dysfunction of orexin in migraine pathophysiology as well [33, 34].

Melatonin also called as sleep hormone is synthesized in epiphysis (pineal gland), and the secretion is regulated by suprachiasmatic nuclei. Thus, having a role in circadian rhythm, melatonin also has an analgesic effect via anti-inflammation, inhibition of dopamine release, and GABAergic and antiglutamatergic effects [35]. Melatonin therapy seemed to be affective in headache treatment even with or without an accompanying sleep-wake disorder [36].

Locus cereleus (LC), periaquaductal gray matter (PAG), and dorsal raphe nucleus (DPN) are important brain structures for both headache and pain [34]. Ventrolateral part of PAG that is activated by lateral orexinergic neurons of hypothalamus is a section of REM off area, and it is active during wakefullness while silent at REM sleep stage [37]. LC takes role in stabilizing the switches from sleep to wake, and DPN mainly takes role in switches from NonREM to REM sleep [38, 39]. The decrease in cyclic alternating pattern in REM sleep of migraineous patients is suspected to be a dysfunction of serotonergic system [40]. Dopaminergic dysfunction is associated with prodromal symptoms of migraine, e.g. yawning and mood changes and also with enhanced risk of restless legs syndrome (RLS). The prodromal symptoms being more in patients with RLS may indicate the role of affected dopaminergic system in both of the diseases [41].

Another suspected mechanism is the role of glymphatic system which is a clearing system of interstitial waste products from central nervous system [42]. In an experimental model, it has been shown that glymphatic system was temporarily disturbed during cortical spreading depression and improved 30 minutes later, indicating the role of glymphatic system in migraine pathophysiology [43]. As during sleep glymphatic system is active, the ameliorating effect of sleep on migraine attacks may be explained by the glymphatic system [4].

3. Sleep disorders and headache

During the evaluation of sleep macrostructure by polysomnography, sleep latency, REM sleep latency, percentages of every sleep stages, total sleep time, and sleep efficiency are mainly analyzed. Polysomnography is a diagnostic test for the detection of both the existence and the type of the sleep disorder. Channels of electroencephalography, electrooculography, and electromyography are used to evaluate all of the sleep stages and wakefullnes. Thermistor, nasal canule, and sensors located on thorax and abdomen belts are used to detect the existence and type of abnormal sleep-related breathing disorders. At the same time, electromyographic channels are used to observe sleep-related movement disorders [44].

The time that the first sleep stage epoch seen is called as sleep latency (minute), and the time that first REM sleep stage seen is named as REM latency (minute) [45]. Sleep effciency (%) is the ratio of total sleep time (TST) to total recording time during polysomnography all night [46]. Microstructure of sleep is evaluated by the detection of arousals and the analysis of cyclic alternating pattern [CAP]. Either macrostructure or microstructure of sleep can be effected by primary headaches, and changes can be observed by polysomnography [47].

Sleep disorders are classified as sleep-related breathing disorders, insomnia, hypersomnia, circadian sleep-wake disorders, parasomnia, sleep-related movement disorders, and other sleep diseases [18]. Polysomnography (PSG), multiple sleep latency test (MSLT), wakefullness maintenance test (WMT), and actigraphy are the main diagnostic tests for sleep disorders. Sleep-related breathing disorders, parasomnias, sleep-related movement disorders, and some types of insomnia can be diagnosed by polysomnography. Actigraphy is a diagnostic test that detects limb movements by using a device worn either on ankle or wrist or both. Whether the patient is awake or asleep can be detected due to the limb movements, but neither sleep stages nor breathing disorders can be evaluated [48]. Actigraphy helps the detection of circadian sleep-wake disorders and insomnia, while MSLT and WMT are used for the diagnosis of hypersomnias [49].

We will discuss the effects of sleep disorders on primary headaches separately as follows:

3.1 Insomnia

Though the high incidence of comorbidity, insomnia is underestimated in patients with headache [19]. Insomnia is identified as difficulty to fall asleep, maintain sleep, or wake up earlier than planned, though all conditions, circumtances, and possibilities are sufficient to sleep [18]. Insomnia may be in relation with several painful symptoms like headache, especially with chronic forms rather than episodic [50, 51]. In patients with the diagnosis of fibromyalgia, insomnia related to pain was thought to be a result of increases in amount of arousals during sleep [52]. During night sleep, frequently seen awakenings may lead to a missing of pain inhibiton and patients may have decreased pain tresholds [24].

Insomnia is the most frequent disorder associated with chronic headache [53]. Nearly 50% of migraneurs may have insomnia symptoms, and insomnia is 1.8 times more in patients with tension-type headache (TTH) [54, 55]. Also, insomnia is a risk factor for the high rate of attacks and chronification for both migraine and TTH [56, 57].

Cognitive behavioral therapy widely used in the treatment of insomnia also decreases the frequency of accompanying headache [58].

3.2 Obstructive sleep apnea

Apnea-hypopnea index (AHI) is calculated by the detection of apneas (cessation of sleep for at least 10 seconds) and hypopneas (shallowing of breath for at least 10 seconds) by polysomnography. Obstructive sleep apnea syndrome (OSAS) is diagnosed when AHI is bigger than 5. AHI in the range of 5–15 is called mild, >15–30 as moderate, and >30 as severe OSAS. Obstructive sleep apnea headache is a secondary type of headache which is characterized by a headache attack that lasts upto 4 hours after awakening [18]. Morning headache is related to nocturnal hypoxemia (oxygen saturation ≤90 %) [59] and/or hypercapnia ($PaCO_2$> 45 mmHg) [60] due to recurrent apneas and/or hypopneas [61]. It has a good response to treatment and decreases or totally diminishes after an effective treatment of OSAS [5, 61, 62].

There may be a comorbidity with migraine, tension-type headache (TTH), cluster and hypnic headache, and OSAS [4, 63, 64]. Snoring which may be a component of sleep apnea is found to be more frequent in chronic types of headache than episodic forms [65]. Treatment of OSAS has favorable effects on the accompanying primary headache [66].

3.3 Parasomnia

Parasomnias are divided into two according to sleep stages as REM parasomnias and NonREM parasomnias. Somnambulism (sleepwalking), sleep terror, sleep-related eating disorders, and confusional arousal are NonREM parasomnias. Nightmare, recurrent isolated sleep paralysis, and REM sleep behavior disorder (RBD) are among REM parasomnias [18].

NonREM parasomnias are common in childhood and adolescent periods and mainly disappear in adulthood. Migraine with aura is more common than migraine without aura in sleepwalking patients [67]. The percentage of a history of sleepwalking in parents of the children with migraine is higher than the other types of headache [68]. As there is a circadian periodicity in NonREM parasomnias, serotonergic and orexinergic systems that have roles in migraine attacks were suspected to be responsible for the underlying comorbidity [69].

The frequency of nightmare, a REM parasomnia, is increased in migraneurs with an increase in awakenings during REM sleep [4, 70]. REM sleep behavior disease was also found to be in association with migraine headache and in relation with disability due to headache [71].

3.4 Sleep-related movement disorders

Restless legs syndrome (RLS) is the most common disorder in this group. It is an upleasant sensation that develops when patient lies to sleep or during inactivity or even resting and makes patients to be in need of moving the legs. The symptoms are mainly seen after evening, but in time they may also develop earlier in a day during resting. Sleep-related periodic limb movements also accompany in a majority of these patients [18]. Prevalance of migraine is high in RLS patients with a high frequency of headache attacks and more disability [72, 73]. Migraine prevalance in RLS is more than RLS in migraine [74].

Also there is an increased association with TTH and RLS [73, 75]. Dopaminergic dysfunction and iron metabolism are claimed for the common pathophysiology of migraine and RLS, while in TTH dopaminergic dysregulation is suspected to connect depression with RLS [76, 77].

Bruxism can occur both in sleep and wakefullness. Sleep bruxism is characterized by recurrent tightening or clenching of teeth with an increased jaw muscle activity during night [78]. Headache comorbidity was found to be higher in adults than children [79]. Bruxism is found to be mainly associated with chronic migraine; nevertheless, the combination of bruxism with temporomandibular dysfunction is associated with both episodic and chronic migraine as well as TTH [53].

3.5 Hypersomnia

In this group, narcolepsy with cataplexia (type 1), narcolepsy without cataplexia (type 2), idiopathic hypersomnia, and recurrent hypersomnia have the main complaint of sleepiness during day [18]. As the disrupted orexinergic system is the main pathology in narcolepsy, migraine is claimed to be in association [80], but the results are conflicting. In a study, no association was found between narcolepsy and migraine, while the others found an increase of migraine attacks in narcoleptic patients and have suspected migraine as an independent risk factor [10, 33, 80].

4. Effects of headache on sleep

4.1 Migraine

Migraine is one of the most disabling primary headache with a pulsating quality and moderate to severe intensity, mainly located unilaterally and lasting for 4 to 72 hours. It may be aggrevated with physical activity and nausea and/or phonophobia, and photophobia may accompany [81]. Migraneurs have higher scores of Pittsburg Sleep Quality Index (PSQI) showing a poor sleep quality [13]. Frequency of attacks were found to be related with decrease in sleep quality, and prevalance of poor sleeper is high in migraine [70]. Poor sleep quality may be in association with chronic migraine, and patients with chronic migraine may have more sleep disorders comparing episodic migraine [8, 54]. Bertisch et al have investigated sleep efficiency of patients with episodic migraine with actigraphy. They have found a relation with poor sleep efficiency characterized by fragmentations during sleep and a migraine attack on the following day. No temporal association was found between poor sleep quality and shorter duration of sleep. They have concluded that fregmantations rather than duration of sleep have a role in inducing frequent attacks in episodic migraine [82]. On the other hand, some studies have not found an association between headache and fregmantations in sleep during night [25, 83].

Obesity is a common risk for both migraine and OSAS, and migraineous patients may be sensitive to hypoxemia which is also associated with OSAS headache [4]. Comorbidity of OSAS increases the frequency of migraine attacks, and chronicity of migraine can be due to sleep apnea [65]. Morning headache attacks in migraineurs can be associated with sleep apne or snoring [84].

The polysomnographic features of migraneurs may show alterations compared to nonheadache population. Sleep latency is found to be normal or longer, and the percentage of NonREM1 sleep is slightly higher, while NonREM3 sleep stage is decreased interictally [40, 84]. The latency of rapid eye movement (REM) sleep may be longer. The total amount of REM sleep may decrease, and sleep efficiency may also be low [13, 84]. In the microstructure of sleep in migraneous patients, rate of CAP and amount of CAP cycles and arousal index in REM and NonREM sleep stages may be low [28, 40], or arousals may be frequent especially in patients with aura [38].

Pediatric patients with chronic migraine or migraine attacks with severe intensity may have short duration of sleep with an increased sleep latency as well as decreased percentages of NonREM 3 and REM sleep stages [84].

Migraneous patients have more complaints of insomnia [85], and the rate increases with the comorbidity of psychiatric diseases such as anxiey disorder and/or depression [86]. Sleep may be fragmentated cause of headache. The patients may also complain of symptoms of insomnia previous night before a morning headache attack with a decrease in the amount of NonREM3 sleep stage [83].

Treatment of associated sleep disorder has beneficial effects on migraine therapy. It has been shown that if the sleep disorder is treated, accompanying migraine may turn from chronic form to episodic one (**Case 1**) [9, 13, 87].

Thirty years old male patient was a shift-worker in a factory. He was admitted to the headache outpatient clinic. Since childhood, he had a pulsatile and throbbing unilateral headache mainly on the right orbitotemporal side. Localization might change at different attacks. Pain duration wass 8 to 10 hours, and pain frequency has increased after he began to work as a shift-worker [>15 days/month]. Pain intensity was severe (with a visual analog score of 10). Phonophobia, photofobia, and nausea and sometimes vomiting accompanied the headache, and pain was elevating by climbing stairs or walking. There was no preceeding aura. Stress and inadequate sleep triggered the pain, while pain alleviated if the patient could sleep during the attack. His mother had a diagnosis of migraine without aura.

As he also complained of snoring without any witnessed apnea and paresthesia on his legs when he lied down for sleep which alleviates as he got up and walked around, he underwent a polysomnographic investigation at the sleep center of University of Health Sciences Erenkoy Mental Health and Neurological Diseases Training and Research Hospital.

On his polysomnographic investigation, we have detected slightly reduced sleep efficiency. NonREM1 percentage was high, while NonREM3 and REM sleep percentages were found to be low during all night. Arousal index in NonREM sleep stages and snoring index both in REM and nonREM sleep stages were increased. Apne-hypopnea index was 2.1 (not pointing out an obstructive sleep apnea syndrome).

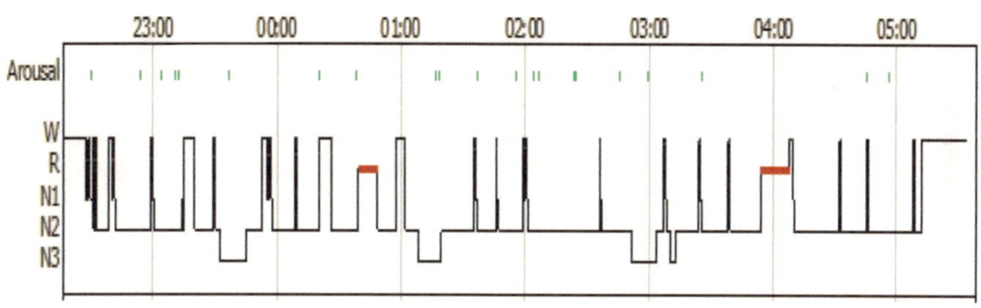

Hypnogram of the patient that has been recorded by polysomnography. Gray horizantal lines show NonREM sleep stages (N1, N2, and N3) and wakefullness [W], red lines show REM sleep stage [R], and green vertical lines show arousals.

CASE 1.
Chronic migraine.

4.2 Tension-type headache

Tension-type headache is the most common primary headache, mainly bilaterally with a pressing or tightening quality and a mild to moderate intensity. The duration may be 30 minutes to 7 days. Routine physical activity does not worsen the pain.

Though photophobia or phonophobia may be present, nausea does not accompany [81]. Several sleep disturbances like hypersomnia, insomnia, or circadian sleep-wake disorders may accompany TTH [8, 88]. The decrease in sleep quantity is one of the most important triggers, and many of the patients with TTH report unsatisfied sleep [8].

Eppworth sleepiness scores and PSQ scores are higher in TTH patients indicating poor sleep quality and daytime sleepiness [83]. Decreased sleep quality is in relation with higher intension of headache attacks and can be a factor for the chronification of TTH [57, 89]. Patients with sleep disorders tend to have lower threshold of pain [83]. Accompanying depression may contribute to decrease the pain treshold [90].

On polysomnographic investigation, NonREM 1 [N1] latency was found to be decreased with an increase in NonREM 3 [N3] sleep, while the structure of REM (both the latency and total amount) was not effected. Decrease in total sleep time and poor sleep efficiency due to increased sleep fregmentation with arousals may also be observed [83].

Sleep-related movements and restless legs syndrome can also accompany TTH [8, 73]. Association with OSAS is not clear, but if OSAS is diagnosed, it must be treated properly [7, 9]. In children, a relation between TTH and bruxism was also found (**Case 2**) [84].

Twenty-three years old female student was admitted to the headache outpatient clinic with a bilateral headache on frontal regions with a pressing quality since 3 years. Pain duration was 3 to 72 hours with a pain frequency more than 15 days in a month. She had no phonophobia or photophobia. Nausea might be present. Short duration of sleep and working with computer for long hours triggered the attacks. Pain intensity was moderate (with a visual analogue score of 6). As she complained of difficulty in maintaining sleep more than 3 times a week and daily sleepiness and snoring, she underwent a polysomnographic investigation at the sleep center of University of Health Sciences Erenkoy Mental Health and Neurological Diseases Training and Research Hospital

On her polysomnographic test, we have found elongation of sleep latency with a reduced sleep efficiency. NonREM1 and NonREM3 percentages are slightly increased. Frequent awakenings and short arousals are observed in the microstructure. Abnormal sleep-related breathing events were not detected.

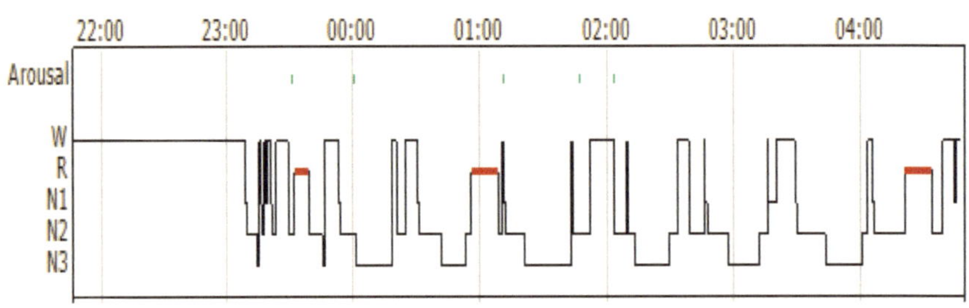

Hypnogram of the patient that has been recorded by polysomnography. Gray horizantal lines show NonREM sleep stages (N1, N2, and N3) and wakefullness (W), red lines show REM sleep stage (R), and green vertical lines show arousals.

CASE 2.
Chronic tension-type headache.

4.3 Cluster headache

Cluster headache is one of the trigeminal autonomic cephalalgias. Pain is mainly located at orbital/supraorbital/temporal regions unilaterally. The duration of severe pain is 15 to 180 minutes with accompanying autonomic symptoms (ICD). The relation with sleep and cluster headache [CH] has been shown in previous studies. CH is thought to be provoked during the switching of REM sleep to NonREM sleep. In episodic CH, attacks are mainly related to REM sleep, though this relaton is not clear in chronic form [8].

Sleep apnea either obstructive or central seem to be frequent in patients with CH. Accompaying sleep apnea may induce CH by leading to nocturnal hypoxemia [91], and effective treatment of sleep apnea either with CPAP or dental device can also be effective to diminish the severity of clusters [9, 64, 92].

Poor sleep quality and short duration of sleep may be seen in both episodic and chronic CH [93]. During the bouts of CH, patients may suffer from transient insomnia which usually resolves after the end of the bout and may recur at the next cluster period in episodic CH [91].

Among shift-workers, episodic cluster headache incidence was found to be higher. This suggested that disturbed sleep due to the work schedule could trigger cluster headache [93, 94]. As shift-working is suspected to induce the attacks, patients with CH can be advised to have a steady daily working plan [93].

During a cluster period, elongated REM latency and diminished total percentage of REM sleep can be detected by polysomnography. Sleep-wake cycle may also be disturbed during this period (**Case 3**) [28, 95].

Fifty-two years old male patient was admitted to the headache outpatient clinic with a complaint of a severe headache on right orbitofrontal region for 3 years. Attacks mainly began at autumn and lasted for 3 to 4 weeks. Each attack has begun early in the morning and awakened him with a duration of 45 to 60 minutes. Pitosis, lacrimation, rinorrhea, and redness on the right eye convoyed the headache, which was very severe (with a visual analog score of 10). Pain was resistant to analgesics, and resting did not alleviate the headache. He did not have a history of any other diseases. Neurological examination and cranial magnetic resonance imaging were normal. He also complained of snoring and witnessed apnea.

His polysomnographic investigation recorded at the sleep center of University of Health Sciences Erenkoy Mental Health and Neurological Diseases Training and Research Hospital shows a slightly increased sleep latency. Awakenings are seen during all night, and sleep efficiency is slightly reduced. He had an attack at 05:00 AM that awakened him from sleep (shown by an arrow). Apne-hypopnea index is 6.2 (mild obstructive sleep apnea syndrome).

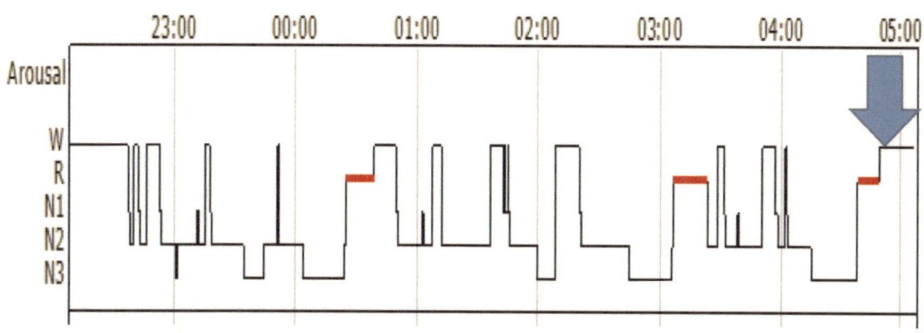

Hypnogram of the patient that has been recorded by polysomnography. Gray horizantal lines show NonREM sleep stages (N1, N2, and N3) and wakefullness (W), and red lines show REM sleep stage)

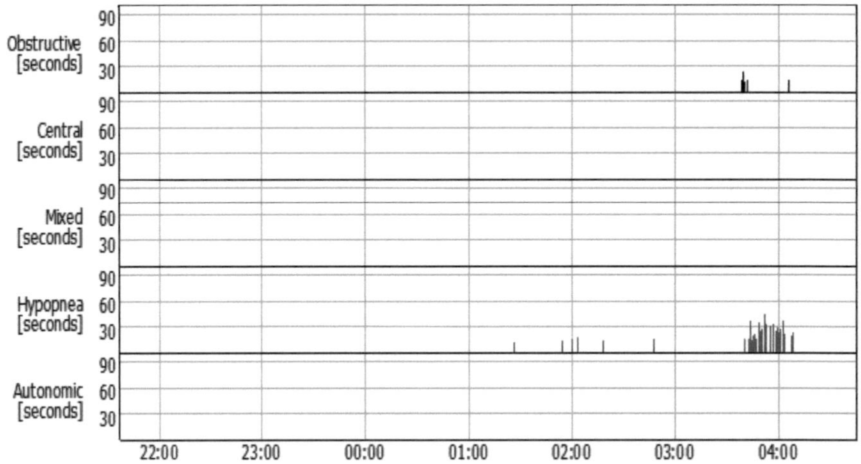

Abnormal sleep-related breathing events (obstructive apneas and hypopneas) are shown by green vertical lines

CASE 3.
Cluster headache.

4.4 Hypnic headache

Hypnic headache is a rare headache disorder that mainly occurs during night sleep as well as at naps during the day [96]. Headache causes wakening and lasts for up to 4 hours without associating characteristic symptoms [81]. Attacks may occur in every stages of sleep [97, 98]. Just before and during the headache attack, elevation in arterial blood pressure has been detected in some patients. It has been hypothesized that these alterations could typify an association with sleep apnea [63]. Sleep apnea may be a trigger for the attack, and the treatment of OSAS also reduces the hypnic headaches [9, 98]. Attacks arising either from REM sleep or NonREM sleep can be documented using polysomnography (**Case 4**) [96].

Fifty-eight years old female patient was admitted to the headache outpatient clinic. She had a pressing headache on vertex which awakened her from sleep nearly every night. Attacks occured in the first half of the night with a duration of 60–120 minutes. Neither phonophobia/photofobia nor nausea/vomiting were present. Autonomic symptoms did not accompany. Pain intensity was mild (with a visual analog score of 6). She did not have any other types of headache. She also complained of snoring, and there was witnessed apnea detected by her husband. She had a history of hypertension which was under control with ramipril.

On her polysomnographic investigation recorded at the sleep center of University of Health Sciences Erenkoy Mental Health and Neurological Diseases Training and Research Hospital, we have found an elongated sleep latency and a reduced sleep efficiency. On the first half of sleep, she has awakened after first REM sleep stage (shown by an arrow on hypnogram) with a headache attack. Apnea-hypopnea index was 8.7 (mild obstructive sleep apnea syndrome). Snoring index both in REM and NonREM sleep stages were increased.

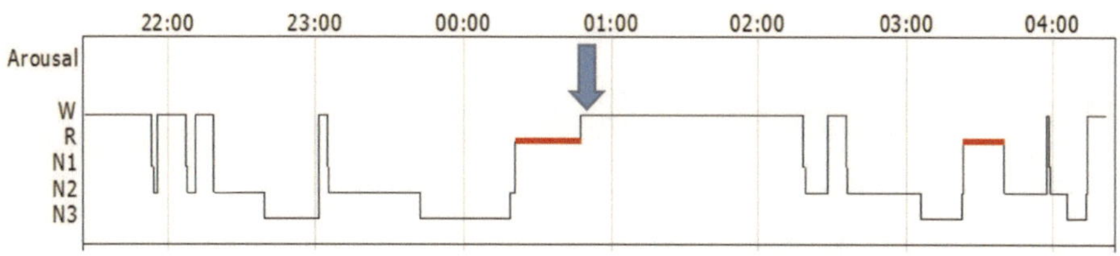

Hypnogram of the patient that has been recorded by polysomnography. Gray horizantal lines show NonREM sleep stages (N1, N2, and N3) and wakefullness (W), and red lines show REM sleep stage)

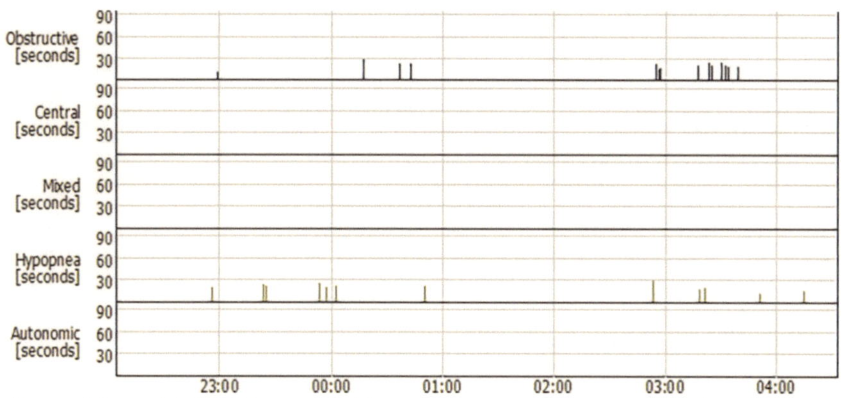

Abnormal sleep-related breathing events (obstructive apneas and hypopneas) are shown by green vertical lines.

CASE 4.
Hypnic headache.

As a conclusion, headache prevalence is high in sleep disorders, and sleep disorders are highly seen in primary headaches. This comorbidity may induce the chronification of both of the syndromes. A detailed history of both disturbances must be taken, and clinicians should consider and behold the treatment of accompanying sleep complaints for an effective management of headache and a better quality of life.

Author details

Füsun Mayda Domaç[1], Derya Uludüz[2*] and Aynur Özge[3*]

1 Department of Neurology and Sleep Medicine Center, University of Health Sciences, Erenkoy Mental Health and Neurological Diseases Training and Research Hospital, Istanbul, Turkey

2 Brain 360 Neurology Health Institute, Istanbul, Turkey

3 Department of Neurology, Mersin University School of Medicine, Algology and Clinical Neurophysiology, Mersin, Turkey

*Address all correspondence to: deryaulu@yahoo.com and aynurozge@gmail.com

References

[1] Stovner LJ, Hagen K, Linde M, Steiner TJ. The global prevalence of headache: An update, with analysis of the influences of methodological factors on prevalence estimates. Journal of Headache Pain. 2022;**23**(1):34

[2] Ha H, Gonzales A. Migraine headache prophylaxis. American Family Physician. 2019;**99**(1):17-24

[3] Buenaver LF, Quartana PJ, Grace EG, et al. Evidence for indirect effects of pain catastrophizing on clinical pain among myofascial temporomandibular disorder participitants: The mediating role of sleep disturbance. Pain. 2012;**153**(6): 1159-1166

[4] Vgontzas A, Cui L, Merikangas KR. Are sleep difficulties associated with migraine attributable to anxiety and depression? Headache. 2008;**48**: 1451-1459

[5] Paiva T, Batista A, Martins P, Martins A. The relationship between headaches and sleep disturbances. Headache. 1995;**35**(10):590-596

[6] Yeung WF, Chung KF, Wong CY. Relationship betwen insomnia and headache in community-based middle-aged Hong Kong Chinese women. The Journal of Headache and Pain. 2010; **11**(3):187-195

[7] Cho SJ, Chu MK. Risk factors of chronic daily headache or chronic migraine. Current Pain and Headache Report. 2015;**19**(1):465

[8] Jennum P, Jensen R. Sleep and headache. Sleep Medicine Rev. 2002; **2002**:471-479

[9] Singh NN, Sahota P. Sleep-related headache and its management. Current Treatment Options in Neurology. 2013; **15**(6):704-722

[10] Evers S. Special issue on headache and sleep. Cephalalgia. 2014;**34**(10):723-724

[11] Korabelnikova EA, Danilov AB. Sleep disorders and headache: A review of correlation and mutual influence. Pain and therapy. 2020;**9**:411-425

[12] Bigal ME, Lipton RB. What predicts the change from episodic to chronic migraine? Current Opinion in Neurology. 2009;**22**(3):269-276

[13] Stanyer EC, Creeney H, Nesbitt AD, Holland PR, Hoffmann J. Subjective sleep quality and sleep architecture in patients with migraine: A meta-analysis. Neurology. 2021;**97**(16):e1620-e1631

[14] Beiske KK, Russell MB, Stavem K. Prevalance and predictors of headache in patients referred to polysomnography. The Journal of Headache and Pain. 2013; **14**:90

[15] Ferini-Strambi L, Galbiati A, Combi R. Sleep disorder-related headaches. Neurological Sciences. 2019; **40**(Suppl. 1):107-113

[16] Irwin MR. Why sleep is important for health: A psychoneuroimmunology perspective. Annual Reviews in Psychology. 2015;**66**(3):143-172

[17] Mayda Domaç F. Etiopathogenesis of circadian sleep-wake rhythm disorders. In: Watson RR, Preedy PR, editors. Neurological Modulation of Sleep: Mechanisms and Function of Sleep Health. United States: Elsevier; 2020. pp. 87-94

[18] Sateia MJ. International classification of sleep disorders-third edition:

Highlights and modifications. Chest. 2014;**146**(5):1387-1394

[19] Zhu Z, Fan X, Li X, Tan G, Chen L, Zhou J. Prevalance and predictive for poor sleep quality among migraneurs in a tertiary hospital headache clinic. Acta Neurologica Belgica. 2013;**113**: 229-235

[20] Burstein R, Jakubowski M, Garcia-Nicas E, et al. Thalamic sensitization transforms localized pain into widespread allodynia. Annals of Neurology. 2010;**68**:81-91

[21] Özge A, Genç H, Güler Aksu G, Uludüz D. Migraine and frontostriatal circuit disorders: What have we learned until now? Neurological Science Neurophysiology. 2021;**38**(2):81-89

[22] Manganotti P, Bongiovanni LG, Fuggetta G, Zanette G, Fiaschi A. Effects of sleep deprivation on cortical excitability in patients affected by juvenile myoclonic epilepsy: A combined transcranial magnetic stimulation and EEG study. Journal of Neurology, Neurosurgery, and Psychiatry. Jan 2006; **77**(1):56-60

[23] Onen SH, Alloui A, Gross A, Eschallier A, Dubray C. The effects of total sleep deprivation, selective sleep interruption and sleep recovery on pain tolerance thresholds in healthy subjects. Journal of Sleep Research. 2001;**10**(1): 35-42

[24] Smith MT, Edwards RR, McCann UD, Haythornthwaite JA. The effects of sleep deprivation on pain inhibition and spontaneous pain in women. Sleep. 2007;**30**:494-505

[25] Goder R, Fritzer G, Kopsokalyvas A, et al. Polysomnographic findings in nights preceding a migraine attack. Cephalalgia. 2001;**21**:31-37

[26] Schulte LH, May A. The migraine generator revisited: Continuous scanning of the migraine cycle over 30 days and three spontaneous attacks. Brain: A Journal of Neurology. 2016;**139**(Pt 7): 1987-1993

[27] Strother LC, Srikiatkhachorn A, Supronsinchai W. Targeted orexin and hypothalamic neuropeptides for migraine. Neurotherapeutics. 2018; **15**(2):377-390

[28] Della MG, Vollono C, Rubino M, et al. A sleep study in cluster headache. Cephalalgia. 2006;**26**:290-294

[29] Caminero-Rodriguez AB, Pareja JA. Anatomical and neurochemical bases accounting for the frequent association between headaches and sleep: The hypnic headache paradigm. Revista de Neurologia. 2008;**47**(6):314-320

[30] May A, Bahra A, Büchel C, et al. PET and MRA findings in cluster headache and MRA in experimental pain. Neurology. 2000;**55**:1328-1335

[31] Arkink EB, Schmitz N, Schoonman GG, et al. The anterior hypothalamus in cluster headache. Cephalalgia. 2017;**37**:1039-1050

[32] Sarchielli P, Rainero I, Coppola F, et al. Involvement of corticotrophin-releasing factor and orexin-A in chronic migraine and medication-overuse headache: Findings from cerebrospinal fluid. Cephalalgia: An International Journal of Headache. 2008;**28**(7): 714-722

[33] Dahmen N, Kasten M, Wieczorek S, Gencik M, Epplen JT, Ullrich B. Increased frequency of migraine in narcoleptic patients: A confirmatory study. Cephalalgia. 2003;**23**(1): 14-19

[34] Holland PR. Headache and sleep: Shared pathophysiological mechanisms. Cephalalgia. 2014;**34**:725-744

[35] Peres MPF. Melatonin, the pineal gland and their implications for headache disorders. Cephalalgia. 2005;**25** (6):403-411

[36] Gagnier JJ. The therapeutic potential of melatonin in migraines and other headache types. Alternative Medicine Review. 2001;**6**:383-389

[37] Sakurai T. The neural circuit of orexin [hypocretin]: Maintaining sleep and wakefulness. Nature Reviews. Neuroscience. 2007;**8**:171-181

[38] Nitz D, Siegel J. GABA release in the dorsal raphe nucleus: Role in the control of REM slee. American Journal of Phsiology. 1997;**273**:451-455

[39] Peyron C, Tighe DK, van den Pol AN, et al. Neurons containing hypocretin [orexin] project to multiple neuronal systems. The Journal of Neuroscience. 1998;**18**:9996-10015

[40] Nayak C, Sinha S, Nagappa M, Nagaraj K, Kulkarni GB, Thennarasu K, et al. Study of sleep microstructure in patients of migraine without aura. Sleep & Breathing. 2016;**20**(1):263-269

[41] Cologno D, Cicarelli G, Petretta V, d'Onofrio F, Bussone G. High Prevalance of dopaminergic premonitory symptoms in migraine patients with restless legs syndrome: A pathogenetic link? Neurological Sciences. 2008;**29**(1): 166-168

[42] Lliff JJ, Wang M, Liao Y, et al. A paravascular pathway facilitiates CSF flow through the brain parenchyma and the clearance of interstitial solutes, including amyloid beta. Science Translational Medicine. 2012;**4**:147

[43] Schain AJ, Mello-Carrillo A, Strassman AM, Burstein R. Cortical spreading depression closes paravascular space and impairs glymphatic flow: Implications for migraine headache. The Journal of Neuroscience. 2017;**37**: 2904-2915

[44] Bloch KE. Polysomnography: A systematic review. Technology and Health Care. 1997;**5**(4):285-305

[45] Berry RB, Quan SF, Abreu AR, et al. The AASM Manual for the Scoring of Sleep and Associated Events: Rules, Terminology and Technical Specifications. Darien, IL: American Academy of Sleep Medicine; 2020

[46] Kushida CA, Chang A, Gadkary C, Guilleminault C, Carrillo O, Dement WC. Comparison of actigraphic, polysomnographic, and subjective assessment of sleep parameters in sleep-disordered patients. Sleep Medicine. 2001;**2**:389-396

[47] Feige B, Baglioni C, Spiegelhalder K, Hircher V, Nissen C, Riemann D. The microstructure of sleep in primary insomnia: An overview and extension. International Journal of Psychophysiology. 2013;**89**(2):171-180

[48] Smith MT, McCrae CS, Cheung J, Martin JL, Harrod CG, Heald JL, et al. Use of actigraphy for the evaluation of sleep disorders and circadian rhythm sleep-wake disorders: An American Academy of Sleep Medicine Systematic Review, Meta-Analysis, and GRADE Assessment. Journal of Clinical Sleep Medicine. 2018;**14**(7):1209-1230

[49] Bollu PC, Goyal MK, Thakkar MM, Sahota P. Sleep medicine: Parasomnias. Missouri Medicine. 2018;**115**(2):169-175

[50] Spierindg EL, von Hoof MJ. Fatique and sleep in chronic headache sufferers:

An age-and sex-controlled questionnaire study. Headache. 1997;**37**(9):549-552

[51] Maizels M, Burchette R. Somatic symptoms in headache patients: The influence of headache diagnosis, frequency, and comorbidity. Headache. 2004;**44**(10):983-993

[52] Drewes AM, Gade K, Nielsen KD, Bjerregard K, Taagholt SJ, Svendsen L. Clustering of sleep electroencephalographic patterns in patients with the fibromyalgia syndrome. British Journal of Rheumatology. 1995;**34**:1151-1156

[53] Fernandes G, Franco AL, Gonçalves DA, Speciali JG, Bigal ME, Camparis CM. Temporomandibular disorders, sleep bruxism, and primary headaches are mutually associated. Journal of Orofacial Pain. 2013;**27**(1):14-20

[54] Kelman L, Rains J. Headache and sleep: Examination of sleep patterns and complaints in a large clinical sample of migraineurs. Headache. 2005;**45**: 904-910

[55] Uhlig BL, Engstrøm M, Ødegård SS, et al. Headache and insomnia in population-based epidemiological studies. Cephalalgia. 2014;**34**:745-751

[56] Tran DP, Spierings E. Headache and insomnia: Their relation reviewed. Cranio. 2013;**31**:165-170

[57] Rains JC, Davis RE, Smitherman TA. Tension-type headache and sleep. Current Neurology and Neuroscience Reports. 2015;**15**:520

[58] Smitherman TA, Walters AB, Davis RE, et al. Randomized controlled pilot trial of behavioral insomnia tratment for chronic migraine with comorbid insomnia. Headache. 2016;**56**: 276-291

[59] Loh NK, Dinner DS, Foldvary N, Skobieranda F, Yew WW. Do patients with obstructive sleep apnea wake up with headaches? Archieves in Internal Medicine. 1999;**159**:1765-1768

[60] Huamani C, de Castro JR. Sleepiness and nocturnal hypoxemia in Peruvian men with obstructive sleep apnea. Sleep Breath. 2014;**18**:467-473

[61] Damiani MF, Falcone VA, Carratu P, Scoditti C, Bega E, Dragonieri S, et al. Using PaCO2 values to grade obesity-hypoventilation syndrome severity: A retrospective study. Multidisciplinary Respiratory Medicine. 2017;**12**:14

[62] Ulfberg J, Carter N, Talback M, Edling C. Headache, snoring and sleep apnea. Journal of Neurology. 1996;**243**: 621-625

[63] Crignotta F, Coccagna G, Sacquegna T, Sfiorza E, Lamontahara G, Zucconi M, et al. Nocturnal headache: systemic arterial pressure and heart rate during sleep. Cephalalgia. 1983;**3**(1): 54-5721

[64] Cherwin RD, Zallej SN, Lin X, Hall JM, Sharma N, Hedger KM. Sleep disordered breathing in patients with cluster headache. Neurology. 2001;**56**: 984

[65] Scher AI, Midgette LA, Lipton RB. Risk factors for headache chronification. Headache. 2008;**48**:16-12

[66] Kallweit U, Hidalgo H, Uhl V, Sandor PS. Continuous positive airway pressure therapy is effective for migraines in sleep apnea syndrome. Neurology. 2011;**76**:1189-1191

[67] Bruni O, Fabrizi P, Ottaviano S, Cortesi F, Giannotti F, Guidetti V. Prevalence of sleep disorders in childhood and adolescence with

headache: A case-control study. Cephalalgia. 1997;**17**:492-498

[68] Casez O, Dananchet Y. Besson G Migraine and somnambulism. Neurology. 2005;**65**(8):1334-1335

[69] Messina A, Bitetti I, Precenzano F, Iacono D, Messina G, Roccella M, et al. Non-rapid eye movement sleep parasomnias and migraine: A role of orexinergic projections. . Frontiers in Neurology. 2018;**9**:95

[70] Headache Classification Committee of the International Headache Society (IHS). The international classification of headache disorders. Cephalalgia. 2008; **38**(1):1-211

[71] Lin YK, Lin GY, Lee JT, et al. Associations between sleep quality and migraine frequency: A cross-sectional case-control. Study Medicine. 2016;**95**: e3554

[72] Suzuki K, Miyamoto T, Miyamoto M, Suzuki S, Watanabe Y, Takashima R, et al. Dream-enacting behaviour is associated with impaired sleep and severe headache-related disability in migraine patients. Cephalalgia. 2013;**33**(10):868-878

[73] Cho SJ, Chung YK, Kim JM, Chu MK. Migraine and restless legs syndrome are associated in adults under age fifty but not in adults over fifty: A population-based study. The Journal of Headache and Pain. 2015;**16**:554

[74] Sevindik MS, Demirci S, Göksan B, Özge A, Savrun FK, Onur H, et al. Accompanying migrainous features in pediatric migraine patients with restless legs syndrome. Neurological Sciences. 2017;**38**:1677

[75] Schürks M, Winter A, Berger K, Kurth T. Migraine and restless legs syndrome: A systematic review. Cephalalgia. 2014;**34**(10):777-794

[76] Yang FC, Lin TY, Chen HJ, Lee JT, Lin CC, Kao CH. Risk of restless legs syndrome following tension-type headache: A nationwide population-based cohort study. Medicine [Baltimore]. 2015;**94**(46):e2109

[77] Earley CJ, Connor J, Garcia-Borreguero D, Jenner P, Winkelman J, Zee PC, et al. Altered brain iron homeostasis and dopaminergic function in restless legs syndrome [Willis-Ekbom disease]. Sleeping Medicine. 2014;**15**(11): 1288

[78] Chung PW, Cho SJ, KİM WJ, Yang KI, Yun CH, Chu MK. Restless legs syndrome and tension-type headache: A population-based study. Journal of Headache Pain. 2017;**18**(1):47

[79] Castrillon EE, Exposto FG. Sleep bruxism and pain. Dental Clinics of North America. 2018;**62**(4):657-663

[80] Canto DL, Singh V, Bigal ME, Major PW, Flores-Mir C. Association between tension-type headache and migraine with sleep bruxism: A systematic review. Headache. 2014; **54**(9):1460-1469

[81] Yang CP, Hsieh ML, Chiang JH, Chang HY, Hsieh VCR. Migraine and risk of narcolepsy in children: A nationwide longitudinal study. PLoS One. 2017;**12**(12):e0189231

[82] Bertisch SM, Wenyuan L, Buettner C, Mostofsky E, Ruescman M, Kaplan ER, et al. Nightly sleep duration, fragmentation, and quality and daily risk of migraine. Neurology. 2020;**94**: 489-496

[83] Engstrøm M, Hagen K, Bjørk M, et al. Sleep quality, arousal and pain

thresholds in tension-type headache: A blinded controlled polysomnographic study. Cephalalgia. 2014;**34**:455-463

[84] Vendrame M. Polysomnographic findings in children with headaches. Pediatric Neuroloy. 2008;**39**(1):6-11

[85] Finan PH, Goodin BR, Smith MT. The association of sleep and pain: An update and a path forward. Pain. 2013; **14**(12):1539-1552

[86] Gupta R, Bhatia MS, Dahiya D, et al. Impact of primary headaches on subjective sleep parameters among adolescents. Annals of Indian Academy of Neurology. 2008;**11**:164-171

[87] Rains JC. Sleep and migraine: Assessment and treatment of comorbid sleep disorders. Headache. Jul 2018;**58** (7):1074-1091

[88] Fernandez-de-las-Penas C, Fernandez-Munoz F, Palacious-Cena M, Paras-Bravo P, Cigaran-Mendez M, Navarro-Pardo E. Sleep disturbances in tension-type headache and migraine. Therapeutic Advanced Neurological Disorders. 2018;**11**:1-6

[89] Kikuchi H, Yoshiuchi K, Yamamoto Y, et al. Does sleep aggravate tension-type headache? An investigation using computerized ecological momentary assessment and actigraphy. Biopsychosocial Medicine. 2011;**5**:10

[90] Chiu YH, Silman AJ, Macfarlane GJ, et al. Poor sleep and depression are independently associated with a reduced pain threshold: Results of a population based study. Pain. 2005;**115**: 316-321

[91] Sahota PK, Dexter JD. Transient recurrent situational insomnia asociated with cluster headache. Sleep. 1993;**16**: 255-257

[92] Graff-Radford SB, Newman A. Obstructive sleep apnea and cluster headache. Headache. 2004;**44**:607-710

[93] de Coo IF, van Oosterhout WPJ, Wilbrink LA, van Zwet EW, Ferrari MD, Fronczek R. Chronobiology and sleep in cluster headache. Headache. 2019;**59**(7): 1032-1041

[94] Ofte HK, Berg DH, Bekkellund SI, et al. Insomnia and peirodicity of headache in an arctic cluster headache population. Heaache. 2003;**53**:1602-1612

[95] Barloese MC, Jennum PJ, Lund NT, et al. Sleep in cluster headache – Beyond a temporal rapid eye movement relationship? European Journal of Neurology. 2015;**22**:656-e40

[96] Manni R, Sances G, Terzaghi M, Ghiotto N, Napsi G. Hypnic headache Psg evidence of both REM and NREM-related Attacks. Neurology 2004

[97] Evers S, Rahmann A, Schwaag S, Ludemann P, Husstedt IW. Hypnic headache—the first German cases including polysomnography. Cephalalgia. 2003;**23**:20-23

[98] Pinessi L, Rainero I, Cicolin A, Zibetti M, Gentile S, Mutani R. Hypnic headache syndrome: Association of the attacks with REM sleep. Cephalalgia. 2003;**23**:150-154

Chapter 9

The Role of Cognitive Reserve in Executive Functioning and Its Relationship to Cognitive Decline and Dementia

Gabriela Álvares-Pereira, Carolina Maruta and Maria Vânia Silva-Nunes

Abstract

In this chapter, we explore how cognitive reserve is implicated in coping with the negative consequences of brain pathology and age-related cognitive decline. Individual differences in cognitive performance are based on different brain mechanisms (neural reserve and neural compensation), and reflect, among others, the effect of education, occupational attainment, leisure activities, and social involvement. These cognitive reserve proxies have been extensively associated with efficient executive functioning. We discuss and focus particularly on the compensation mechanisms related to the frontal lobe and its protective role, in maintaining cognitive performance in old age or even mitigating the clinical expression of dementia.

Keywords: cognitive reserve, executive functions, aging, cognitive decline, dementia

1. Introduction

The impact of brain aging on cognition is far from uniform, ranging from perfect fitness to cognitive impairment. The prevalence of dementia is estimated to increase from 57.4 million in 2019 to 152.8 million in 2050 [1], thus representing a major public health problem. Still, evidence shows that more than one-third of all cases of dementia could be prevented or modified by changes in lifestyle, correction of risk factors, and specific therapeutic interventions [2–5]. In fact, despite the absence of pharmacological treatment for degenerative diseases, such as Alzheimer's disease (AD), it is known that vascular risk factors increase the likelihood of cognitive decline. Simple measures, the control of hypertension, for instance, may revert cognitive impairment and reduce conversion to dementia [6]. Likewise, healthy lifestyle patterns, physical exercise, and intellectual and social enrichment may improve performance and change the biomarker trajectories of individuals classified as cognitively impaired [7–9].

Promoting the presence of protective factors throughout life may help to cope with the negative consequences of pathology through resilience or resistance mechanisms. The term "resistance" refers to the notion of avoiding pathology (i.e., being free from significant AD pathology) in the sense that it is inferred from an observed absence or lower level of dementia-associated brain injury, relative to an expected greater frequency or severity based on age, genetic factors, or other individual characteristics. On the other hand, "resilience" is mostly used in the sense of coping with pathology (i.e., remaining cognitively intact despite significant AD pathology) and is inferred from the observation of a higher than expected cognitive functioning related to the level of brain injury [10, 11]. While the first is linked to an absence or delay of brain changes ("brain maintenance"), the latter is closely associated with the concept of the reserve, which can be measured or inferred either as brain structural and/or physiological premorbid capacity [11, 12].

The construct of reserve firstly emerged to describe patients with extensive destruction of nervous tissue following brain damage but not the expected level of functional changes [13]. It was then proposed that larger brains, with greater weight and a larger number of neurons, could have protective effects on the cognitive decline due to a higher "brain reserve" capacity [14]. Years later, Stern [15] defined the concept of Cognitive Reserve (CR) as the brain's ability to optimize and maximize performance through the differential recruitment of brain networks or the use of alternative cognitive strategies to cope with brain dysfunction. Stern's proposal claims that the mechanisms underlying CR are active processes by which the brain tries to compensate for the neural loss. These processes can be influenced by the interaction between innate factors (e.g., in utero or genetically determined) and, mainly, lifelong experience (e.g., intelligence, education, occupation, physical exercise, leisure activities, or social involvement). In contrast, the passive models propose that response to neural damage is related to brain size or the number of synapses (brain reserve), which can affect the threshold for clinical expression [16]. Brain reserve and CR are not mutually exclusive in the sense that brain reserve does not protect against the accumulation of pathology, but it does protect against its negative effects [17]. Instead, they influence each other—life experiences and the involvement in stimulating cognitive activities can modify brain anatomy (i.e., neurogenesis, angiogenesis, and resistance to apoptosis) and positively regulate compounds that promote neural plasticity [18].

The concept of CR has progressively evolved in such a way that it is now central in the literature on normal and pathological aging, notwithstanding the theoretical pitfalls and methodological controversies generated by years of studies and reserve-associated concepts. The most striking challenges are the absence of an operational definition of CR and the lack of clarification of its neural bases, the relationship between the brain and CR, and which factors affect brain reserve [19]. Making an effort to overcome these difficulties, a consensus report tried to clarify CR terminology [17] by claiming "resilience" as an umbrella concept that describes the process of coping with age- and disease-related changes, which includes multiple reserve related-concepts, such as brain reserve, brain maintenance, and CR.

Normal aging is characterized by several brain changes at the morphometric and functional level, and associated neuropsychological changes, that are particularly relevant in the frontal lobes on which Executive Functions (EF) heavily depend upon [20]. Among other areas of cognition, EF play a critical role in everyday life, allowing individuals to plan ahead, focus attention, and switch between tasks, hence maintaining effective levels of independent functioning. Variable EF trajectories include development into early adulthood and decline into older age, associated with structural and functional changes in the prefrontal cortex [21]. Despite this age-related decline, EF also assumes an important role in maintaining global cognitive efficiency in the late period of life, thanks to a higher CR [22].

There is considerable interest in understanding the processes underlying cognitive decline (and whether they result from specific or general impairments that reflect different patterns and different pathological processes) but also in how the brain actively copes with these deleterious effects on EF so functional independence can be maintained. Next in this chapter, we will review evidence that focuses on certain socio-behavioral CR proxies (e.g., education, occupational complexity, leisure activities, and social involvement), how they may help to cope with age-related changes and brain pathology, and how they relate with EF. Further, differences between "active" versus "passive" models of reserve and the underlying CR mechanism ("neural reserve" vs. "neural compensation") are described.

2. Socio-behavioral proxies in the prevention of cognitive decline and dementia

One of the major limitations of the CR construct is that it can hardly be measured directly. Three methods are usually used to quantify and measure it—(a) socio-behavioral indicators, (b) residual approach, and (c) functional neuroimaging studies [17]. Hence, studies should include not only measures of the status of the brain (reflecting brain alteration or pathology), but also clinical or cognitive performance (consequences of brain damage), and socio-behavioral indicators (e.g., an index of life-long experience/premorbid capacity) when assessing the role of CR. The goal is to be able to predict an individual's cognitive performance through the interaction between the proposed CR factors and the state of the brain/pathology.

Several studies have shown that CR proxies may decrease the rate of conversion to dementia in subjects with identical degrees of the pathological burden of AD [23, 24], and even have a protective role against the cognitive impairment associated with brain white-matter changes (WMC) or higher ventricular volume [25, 26] delaying the onset of clinical deficits [27]. Understanding the role of these proxies on the prediction of cognitive trajectories serves a two-fold objective, either it is disease prevention or disease diagnosis.

Different CR proxies have been identified [28], but recent systematic reviews indicate that education, occupational attainment, leisure activities, and social involvement are the most common indicators [29–31].

The number of years of formal education is the most consistently used across studies. A protective effect of education for age-associated cognitive decline appears to result in higher levels of CR [30, 32]. This is supported by strong positive associations between the number of years of formal education and crystallized measures (e.g., vocabulary) and EF, explaining, in the latter case, even more variance than age itself [33–35], compared with fluid abilities, such as processing speed, memory, or visuospatial abilities [34, 36, 37]. Robust scientific evidence also supports that lower-educated individuals are more likely to suffer from dementia in a wide range of settings [38]. For example, Contador et al. [39] found that living in a rural area (early and mid-life stages) increased the likelihood of dementia, with the risk being particularly high in people with low education. However, the effect of education on age-related cognitive changes remains controversial [40]. Kremen et al. [41] sought to demonstrate that the impact of CR factors is primarily downstream of intellectual capacity. These researchers concluded that brain development is substantial during childhood and adolescence and that further education from the age of 20 years would contribute much less to brain development. Moreover, it should be considered that the quality of the educational experience is not the same for all individuals, which may influence its potential impact as a CR proxy.

The protective effect of education not only mediates the transition between normal and pathological aging but also between stages of cognitive impairment. Based on the hypothesis that less automatized cognitive domains (or those that did not achieve proper consolidation throughout life) may deteriorate sooner than more consolidated ones, a recent retrospective study aimed to investigate whether education modifies the profile of cognitive/executive performance (i.e., sustained and divided attention, inhibitory control, working memory, verbal, motor and graphomotor fluency, planning, abstract reasoning, and episodic memory) in Mild Cognitive Impairment (MCI). It was found that despite a similar pattern of cognitive decline in both higher and lower education groups, patients with higher education revealed a trend toward a higher proportion of abnormal performances (≤ -1.5 standard deviation on age- and education-adjusted normative scores) and a steeper decline in measures of sustained attention and episodic memory [42]. These findings suggest that patients with higher levels of education have a higher CR because they show a more pronounced decline in executive control that does not reflect differences in clinical disease staging [35, 43, 44]. On the opposite extreme of educational level, elderly illiterate subjects may be more vulnerable to cognitive decline and dementia, due to the lack of the protective effect of education [45–49].

It is worth noting that, although education is usually measured by the number of years of formal education, there may be other indicators that better capture its true impact. In a recent prospective longitudinal cohort study on aging and cognition, which recruited and followed 275 healthy community subjects seen in primary care settings, with 50 years or older, over a 5-year period, investigators found that being male, older, and with a lower age- and education-adjusted z-scores on divided attention/mental flexibility measures were significant independent predictors of cognitive impairment 5 years later. Moreover, vocabulary emerged as a stronger predictor of cognitive stability or decline than education, independently of their correlation [50]. This highlights the relevance of this measure by reflecting more accurately the degree of cognitive stimulation and intellectual enrichment that may account for subtle differences between subjects at the same educational level, particularly relevant in overall low-literacy populations.

Occupational and leisure activities may also have markedly significant protective effects on cognitive decline and dementia, especially for individuals whose jobs involve social interaction [51]. In fact, it is known that engagement in mentally stimulating activities throughout life may promote neural connectivity [52]. With respect to occupational activity, cognitively demanding work conditions are associated with a decreased risk of cognitive decline in older adults [53]. Middle-aged people at risk for AD (decreased hippocampal volume and increased brain atrophy) with greater occupational complexity (e.g., involving complex social interactions) maintained a similar level of cognitive performance as those with less pathology [51, 54]. However, since higher levels of education are usually associated with jobs that are more cognitively demanding, whether or not the protective effect of education is independent of the levels of work complexity in middle age remains controversial [55, 56]. Moreover, a synergistic effect of low education and occupation on the risk of developing AD was described by Stern et al. [57], particularly when it is combined with cognitively demanding work activity in adulthood [55, 57]. For instance, some studies indicate that level of literacy is a more accurate measure of CR than years of education [58], especially in those individuals from disadvantaged groups or with null/low educational attainment [39, 59]. Regarding involvement in leisure activities, it has also been associated with a reduced risk of AD [60, 61] and protective effects against cognitive decline [62, 63].

There seems to be evidence that lifestyle and the environment effectively regulate cognitive aging and that this regulation may be particularly relevant in the hippocampal-mediated memory functions in mammals. Although the causal nature of this relationship has not yet been established [64], studies in animal models seem to indicate that it may exist, but more clinical studies are needed to specifically understand how social involvement and integration can be used to prevent cognitive decline. Additionally, the mechanisms underlying this relationship seem to indicate a relevant role for growth factors, neuroinflammation, and neurogenesis processes. In this context, physical activity, for instance, has been identified as inducing neurogenesis due to its benefits on structural and functional plasticity in hippocampus-dependent learning and memory tasks. Accordingly, maintaining an active social life at older ages can improve CR and benefit cognitive function. This is especially relevant since some aspects, such as education or occupational complexity, developed at a young age and middle age cannot be modified. Social activity can contribute to an increased reserve even in a more advanced stage of life, with gains in cognitive performance. In fact, living alone was significantly associated with an increase in cognitive complaints and a significant predictor of future cognitive decline in specific linguistic/executive measures, such as verbal fluency over a 5-year follow-up [33, 65, 66]. Social interactions can be viewed as natural forms of cognitive stimulation and may play a relevant part in the stimulation of language skills, whereas living alone would represent a relative cognitive deprivation situation, with reduced cognitive stimulation and lower CR [67, 68]. Interventions that reduce social isolation at a more advanced stage can benefit cognitive function both directly and indirectly by building reserve, especially in individuals with low CR in middle age. This aspect has important implications for interventions suggesting that combating social isolation can contribute to a healthier cognition [69].

3. Compensatory mechanisms of CR and EF

Traditionally, late-mature regions, such as the frontal lobes, are considered especially vulnerable to normal age changes, inspiring theories of cognitive aging, such as the “last in, first-out” or “retrogenesis” hypothesis. This hypothesis considers an anteroposterior gradient of age vulnerability, which explains the decline in EF often observed in healthy older adults [70].

Executive functions, such as processing speed, working memory, inhibitory control, top–down suppression, or shifting ability, are shaped by education and by other CR proxies. A decline in executive performance has been shown to be associated with low performance in activities of daily living and to predict conversion from MCI to dementia [71, 72]. Moreover, EF are known to be sensitive to damage in other parts of the brain, such as subcortical white matter changes [34], thalamic nuclei, the limbic system, and basal ganglia [73] apart from prefrontal lobe damage.

The perspective that age-related cognitive decline emerges when a person is no longer able to compensate for the reduced functioning of the primary brain structures and circuits, is largely supported in the literature. Relevant conceptual models have emerged over the last 20 years, aiming to describe and understand brain reorganization in response to age-related changes and brain injury. Older adults may use alternative networks to aim for the same level of functioning as younger individuals, which can represent a mechanism of neural compensation [74, 75]. The "Scaffolding Theory of Aging and Cognition" (STAC) model proposed by Park and Reuter-Lorenz [76] claims the recruitment of additional circuits as a way to strengthen the declining structures whose functioning has become inefficient. These strategies lose efficiency in the aged brain and are eventually no longer accessible when there is cerebral pathology, as in the case of AD. The "normalcy-pathology homology" phenomenon suggests that there are regions more vulnerable to age-related changes and that this age vulnerability renders them more susceptible to additional pathological AD-related changes. This is particularly clear in frontotemporal regions where the elderly, even with a low risk of AD, present prominent cortical reductions [70].

The Cognitive Reserve framework suggests that individual differences in cognitive performance are based on more efficient recruitment of brain networks (neural reserve) or the enhanced ability to recruit alternate (compensation) brain networks [15, 77]. Regarding neural reserve, it is postulated that inter-individual variability related to the efficiency, capacity, or flexibility of the brain networks will influence how the healthy brain can deal with the demands imposed by the emergence of brain injuries or pathologies. The neural reserve allows healthy young individuals with greater CR to solve tasks more efficiently and more capably and, in turn, may better confront the disruptions imposed by brain pathology due to the increased flexibility of brain networks. Neural compensation concerns task-related activation, a mechanism that only appears when new resources are needed to maintain or improve performance due to changes in the brain structure. Hence, neural compensation is a mechanism usually referring to people who have brain pathology [15, 77]. The degree of compensation can also vary in individuals in terms of expression and effectiveness. In fact, neural compensation refers to inter-individual variability to compensate for the disruption of standard processing networks. In this situation, brain structures or networks that are not normally used by individuals with intact brains become activated. Both neural reserve and neural compensation support CR, with compensation being the most common mechanism in more advanced stages of the aging spectrum [78].

As previously stated, individuals with higher CR can maintain a more efficient and capable network or compensate advantageously in the face of a comparable amount of brain pathology [79]. Accordingly, Scarmeas et al. [80], using a set of memory tasks, identified brain regions where systematic relationships between CR and brain activation differed as a function of aging. Thus, when facing certain tasks, young and older people activate similar brain regions but as the difficulty of the memory task increased the magnitude of activation was often higher in older individuals, suggesting more efforts to achieve a comparable level of performance, which can be related to network efficiency. In addition, the older adults recruited additional regions of the brain not used by young people while performing certain memory tasks, which can represent a form of active neural compensation [80]. A similar pattern of compensation was also found when comparing old adults schooled later in life with old adults schooled at the proper age, in a memory recognition task using Magnetoencephalography (MEG), and the first ones displayed additional activations to keep the level of performance [81].

In the last few decades, scientific studies have tried to capture the "neural implementation" of CR through functional neuroimaging [78]. This approach seeks to identify resting state or task-related functional activation brain networks that may underlie CR. Potentially, the expression of these networks may be associated with the influence of CR proxies, moderating the effect of brain changes on cognition. If these networks were identified through functional neuroimaging research methods (not properly used in clinical practice), their degree of manifestation would be a more direct measure of CR than other types of proxies [17]. Tucker and Stern [37] suggested that there may be at least one "generic CR network" that can be activated during the performance of many tasks, explaining how CR protects against brain pathology, which seems to be a promising line of future investigation [17].

A recent systematic review indicates that a resting-state network, implicating medial temporal regions and cingulate cortex (anterior or posterior), is associated with neural reserve, whereas frontal regions and the dorsal attentional network (DAN), activated during the cognitive engagement, are related to neural compensation [78]. Task-related studies have found a positive correlation between CR proxies (mostly premorbid IQ and education) and higher frontal activity in healthy older adults compared to young adults [82–85]. Moreover, a positive association between CR (i.e., education-occupation attainment, premorbid IQ, and leisure activities) and frontal activity in MCI and AD patients compared to healthy older adults has been found [86, 87].

The mechanisms on which the function and resilience of large-scale brain networks are based are still poorly understood. Early lifespan environmental influences can contribute to understanding phenomena such as reserve, as, at least partially, to determine the variance of the underlying structural network. This may have implications for global and regional network controllability. A dynamic network theory can be crucial for advancing the understanding of the resilience of the human brain, reinforcing the need for a spatiotemporal analysis in complex systems [88]. In fact, the human capacity to perform a variety of tasks seems to be associated with cognitive control networks, specifically the frontoparietal control network (FPN) in the left posterior parietal cortex. The adaptability of this network, whose global connectivity pattern seems to change more than other networks, and the connectivity patterns that can be used to identify task performance, point to the importance of this network in cognitive control and task performance. It seems to be possible through "flexible hubs," that is, regions that quickly update their connectivity pattern according to task demands [89].

This greater variability in FPN connectivity, both between networks and between tasks, supports the notion that this network implements core flexible hubs, allowing cognitive control across various and distinct tasks [89]. This is especially relevant for this chapter's purpose as the existence of this control network appears to be crucial for reserve. Specifically, one of its hubs, the left frontal cortex (LFC, covering BA 6/44) [90, 91], is a likely candidate for the neural implementation of CR. The resting-state connectivity of that LFC hub region had previously been associated with protective factors such as high IQ and high cognitive performance. Concretely, it had already been demonstrated that the lateral prefrontal cortex (LPFC) is a hub region with an especially high global connectivity but, more than that, it has been shown that this global connectivity could predict the fluid intelligence of individuals, appearing to be a global hub connector [92]. This level of the organization thus appears to be especially relevant for understanding the brain and CR that involve distributed circuits and complex psychological constructs.

Global connectivity of the LFC hub (close to the Broca area), in resting-state fMRI, is associated with more years of education (CR proxy) and with milder effects of FDG-PET hypometabolism on memory performance in prodromal AD [91]. This can be important for instance in the selection of participants for intervention trials since MCI patients with higher CR seem to have a higher likelihood to benefit from a cognitive intervention [93].

Increased frontoparietal activation may reflect a compensatory mechanism, helping to protect memory task performance in early-stage AD. Additionally, increased global connectivity of LFC can support frontoparietal increased activation and that is associated with CR, moderating the association between AD neuropathology and cognitive decline, and helping to maintain better memory performance [90]. In a task-related fMRI study, the authors tried to understand if LFC hub connectivity during an episodic memory task was associated with a reserve in aging and MCI. More years of education were associated with increased LFC connectivity during memory processing, and increased LFC connectivity was associated with a higher reserve in the memory domain. This result pattern was found in controls and MCI groups, which was interpreted as suggestive that connectivity of a key hub of the frontoparietal control network contributes to reserve in both normal and pathological aging. This conclusion reinforces that LFC is a good candidate for the neural basis of reserve and that a higher LFC connectivity may be a long-lasting trait that is influenced by environmental stimuli, namely education [91]. In fact, CR, being the result of multiple and distinct stimulations, involves connectivity between different tasks and domains. Consistently with this, the LFC (BA 6/44) ranks among the top 5% of brain regions in terms of number of connections in the brain, being high and globally connected to the rest of the brain and is a key connector hub between different functional networks [91]. Taken together the results seem to point out that the cognitive control network, particularly LFC, works as a hub of the frontoparietal control network, which is associated with greater reserve. Later work showed that education is associated with better performance on memory tasks thanks to greater efficiency of functional networks, clearly demonstrating the effects of education on DMN/DAN small-worldness, mediated via LFC connectivity, and reinforcing its role as a neural basis of the reserve [94].

Moreover, evidence also shows that education facilitates the brain's ability to form segregated functional groups of networks, with stronger signals in parietal and occipital regions [95]. This fact reinforces the perspective that more years of schooling trigger a more specialized use of neural processing. However, CR (residual variance in memory and general executive functioning) was also associated with higher global network efficiency (i.e., functional integration). In this sense, this study corroborates that CR is associated both with increased functional connectivity and better organization of the network topology.

The protective role of higher global functional connectivity in the FPN and higher local connectivity between the salience network (anterior cingulate cortex) and medial frontal cortex can significantly mitigate the impact of white matter lesions on EF [96], emphasizing the role of the cognitive control network as a neural substrate for CR. As pointed out by the authors, both the salience and the FPN are important cognitive control networks, that are crucial for appropriate brain functioning, with the FPN flexibly regulating the activity of other networks and the salience network integrating inputs from different sources. Their results reinforce the notion that cognitive control networks may play a role in brain resilience mechanisms with increased connectivity being linked to better cognition.

Overall, these findings suggest that greater activity of frontal regions, namely via LFC connectivity, is a potential component of functional networks underlying neural compensation. Conversely, MTL regions, which are known to be critical for the conversion from MCI to AD, may reflect the capacity of the neural reserve [97, 98].

4. Conclusions

The understanding of the mechanisms involved with successful aging is far from straightforward and the growing number of publications in this field shows the interest of the scientific community to understand the importance of complex related concepts in its promotion. In this chapter, we focused on several socio-behavioral CR proxies identified as protective factors against cognitive decline and dementia and how they impact EF by means of neural compensation mechanisms related to the increased functional activity of the frontal lobe.

The relationship between CR proxies and the maintenance of cognitive efficiency in the context of age-related changes/brain pathology is dynamic. Not only do the skills, social involvement, and occupational attainment developed throughout life have a mediating role in improving neural connections (i.e., in terms of activation, flexibility, and efficiency), but also this enhancement of brain connectivity patterns expresses itself in better cognitive performance. Despite its vulnerability to the effects of senescence, the frontal lobes play a key role in CR allowing for the preservation of the overall cognitive function by means of enrichment of EF (e.g., planning, sequencing, inhibitory control, abstract reasoning) via a higher CR. Indeed, people with high CR show an advantage in the use of these more developed EF, thus increasing frontal lobe activity. The use of alternative task-relevant circuits compensates for effectiveness (e.g., MTL, especially relevant for memory and selectively affected in AD) thus mitigating the clinical expression of dementia. In this compensation mechanism, DAN and FPN networks are particularly relevant, with a sub-region in the LFC being identified as a potential candidate for a neural marker of CR.

Several caveats still, however, need to be fully addressed. First and foremost, it is unclear how CR proxies may specifically influence different aetiologies of dementia and modulate different cognitive trajectories. Second, EF cannot have a double role as an age–/pathology-dependent measure and as a factor that changes the relationship. As a consequence, all EF may not be appropriate measures for CR, since it is not stable throughout life and is vulnerable to age-related changes. Thus, according to the gain/loss hypothesis, one should carefully select aspects of EF that are robust and resistant to aging, in order to include them as components of CR. Stern et al. [99] considered that this approach should be better explored in the future, despite currently presenting some limitations that are difficult to overcome. From the outset, the fact that the brain measures used to predict cognition only partially capture brain pathology or physiology and different lifestyles cannot be explained by known brain predictors can lead to a high risk of including many aspects that are not reserved. Third, the differential impact of CR depending on the demographic characteristics of the population or discrepancies in measuring CR measures or outcomes (i.e., cognitive or functional) needs to be addressed as well. In fact, precise operational definitions of CR and other related theoretical constructs are needed. Advances in multimodal imaging, preferably longitudinal studies, will allow a better understanding of the neural mechanisms underlying CR. Future work should focus on the design of studies that will help to clarify the relationship between CR proxies and brain reserve, as well as improve their measurement. These studies will make it possible to improve and integrate the existing conceptual models of the moderation of CR in cognitive performance. Further, it is expected that the contribution of these investigations could lead to objective guidelines and strategies for the development of differentiated, validated, and accessible intervention programs aimed to provide more functionality and better quality of life in older adults [17].

If the existence of a compensatory capacity in individuals with a high CR seems to be clear, it is consensual that it is still not entirely clear what reserve consists of in neural terms. Potential candidates have been proposed but the discovery of this neural basis is particularly relevant as, in addition to traditional cognitive and psychosocial stimulation techniques, it could also open doors to more direct brain stimulation allowing the use of a whole arsenal of new non-invasive brain stimulation technologies which is predicted to have increasing importance in intervention.

Acknowledgements

This work is financially supported by the National Funds through FCT—Fundação para a Ciência e a Tecnologia, I.P., under the project UIDB/04279/2020. The authors thank the Centre for Interdisciplinary Research in Health and Católica Research Centre for Psychological, Family & Social Wellbeing, Universidade Católica Portuguesa for their support.

Author details

Gabriela Álvares-Pereira[1*], Carolina Maruta[2] and Maria Vânia Silva-Nunes[1]

1 Institute of Health Sciences, Centre for Interdisciplinary Research in Health, Universidade Católica Portuguesa, Lisbon, Portugal

2 Faculty of Human Sciences, Católica Research Centre for Psychological, Family and Social Wellbeing, Universidade Católica Portuguesa, Lisbon, Portugal

*Address all correspondence to: gpereira@ucp.pt

References

[1] GBD 2019 Dementia Forecasting Collaborators. Estimation of the global prevalence of dementia in 2019 and forecast prevalence in 2050: An analysis for the global burden of disease study 2019. The Lancet Public Health. 2022;**00249-8**(1):S2468-S2667. DOI: 10.1016/S2468-2667(21)00249-8

[2] Livingston G, Sommerlad A, Orgeta V, Costafreda SG, Huntley J, Ames D, et al. Dementia prevention, intervention, and care. Lancet. 2017;**390**(10113):2673-2734. DOI: 10.1016/S0140-6736(17)31363-6

[3] Livingston G, Huntley J, Sommerlad A, Ames D, Ballard C, Banerjee S, et al. Dementia prevention, intervention, and care: 2020 report of the lancet commission. Lancet. 2020;**396**(10248):413-446. DOI: 10.1016/S0140-6736(20)30367-6

[4] Morris MC, Tangney CC, Wang Y, Sacks FM, Bennett DA, Aggarwal NT. MIND diet associated with reduced incidence of Alzheimer's disease. Alzheimer's & Dementia. 2015;**11**(9):1007-1014. DOI: 10.1016/j.jalz.2014.11.009

[5] Zheng G, Xia R, Zhou W, Tao J, Chen L. Aerobic exercise ameliorates cognitive function in older adults with mild cognitive impairment: A systematic review and meta-analysis. British Journal of Sports Medicine. 2016;**50**(23):1443-1450. DOI: 10.1136/bjsports-2015-095699

[6] Sachdev PS, Lipnicki DM, Crawford J, Reppermund S, Kochan NA, Trollor JN, et al. Factors predicting reversion from mild cognitive impairment to normal cognitive functioning: A population-based study. PLoS One. 2013;**8**(3):e59649. DOI: 10.1371/journal.pone.0059649

[7] Arenaza-Urquijo EM, Wirth M, Chételat G. Cognitive reserve and lifestyle: Moving towards preclinical Alzheimer's disease. Frontiers in Aging Neuroscience. 2015;**7**:134. DOI: 10.3389/fnagi.2015.00134

[8] Vemuri P, Lesnick TG, Przybelski SA, Machulda M, Knopman DS, Mielke MM, et al. Association of lifetime intellectual enrichment with cognitive decline in the older population. JAMA Neurology. 2014;**71**(8):1017-1024. DOI: 10.1001/jamaneurol.2014.963

[9] Willey JZ, Gardener H, Caunca MR, Moon YP, Dong C, Cheung YK, et al. Leisure-time physical activity associates with cognitive decline: The northern Manhattan study. Neurology. 2016;**86**(20):1897-1903. DOI: 10.1212/WNL.0000000000002582

[10] Arenaza-Urquijo EM, Vemuri P. Resistance vs resilience to Alzheimer disease: Clarifying terminology for preclinical studies. Neurology. 2018;**90**(15):695-703. DOI: 10.1212/WNL.0000000000005303

[11] Montine TJ, Cholerton BA, Corrada MM, Edland SD, Flanagan ME, Hemmy LS, et al. Concepts for brain aging: Resistance, resilience, reserve, and compensation. Alzheimer's Research & Therapy. 2019;**11**(1):22. DOI: 10.1186/s13195-019-0479-y

[12] Álvares Pereira G, Nunes MVS, Alzola P, Contador I. Cognitive reserve and brain maintenance in aging and dementia: An integrative review. Applied Neuropsychology. Adult. 2021:1-11. DOI: 10.1080/23279095.2021.1872079

[13] Anonymous. Reserve capacity of the brain. British Medical Journal. 1940;**2**(4167):673-674

[14] Katzman R, Terry R, DeTeresa R, Brown T, Davies P, Fuld P, et al. Clinical, pathological, and neurochemical changes in dementia: A subgroup with preserved mental status and numerous neocortical plaques. Annals of Neurology. 1988;**23**(2):138-144. DOI: 10.1002/ana.410230206

[15] Stern Y. What is cognitive reserve? Theory and research application of the reserve concept. Journal of the International Neuropsychological Society. 2002;**8**(3):448-460. DOI: 10.1017/S1355617702813248

[16] Satz P. Brain reserve capacity on symptom onset after brain injury: A formulation and review of evidence for threshold theory. Neuropsychology. 1993;**7**(3):273-295. DOI: 10.1037/0894-4105.7.3.273

[17] Stern Y, Arenaza-Urquijo EM, Bartrés-Faz D, Belleville S, Cantilon M, Chételat G, et al. Whitepaper: Defining and investigating cognitive reserve, brain reserve, and brain maintenance. Alzheimer's & Dementia. 2020;**16**(9):1305-1311. DOI: 10.1016/j.jalz.2018.07.219

[18] Stern Y. Cognitive reserve. Neuropsychologia. 2009;**47**(10):2015-2028. DOI: 10.1016/j.neuropsychologia.2009.03.004

[19] Nilsson J, Lövdén M. Naming is not explaining: Future directions for the "cognitive reserve" and "brain maintenance" theories. Alzheimer's Research & Therapy. 2018;**10**:34. DOI: 10.1186/s13195-018-0365-z

[20] Buckner RL. Memory and executive function in aging and AD: Multiple factors that cause decline and reserve factors that compensate. Neuron. 2004;**44**(1):195-208. DOI: 10.1016/j.neuron.2004.09.006

[21] Raz N, Lindenberg U, Rodrigue KM, Kennedy KM, Head D, Williamson A, et al. Regional brain changes in aging healthy adults: General trends, individual differences and modifiers. Cerebral Cortex. 2005;**15**(11):1676-1689. DOI: 10.1093/cercor/bhi044

[22] Oosterman JM, Jansen MG, Scherder EJA, Kessels RPC. Cognitive reserve relates to executive functioning in the old-old. Aging Clinical and Experimental Research. 2021;**33**(9):2587-2592. DOI: 10.1007/s40520-020-01758-y

[23] Bennett DA, Wilson RS, Schneider JA, Evans DA, Mendes de Leon CF, Arnold SE, et al. Education modified the relation of AD pathology to level of cognitive function in older persons. Neurology. 2003;**60**(12):1909-1915. DOI: 10.1212/01.wnl.0000069923.64550.9f

[24] EClipSE Collaborative Members, Brayne C, Ince PG, Keage HAD, IG MK, Matthews FE, et al. Education, the brain and dementia: Neuroprotection or compensation? Brain. 2010;**133**(Pt8):2210-2216. DOI: 10.1093/brain/awq185

[25] Brickman AM, Muraskin J, Zimmerman ME. Structural neuroimaging in Alzheimer's disease: Do white matter hyperintensities matter? Dialogues in Clinical Neuroscience. 2009;**11**(2):181-190. DOI: 10.31887/DCNS.2009.11.2/ambrickman

[26] Schmidt R, Schmidt H, Haybaeck J, Loitfelder M, Weis S, Cavalieri M, et al. Heterogeneity in age-related white matter changes. Acta Neuropathologica. 2011;**122**(2):171-185. DOI: 10.1007/s00401-011-0851-x

[27] Sundermann EE, Biegon A, Rubin LH, Lipton RB, Mowery W, Landau S, et al. Better verbal memory in women than men in MCI despite

similar levels of hippocampal atrophy. Neurology. 2016;**86**(15):368-1376. DOI: 10.1212/WNL. 00000000000 02570

[28] Farina M, Paloski LH, Oliveira CR, Argimon IIL, Irigaray TQ. Cognitive reserve in elderly and its connection with cognitive performance: A systematic review. Ageing International. 2018;**43**(4):496-597. DOI: 10.1007/ s12126-017-9295-5

[29] Chapko D, McCormack R, Black C, Staff R, Murray A. Life-course determinants of cognitive reserve (CR) in cognitive aging and dementia—A systematic literature review. Aging & Mental Health. 2018;**22**(8):915-926. DOI: 10.1080/13607863.2017.1348471

[30] Chen Y, Lv C, Li X, Zhang J, Chen K, Liu Z, et al. The positive impacts of early-life education on cognition, leisure activity, and brain structure in healthy aging. Aging. 2019;**11**(14):4923-4942. DOI: 10.18632/aging.102088

[31] Harrison TM, Maass A, Baker SL, Jagust WJ. Brain morphology, cognition, and beta-amyloid in older adults with superior memory performance. Neurobiology of Aging. 2018;**67**:162-170. DOI: 10.1016/j. neurobiolaging.2018.03.024

[32] Chapko D, McCormack R, Black C, Staff R, Murray A. Life course determinants of cognitive reserve (CR) in cognitive aging and dementia—A systematic literature revie. Aging and Mental Health. 2017;**13**:1-12. DOI: 10.1080/13607863.2017.1348471

[33] Martins IP, Maruta C, Silva C, Rodrigues P, Chester C, Ginó S, et al. The effect of education on age-related changes in three cognitive domains: A cross-sectional study in primary care. Applied Neuropsychology: Adult. 2012;**19**(4):287-298. DOI: 10.1080/ 09084282.2012.670145

[34] Martins IP, Maruta C, Freitas V, Mares I. Executive performance in older Portuguese adults with low education. The Clinical Neuropsychologist. 2013;**27**(3):410-425. DOI: 10.1080/ 13854046.2012.748094

[35] Sundermann EE, Biegon A, Rubin LH, Lipton RB, Mowrey W, Landau S, et al. Better verbal memory in women than men in MCI despite similar levels of hippocampal atrophy. Neurology. 2016;**86**(15):1368-1376. DOI: 10.1212/WNL.0000000000002570

[36] Salthouse TA. Mental exercise and mental aging: Evaluating the validity of the "use it or lose it" hypothesis. Perspectives on Psychological Science. 2006;**1**(1):68-87. DOI: 10.1111/j. 1745-6916.2006.00005.x

[37] Tucker AM, Stern Y. Cognitive reserve in aging. Current Alzheimer Research. 2011;**8**(4):354-360. DOI: 10.2174/156720511795745320

[38] Meng X, D'Arcy C. Education and dementia in the context of the cognitive reserve hypothesis: A systematic review with meta-analyses and qualitative analyses. PLoS One. 2012;**7**(6):e38268. DOI: 10.1371/journal.pone.0038268

[39] Contador I, Bermejo-Pareja F, Del Ser T, Benito-León J. Effects of education and word reading on cognitive scores in a community-based sample of Spanish elders with diverse socioeconomic status. Journal of Clinical and Experimental Neuropsychology. 2015;**37**(1):92-101. DOI: 10.1080/13803395.2014.989819

[40] Seblova D, Berggren R, Lövdén M. Education and age-related decline in cognitive performance: Systematic review and meta-analysis of longitudinal

cohort studies. Ageing Research Reviews. 2020;**58**:101005. DOI: 10.1016/j.arr.2019.101005

[41] Kremen WS, Beck A, Elman JA, Gustavson DE, Reynolds CA, Tu XM, et al. Influence of young adult cognitive ability and additional education on later-life cognition. Proceedings of the National Academy of Sciences of the United States of America. 2019;**116**(6):2021-2026. DOI: 10.1073/pnas.1811537116

[42] Godinho F, Maruta C, Borbinha C, Pavão MI. Effect of education on cognitive performance in patients with mild cognitive impairment. Applied Neuropsychology. Adult. 2021;**15**:1-10. DOI: 10.1080/23279095.2021.1887191

[43] Scarmeas N, Stern Y, Mayeux R, Luchsinger JA. Mediterranean diet, Alzheimer disease, and vascular mediation. Archives of Neurology. 2006;**63**(12):1709-1717. DOI: 10.1001/archneur.63.12.noc60109

[44] Wilson RS, Li Y, Aggarwal NT, Barnes LL, McCann JJ, Gilley DW, et al. Education and the course of cognitive decline in Alzheimer disease. Neurology. 2004;**63**(7):1198-1202. DOI: 10.1212/01.wnl.0000140488.65299.53

[45] Brayne C, Ince PG, Keage HAD, McKeith IG, Matthews FE, Polvikoski T, et al. Education, the brain and dementia: Neuroprotection or compensation? EClipSE Collaborative Members. Brain. 2010;**133**(8):2210-2216. DOI: 10.1093/brain/awq185

[46] Brickman AM, Siedlecki KL, Muraskin J, Manly JJ, Luchsinger JA, Yeung L-K, et al. White matter hyperintensities and cognition: Testing the reserve hypothesis. Neurobiology of Aging. 2011;**32**(9):1588-1598. DOI: 10.1016/j.neurobiolaging.2009.10.013

[47] Koepsell TD, Kurland BF, Harel O, Johnson EA, Zhou XH, Kukull WA. Education, cognitive function, and severity of neuropathology in Alzheimer disease. Neurology. 2008;**70**(19 Pt 2): 1732-1739. DOI: 10.1212/01.wnl.0000284603.85621.aa

[48] Lane EM, Paul RH, Moser DJ, Fletcher TD, Cohen RA. Influence of education on subcortical hyperintensities and global cognitive status in vascular dementia. Journal of the International Neuropsychological Society. 2011;**17**(3):531-536. DOI: 10.1017/S1355617711000324

[49] Valenzuela MJ, Sachdev P. Brain reserve and dementia: A systematic review. Psychological Medicine. 2006;**36**(4):441-454. DOI: 10.1017/S0033291705006264

[50] Martins IP, Maruta C, Morgado J, Loureiro C, Tavares J, Freitas V, et al. Predictors of cognitive stability or decline during aging: A longitudinal study in primary care. Applied Neuropsychology: Adult. 2020;**27**(1):22-34. DOI: 10.1080/23279095.2018.1476866

[51] Boots EA, Schultz SA, Almeida RP, Oh JM, Koscik RL, Dowling MN, et al. Occupational complexity and cognitive reserve in a middle-aged cohort at risk for Alzheimer's disease. Archives of Clinical Neuropsychology. 2015;**30**(7):634-642. DOI: 10.1093/arclin/acv041

[52] Serra L, Bruschini M, Di Domenico C, Gabrielli GB, Marra C, Caltagirone C, et al. Memory is not enough: The neurobiological substrates of dynamic cognitive reserve. Journal of Alzheimer's Disease. 2017;**58**(1):171-184. DOI: 10.3233/JAD-170086

[53] Then FS, Luck T, Luppa M, Arélin K, Schroeter ML, Engel C, et al. Association between mental demands at work and cognitive functioning in the general population—Results of the health study of the Leipzig research center for civilization diseases (LIFE). Journal of Occupational Medicine and Toxicology. 2014;**9**:23. DOI: 10.1186/1745-6673-9-23

[54] Kröger E, Andel R, Lindsay J, Benounissa Z, Verreault R, Laurin D. Is complexity of work associated with risk of dementia? The Canadian study of health and aging. American Journal of Epidemiology. 2008;**167**(7):820-830. DOI: 10.1093/aje/kwm382

[55] Dekhtyar S, Wang H-X, Scott K, Goodman A, Koupil I, Herlitz A. A life-course study of cognitive reserve in dementia—From childhood to old age. The American Journal of Geriatric Psychiatry. 2015;**23**(9):885-896. DOI: 10.1016/j.jagp.2015.02.002

[56] Karp A, Andel R, Parker MG, Wang HX, Winblad B, Fratiglioni L. Mentally stimulating activities at work during midlife and dementia risk after age 75: Follow-up study from the Kungsholmen project. The American Journal of Geriatric Psychiatry. 2009;**17**(3):227-236. DOI: 10.1097/JGP.0b013e318190b691

[57] Stern Y, Gurland B, Tatemichi TK, Tang MX, Wilder D, Mayeux R. Influence of education and occupation on the incidence of Alzheimer's disease. Journal of the American Medical Association. 1994;**271**(13):1004-1010

[58] Baldivia B, Andrade VM, Bueno OFA. Contribution of education, occupation and cognitively stimulating activities to the formation of cognitive reserve. Dementia & Neuropsychologia. 2008;**2**(3):173-182. DOI: 10.1590/S1980-57642009DN20300003

[59] Whalley LJ, Staff RT, Fox HC, Murray AD. Cerebral correlated of cognitive reserve. Psychiatry Research: Neuroimaging. 2016;**247**:65-70. DOI: 10.1016/j.pscychresns.2015.10.012

[60] Scarmeas N, Levy G, Tang MX, Manly J, Stern Y. Influence of leisure activity on the incidence of Alzheimer's disease. Neurology. 2001;**57**(12):2236-2242. DOI: 10.1212/wnl.57.12.2236

[61] Verghese J, Lipton RB, Katz MJ, Hall CB, Derby CA, Kuslansky G, et al. Leisure activities and the risk of dementia in the elderly. The New England Journal of Medicine. 2003;**348**(25):2508-2516. DOI: 10.1056/NEJMoa022252

[62] Ribeiro AM, Monteiro S, Pereira AS. Leisure activities as a predictor of cognitive decline and dementia in old age. Open Journal of Social Sciences. 2017;**5**(3):254-259. DOI: 10.4236/jss.2017.53023

[63] Wang H-X, Jin Y, Hendrie HC, Liang C, Yang L, Cheng Y, et al. Late life leisure activities and risk of cognitive decline. The Journals of Gerontology. Series A, Biological Sciences and Medical Sciences. 2013;**68**(2):205-213. DOI: 10.1093/gerona/gls153

[64] Dause TJ, Kirby ED. Aging gracefully: Social engagement joins exercise and enrichment as a key lifestyle factor in resistance to age-related cognitive decline. Neural Regeneration Research. 2019;**14**:39-42. DOI: 10.4103/1673-5374.243698

[65] Martins IP, Mares I, Stilwell PA. How subjective are subjective language complaints. European Journal of Neurology. 2012;**19**:666-671. DOI: 10.1111/j.1468-1331.2011.03635.x

[66] Maruta C, Martins IP. May subjective language complaints predict future

language decline in community-dwelling subjects? Frontiers in Psychology. 2019;**10**:1974. DOI: 10.3389/fpsyg.2019.01974

[67] Bennett DA, Schneider JA, Tang Y, Arnold SE, Wilson RS. The effect of social networks on the relation between Alzheimer's disease pathology and level of cognitive function in old people: A longitudinal cohort study. Lancet Neurology. 2006;**5**:406-412. DOI: 10.1016/S1474-4422(06)70417-3

[68] Gow AJ, Corley J, Starr JM, Deary IJ. Which social network or support factors are associated with cognitive abilities in old age? Gerontology. 2013;**59**:454-463. DOI: 10.1159/000351265

[69] Evans ISEM, Llewellyn DJ, Matthews FE, Woods RT, Brayne C, Clare L, et al. Social isolation, cognitive reserve, and cognition in healthy older people. PLoS One. 2018;**13**(8):e0201008. DOI: 10.1371/journal.pone.0201008

[70] Fjell AM. McEvoy L, Holland D, dale AM, Walhovd KB, Alzheimers's disease neuroimaging initiative. What is normal in normal aging? Effects of aging, amyloid and Alzheimer's disease on the cerebral cortex and the hippocampus. Progress in Neurobiology. 2014;**117**:20-40. DOI: 10.1016/j.pneurobio.2014.02.004

[71] Gibbons LE, Carle AC, Mackin RS, Harvey D, Mukherjee S, Insel P, et al. A composite score for executive functioning, validated in Alzheimer's Disease Neuroimaging Initiative (ADNI) participants with baseline mild cognitive impairment. Brain Imaging and Behavior. 2012;**6**(4):517-527. DOI: 10.1007/s11682-012-9176-1

[72] Tabert MH, Manly JJ, Liu X, Pelton GH, Rosenblum S, Jacobs M, et al. Neuropsychological prediction of conversion to Alzheimer disease in patients with mild cognitive impairment. Archives of General Psychiatry. 2006;**63**(8):916-924. DOI: 10.1001/archpsyc.63.8.916

[73] Lezak MD, Howieson DB, Bigler ED, Tranel D. Neuropsychological Assessment. 5th ed. Oxford: Oxford University Press; 2012

[74] Cabeza R. Hemispheric asymmetry reduction in older adults: The HAROLD model. Psychology and Aging. 2002;**17**(1):85-100. DOI: 10.1037//0882-7974.17.1.85

[75] Stern Y, Zarahn E, Hilton HJ, Delapaz R, Flynn J, Rakitin B. Exploring the neural basis of cognitive reserve. Journal of Clinical and Experimental Neuropsychology. 2003;**5**:691-701. DOI: 10.1076/jcen.25.5.691.14573

[76] Park DC, Reuter-Lorenz P. The adaptive brain: Aging and neurocognitive scaffolding. Annual Review of Psychology. 2009;**60**:173-196. DOI: 10.1146/annurev.psych.59.103006.093656

[77] Stern Y. An approach to studying the neural correlates of reserve. Brain Imaging and Behavior. 2017;**11**(2):410-416. DOI: 10.1007/s11682-016-9566-x

[78] Anthony M, Lin F. A systematic review for functional neuroimaging studies of cognitive reserve across the cognitive aging spectrum. Archives of Clinical Neuropsychology. 2018;**33**(8):937-948. DOI: 10.1093/arclin/acx125

[79] Barulli D, Stern Y. Efficiency, capacity, compensation, maintenance, plasticity: Emerging concepts in cognitive reserve. Trends in Cognitive Sciences. 2013;**17**(10):502-509. DOI: 10.1016/j.tics.2013.08.012

[80] Scarmeas N, Zarahn E, Anderson KE, Hilton J, Flynn J, Van Heertum RL, et al. Cognitive reserve modulates functional brain responses during memory tasks: A PET study in healthy young and elderly subjects. NeuroImage. 2003;**19**(3):1215-1227. DOI: 10.1016/s1053-8119(03)00074-0

[81] Silva Nunes MV, Castro-Caldas A, Del Rio D, Maestú F, Ortiz T. The ex-illiterate brain: The critical period, cognitive reserve and HAROLD model. Dementia & Neuropsychologia. 2009;**3**(3):222-227. DOI: 10.1590/S1980-57642009DN30300008

[82] Bartrés-Faz D, Solé-Padullés C, Junqué C, Rami L, Bosch B, Bargalló N, et al. Interactions of cognitive reserve with regional brain anatomy and brain function during a working memory task in healthy elders. Biological Psychology. 2009;**80**(2):256-259. DOI: 10.1016/j.biopsycho.2008.10.005

[83] Springer MV, McIntosh AR, Winocur G, Grady CL. The relation between brain activity during memory tasks and years of education in young and older adults. Neuropsychology. 2005;**19**(2):181-192. DOI: 10.1037/0894-4105.19.2.181

[84] Steffener J, Reuben A, Rakitin BC, Stern Y. Supporting performance in the face of age-related neural changes: Testing mechanistic roles of cognitive reserve. Brain Imaging and Behavior. 2011;**5**(3):212-221. DOI: 10.1007/s11682-011-9125-4

[85] Waiter GD, Fox HC, Murray AD, Starr JM, Staff RT, Bourne VJ, et al. Is retaining the youthful functional anatomy underlying speed of information processing a signature of successful cognitive ageing? An event-related fMRI study of inspection time performance. NeuroImage. 2008;**41**:581-595. DOI: 10.1016/j.neuroimage.2008.02.045

[86] Bosch B, Bartrés-Faz D, Rami L, Arenaza-Urquijo EM, Fernández-Espejo D, Junqué C, et al. Cognitive reserve modulates task-induced activations and deactivations in healthy elders, amnestic mild cognitive impairment and mild Alzheimer's disease. Cortex. 2010;**46**(4):451-461. DOI: 10.1016/j.cortex.2009.05.006

[87] Solé-Padullés C, Bartrés-Faz D, Junqué C, Vendrell P, Rami L, Clemente IC, et al. Brain structure and function related to cognitive reserve variables in normal aging, mild cognitive impairment and Alzheimer's disease. Neurobiology of Aging. 2009;**30**(7):1114-1124. DOI: 10.1016/j.neurobiolaging.2007.10.008

[88] Medaglia JD, Pasqualetti F, Hamilton RH, Thompson-Schill SL, Bassett DS. Brain and cognitive reserve: Translation via network control theory. Neuroscience and Biobehavioral Reviews. 2017;**75**:53-64. DOI: 10.1016/j.neubiorev.2017.01.016

[89] Cole MW, Reynolds JR, Power JD, Repovs G, Anticevic A, Braver TS. Multi-task connectivity reveals flexible hubs for adaptive task control. Nature Neuroscience. 2013;**16**:1348-1355. DOI: 10.1038/nn.3470

[90] Franzmeier N, Araque-Caballero MÁ, Taylor ANW, Simon-Vermot L, Buerger K, Ertl-Wagner B, et al. Resting-state global functional connectivity as a biomarker of cognitive reserve in mild cognitive impairment. Brain Imaging and Behavior. 2017;**11**(2):368-382. DOI: 10.1007/s11682-016-9599-1

[91] Franzmeier N, Duering M, Weiner M, Dichgans M, Ewers M. Alzheimer's Disease Neuroimaging Initiative

(ADNI). Left frontal cortex connectivity underlies cognitive reserve in prodromal Alzheimer disease. Neurology. 2017;**88**(11):1054-1061. DOI: 10.1212/WNL.0000000000003711

[92] Cole MW, Ito T, Braver TS. Lateral prefrontal cortex contributes to fluid intelligence through multinetwork connectivity. Brain Connectivity. 2015;**5**:497-504. DOI: 10.1089/brain.2015.0357

[93] Franzmeier N, Unterauer E, Ewers M, Düring M, Müller C, Ruiescu D, et al. Effects of age, APOE ε4, cognitive reserve and hippocampal volume on cognitive intervention outcome in amnestic mild cognitive impairment. Journal of Alzheimer's Disease. 2016;**6**:1-7. DOI: 10.4172/2161-0460.1000246

[94] Franzmeier N, Düzel E, Jessen F, Buerger K, Levin J, Duering M, et al. Left frontal hub connectivity delays cognitive impairment in autosomal-dominant and sporadic Alzheimer's disease. Brain. 2018;**141**:1186-1200. DOI: 10.1093/brain/awy008

[95] Marques P, Moreira P, Magalhães R, Costa P, Santos N, Zihl J, et al. The functional connectome of cognitive reserve. Human Brain Mapping. 2016;**37**(9):3310-3322. DOI: 10.1002/hbm.23242

[96] Benson G, Hildebrandt A, Lange C, Schwarz C, Köbe T, Sommer W, et al. Functional connectivity in cognitive control networks mitigates the impact of white matter lesions in the elderly. Alzheimer's Research & Therapy. 2018;**10**(1):109. DOI: 10.1186/s13195-018-0434-3

[97] Bozzali C, Cercignani M. The impact of cognitive reserve on brain functional connectivity in Alzheimer's disease. Journal of Alzheimer's Disease. 2015;**44**(1):243-250. DOI: 10.3233/JAD-141824

[98] Lee JC, Kim SJ, Hong S, Kim Y. Diagnosis of Alzheimer's disease utilizing amyloid and tau as fluid biomarkers. Experimental & Molecular Medicine. 2019;**51**:53. DOI: 10.1038/s12276-019-0250-2

[99] Stern Y, Gazes Y, Razlighi Q, Steffener J, Habeck C. A task-invariant cognitive reserve network. NeuroImage. 2018;**178**:36-45. DOI: 10.1016/j.neuroimage.2018.05.033

Chapter 10

Understanding the Neuropathophysiology of Psychiatry Disorder Using Transcranial Magnetic Stimulation

Jitender Jakhar, Manish Sarkar and Nand Kumar

Abstract

Transcranial magnetic stimulation (TMS) is a safe and non-invasive tool that allows researchers to probe and modulate intracortical circuits. The most important aspect of TMS is its ability to directly stimulate the cortical neurons, generating action potentials, without much effect on intervening tissue. This property can be leveraged to provide insight into the pathophysiology of various neuropsychiatric disorders. Using multiple patterns of stimulations (single, paired, or repetitive), different neurophysiological parameters can be elicited. Various TMS protocol helps in understanding the neurobiological basis of disorder and specific behaviors by allowing direct probing of the cortical areas and their interconnected networks. While single-pulse TMS can provide insight into the excitability and integrity of the corticospinal tract, paired-pulse TMS (ppTMS) can provide further insight into cortico-cortical connections and repetitive TMS (rTMS) into cortical mapping and modulating plasticity.

Keywords: TMS, investigation, pathophysiology, psychiatry disorder, non-invasive

1. Introduction

Transcranial magnetic stimulation (TMS) is an experimental tool that allows researchers to noninvasively explore various neural processes and measure a variety of cortical phenomena and different timescales. The most important aspect of TMS is its ability to directly stimulate the cortical neurons, generating action potentials, without much effect on intervening tissue. This property can be leveraged to provide insight into the pathophysiology of various neuropsychiatric disorders. Using multiple patterns of stimulations (single, paired, or repetitive), different neurophysiological parameters can be elicited. In this article, we review the role of TMS as a tool to study motor neurophysiology of major neuropsychiatric disorders. TMS-related parameters reflect underlying cortical excitability changes during any brain motor action. New findings of motor system abnormality through TMS parameters have provided new insight into the pathophysiology of neuropsychiatric disorder.

2. Basic principle of TMS and cortical reactivity

Since its introduction by Barker in the 1980s [1], who first discovered the induction of finger and foot movements through the use of a magnetic coil placed on the motor cortex, TMS has greatly advanced our ability to explore and understand neural circuitry in neurology, psychiatry, and neuropsychological research.TMS uses principles of electromagnetic induction [2]. According to the principle whenever an electric current is passed through a coil, a transient magnetic field is generated, which induces a current in the corresponding neural tissue, consistent with Faraday's law. When the induced current is sufficient (several mA/cm^2), depolarization of neuronal membranes occurs, and hence generation of action potentials, which is recorded peripherally using electromyography (**Figure 1**). Based on which area of the cortex is stimulated, different functions can be assessed. In the case of the stimulation of the primary motor cortex, TMS is thought to predominantly activate the pyramidal cells transynaptically through excitatory intraneuronal elements. The corticospinal tract (CST) is the main descending motor pathway from the cerebral cortex to the spinal cord that can be activated by TMS. The CST originates from large pyramidal cells predominantly in the fifth layer of the cerebral cortex. The descending corticospinal tract is known to make monosynaptic connections with spinal motoneurons in humans. This organized electrical activity in the corticospinal tract is also regulated by the balance of GABAergic inhibitory postsynaptic firing and excitatory glutamate receptor activations. This contrasting cortical modulation by GABAergic vs. Glutamatergic fibers in neural tissue can be studied non-invasively in the human brain through TMS parameters. Cortical inhibition by GABAergic neurons mediates the balance between the excitatory and inhibitory systems of the nervous system. TMS has been used to study a variety of neuropsychiatric disorders including anxiety, obsessive–compulsive disorder, attention deficit hyperkinetic disorder, post-traumatic stress disorder, schizophrenia, and mood disorder by assessing

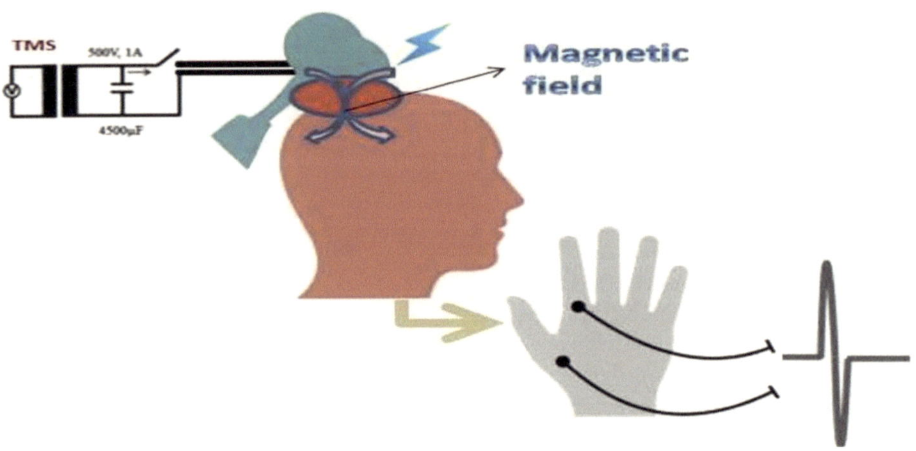

Figure 1.
TMS figure explaining the procedure from TMS machine to recording of motor evoked potential from hand; figure depicts TMS machine, figure of 8 coil (green), generation of the magnetic field through the coil, induced current in the cortex and finally TMS elicited motor potential recording of the event through hand muscle. Magnetic pulses activate cortical pyramidal neurons, leading to a corticospinal output that can be measured peripherally as a motor-evoked potential (MEP) using electromyography.

these cortical reactivity parameters. Gamma-aminobutyric acid (GABA) is one of the most important inhibitory neurotransmitters in the brain, widely distributed, and plays an important role in the modulation of cortical reactivity and neuroplasticity. GABAergic neurons represent between 20–40% of all neurons present in the central nervous system [3, 4] and they are present throughout all levels of the neuraxis, and play an important role to balance and fine-tune excitatory neurotransmission of various other neuronal systems including the cholinergic and monoaminergic projection to the area of the forebrain. Gamma-aminobutyric acid-ergic (GABAergic) deficit pathology is widely studied in various neuropsychiatric disorders [5, 6]. GABA shows its actions by interacting with two different subtypes of receptors a) GABAA receptors (GABAARs)-ionotropic b) GABAB receptor (GABABRs)-metabotropic. GABAARs are predominately responsible for anxiety and mood disorders, due to marked evidence suggesting altered GABAAR signaling in both disorders [7, 8]. Benzodiazepine act as an allosteric modulator of a major subset of GABAARs demonstrating potent anxiolytic activity and playing key control elements of anxiety state [7, 9]. GABAB Rs, Coded by GABA 1 gene and GABA 2 gene are members of the G-protein coupled receptor family and their role in the causation of affective disorder have been recently implicated in mice who demonstrated alerted anxiety and depression-related behavior after subjecting to pharmacological and genetic manipulations of these receptors [10]. The results from this study reflect that future development of therapeutic anxiolytic can be based on modulating GABAB receptors in the experimental study. In 2010, one review suggested that there are compelling evidence of both GABAA and GABAB inhibitory deficit in the pathophysiology of depressive disorder [11]. Both GABA-A and GABA-B receptors are involved in cortical inhibition with GABA-A mediating short interval cortical inhibition (SICI) and GABA B mediating cortical silent period (CSP), long interval cortical inhibition (LICI), interhemispheric inhibition (IHI) [12].

3. TMS paradigm

Various TMS protocol helps in understanding the neurobiological basis of disorder and specific behaviors by allowing direct probing of the cortical areas and their interconnected networks. While single-pulse TMS can provide insight into the excitability and integrity of the corticospinal tract, paired-pulse TMS (ppTMS) can provide further insight into cortico-cortical connections and repetitive TMS (rTMS) into cortical mapping and modulating plasticity. Few paradigms used are mentioned in **Table 1**.

3.1 TMS as a tool to measure cortical excitability

TMS can be used in humans to measure parameters of cortical excitability *in vivo*. Excitation is mainly facilitated by the action of glutamate on N-methyl-d-aspartate (NMDA), and non-NMDA receptors. Single-pulse TMS had been initially employed to test the functional integrity of human corticospinal pathways. When a stimulus of sufficient intensity is applied to the motor cortex, it will produce a motor evoked potential (MEP) in the muscle supplied by the cortical area that can be measured with electromyographic equipment [13]. Cortical excitability can be assessed by either calculation of resting motor threshold (RMT) or by calculation of MEP. RMT is defined

Number	Function assessed	TMS paradigm used	
1.	Cortical excitability	• Motor evoked potential • Resting motor threshold • Intracortical facilitation	
2.	Cortical inhibition	• Cortical silent period • Short interval intracortical inhibition • Long interval intracortical inhibition	
3.	Cortical connectivity	Interhemispheric ~Transcallosal inhibition ~Trans cerebellar inhibition	Intra-hemispheric ~Parietal-Motor networks
4.	Cortical plasticity	Long term potentiation-like: ~High-frequency rTMS ~Intermittent theta burst ~ Paired Associative Stimulation 25	Long term depression-like: ~Low-frequency rTMS ~Continuous theta burst ~ Paired Associative Stimulation 10
5.	Putative mirror neuron system activity	Motor cortical facilitation during action observation relative to rest states	
6.	Cortical mapping	Virtual lesion after-effects	
7.	Cortical connectivity with a higher temporal and spatial resolution	• TMS-EEG • TMS-fMRI	

Table 1.
Different cortical function and paradigm assessed using TMS.

as the minimum TMS intensity (expressed as a percentage of maximum stimulator output) that elicits reproducible MEP responses of at least 50 μV in 50% of 5–10 consecutive trials. The majority of application to date has involved the motor cortex but can be applied to another area of the cortex eg. stimulation in the visual cortex can produce flashes of light known as phosphenes, and stimulation of prefrontal areas can produce TMS-evoked EEG potentials.

Paired pulse stimulation (ppTMS) can also be used to assess cortical excitability and it has been accepted as a tool specifically for corticocortical circuit evaluation, whether interhemispheric, interhemispheric, or interregional. Two reactivity parameters that are commonly assessed are inhibition and excitation [14]. In ppTMS, the baseline single pulse is referred to as the test stimulus (TS), while the priming additional pulse administered a few milliseconds prior to the TS is the conditioning stimulus (CS). Conditioning stimuli strength may vary from less than (subthreshold) to greater than (suprathreshold) the RMT. A high-intensity suprathreshold pulse activates cortical pyramidal neurons directly and indirectly, via excitatory interneurons, leading to a corticospinal output that can be measured peripherally as a MEP. The response to the paired stimuli predominately depends upon interstimulus interval and CS strength. The two most commonly pair pulse paradigms used for facilitatory circuits [15, 16] are Intracortical facilitation (ICF) and short-interval intracortical facilitation (SICF) (**Table 2**). These paradigms typically reflect glutamatergic neurotransmission in the brain.

1	Intracortical facilitation (ICF)	Subthreshold CS (80%RMT) given 6–25 ms before suprathreshold test stimulus (TS) lead to the facilitation of MEP
2	Short-Interval Intracortical Facilitation (SICF)	A suprathreshold stimulus is followed by a subthreshold stimulus when two stimuli near the motor threshold are given consecutively

Table 2.
Cortical excitation parameters ICF and SICF.

3.2 TMS as a tool to measure cortical inhibition

The ability of TMS to measure cortical inhibition depends on the stimulation of interneurons in addition to corticospinal neurons. At low intensities, only intracortical inhibitory and excitatory neurons are stimulated without any effect on the excitability of corticospinal output and, therefore, do not result in an MEP. Thus, by combining a subthreshold pulse with a suprathreshold pulse, one can assess the inhibitory effects of interneurons on cortical output. The paradigms that demonstrate cortical inhibition include paired-pulse TMS (ppTMS), cortical silent period, and transcallosal inhibition (TCI). The cortical silent period is measured as isoelectric EMG (silent period) elicited by delivering a stimulus (110–160% of RMT) in the contralateral motor cortex, while the hand muscle is in a contraction phase. There is evidence that the early and late phase of the silent period may be mediated through different mechanisms with the late phase produced through G-protein coupled $GABA_B$ receptor and the early phase potentially complicated by spinal effects [17]. Next, when paired-pulse TMS is applied to the same cortical location, there are at least two inhibitory corticocortical circuits one can activate: short-interval intracortical inhibition (SICI) and long-interval intracortical inhibition (LICI) (**Table 3**). The cortical inhibition appears to be produced by the GABAergic receptor.

Various studies in the past have linked the role of GABA in the pathophysiology of different neuropsychiatric disorders, most important among them include major depressive disorder, schizophrenia, and obsessive–compulsive disorder (OCD) [18]. The deficit in GABAergic inhibition is noticed widely in psychiatric disorders, however, each illness may have a distinct profile and varied response to treatment. A meta-analysis in 2013 suggested that deficit in SICI– mediated by the $GABA_{(A)}$ergic inhibition is a universal finding in severe psychiatric illnesses including Obsessive–Compulsive disorder, Major depressive disorder, Schizophrenia but enhancement of intracortical facilitation was specific to OCD [19]. A recent review for understanding the neurobiological basis of psychiatric disorders using the TMS-based paradigms points to significant impairment in cortical inhibitory, excitatory, and oscillatory activity, especially in the frontal region [20].

3.3 TMS as a tool to measure connectivity

Measurement of corpus callosum connectivity in illnesses such as schizophrenia, in which the pathophysiology has been closely linked to dysfunctional cerebral connectivity, is helpful. Additionally, while the application of the TS often remains fixed to a given motor cortical region, the CS location may be varied to interhemispheric or interregional location including the contralateral motor cortex, cerebellum, and

1.	Short interval intracortical inhibition(SICI)	Subthreshold CS (80% RMT) given 1-6 ms before suprathreshold test stimulus (TS) leads to inhibition of TS evoked response in the contralateral muscle. This is mediated by the fast-acting, but weaker $GABA_A$ receptor neurotransmission.
2.	Long-Interval Intracortical Inhibition (LICI)	Suprathreshold CS delivered 50–200 ms before suprathreshold TS, lead to inhibition of MEP. This is mediated by the slow-acting, but stronger $GABA_B$ receptor neurotransmission.

Table 3.
Cortical inhibition parameters SICI and LICI.

peripheral nerves. These circuits correspond to pathways of interhemispheric inhibition (IHI), cerebellar inhibition (CBI), and short- (SAI) or long-latency afferent inhibition (LAI), respectively.

3.3.1 Interhemispheric inhibition (IHI)-

The relationship between the two motor cortices can be studied by paired-pulse TMS at both motor cortices. In this stimulation paradigm, MEP in the distal hand muscles by test TMS was inhibited by prior CS on the opposite side at ISI between 6 and 30 ms to investigate the transcallosal route and connectivity between brain regions. IHI requires an intact inhibitory system in the contralateral motor cortex as transcallosal fiber from the motor cortex synapse on GABAergic inhibitory interneuron [21]. A similar technique can be used to investigate connectivity between the motor cortex and the cerebellum. Inter-hemispheric inhibition is thought to be mediated through excitatory axons that cross the corpus callosum to act on local inhibitory (mainly $GABA_B$-mediated) neurons in the contralateral motor cortex. Also, in short-latency afferent inhibition, afferent sensory input through stimulation of the median nerve at the wrist or cutaneous fibers at the index finger can modify the excitability of the motor cortex with a complex time course and thought to be regulated by muscarinic and cholinergic cerebral circuits.

3.3.2 Intrahemispheric inhibition

Functional and anatomical connections between motor and parietal areas have been studied before in Humans and studies have collectively proved distinctly defined connections from parietal (anterior and posterior part) to motor areas [22]. Anatomically, the anterior part of the inferior parietal lobule (IPL) is connected to the ventral premotor and prefrontal regions, whereas the posterior IPL is linked to caudal-lateral prefrontal regions. Hence, we have to use the twin coil TMS (Tc TMS) protocol for investigating these parietal-motor connections in humans. In tcTMS a conditioning stimulus (CS) is delivered to an area of interest and followed by a test stimulus (TS) to the primary motor cortex (M1). Koch and colleagues had shown that this protocol can be used to probe parietal-motor connections and since then widely used to investigate the time course and locality of parietal-motor interaction, both during the task and at rest in studies [23]. Studies have shown the facilitatory connection between the posterior portion of IPL and M1 when a conditioning stimulus is given to the posterior IPL 2-8 ms prior to the test stimulus over M1and the EMG response triggered by M1 pulse is enhanced.

3.4 TMS as a tool to measure cortical plasticity

TMS can be used as a strategic tool to probe plastic changes in humans and this approach was used initially in neurorehabilitation to study cortical reorganization. First demonstrated by Classen to measure the effects of neuro-rehabilitative strategies in stroke patients by applying TMS over an optimal position in the motor cortex. He relates that the size of topographic motor maps in the vicinity of a cortical lesion shrinks following inactivity but often expands after the physical activity of the limb affected by the lesion [24] and areas such as the premotor cortex overtakes the functions typically executed by the primary motor cortex. MS cortical motor maps enlarge after intense motor training in stroke patients and such a plasticity effect can also be demonstrated after a yoga intervention. Similarly, another example of transmodal plasticity can also be elicited in patients who are blind from early life read Braille, where somatosensory information gets routed to the visual cortex and show activation sign in the visual cortex in functional neuroimaging studies [25]. But this did not prove that activity in the visual cortex was being used for actual analysis of the information. Using rTMS during reading showed that this function was impaired when the visual cortex was disrupted. Of potential clinical importance, aberrant synaptic plasticity is a pathophysiological characteristic of schizophrenia. and using in-vivo perturbation protocols like TMS and tDCS, studies have demonstrated diminished LTP and LTD-like motor cortical plasticity [26]. The common paradigms in TMS for assessing cortical plasticity are:

3.4.1 Paired associative stimulation (PAS)-

PAS involves repetitive activation of sensory inputs (mostly median nerve) to the motor cortex using TMS and producing long-term changes in the excitability of the motor cortex that can last for several hours. The effect of PAS on MEP size was found to be dependent on the timing of the TMS pulse with respect to the afferent stimulation. In short, during PAS with an ISI of −10 ms (PAS10), LTD-type effects were induced in the motor cortex as reflected in reduced MEPs. On the other hand, when an ISI of 25 ms (PAS 25) was used, long-term potentiation (LTP-type) effects were induced as evidenced by increases in MEP responses [27].

3.4.2 Repetitive TMS

This can be used to induce sustained changes in cortical reactivity that significantly outlasts the stimulation period. Repetitive TMS can either activate or inhibit cortical activity, depending on stimulation frequency. Low-frequency stimulation results in depression of the target brain area, while high-frequency stimulation induces the facilitation of the region. Low frequency (1 Hz) stimulation for a period of approximately 15 minutes induces a transient inhibition of the cortex. The mechanisms behind such inhibition are unclear, although there are similarities to long-term depression-like synaptic plasticity. In contrast, stimulation at frequencies of 10–20 Hz has been shown to increase cortical activation. Newer theta-burst stimulation technique is a high-frequency stimulation paradigm that can produce either inhibitory (if applied continuously) or facilitatory (if applied intermittently) effects. These effects are thought to be predominantly mediated by NMDA receptors as well as by modulation of GABA receptor functions [28]. Repetitive TMS-induced changes in cortical plasticity can potentially be studied as a marker of brain resilience to neuropathology [29]. It has been demonstrated that individuals with schizophrenia have diminished cortical plasticity as measured by these techniques [26].

3.5 TMS as a tool to measure mirror neuron system (MNS) activity

Among the various networks involved in the pathology of social cognition, the mirror neuron system is most extensively studied. Typically, there is a quantifiable motor cortical reactivity facilitation (increased motor evoked potentials or reduced intracortical inhibition) or motor resonance in the same muscle group that is observed to be in action. This index of motor resonance in the primary motor cortex is likely to be driven by premotor MNS-activity and is used as a putative or indirect marker of the premotor mirror neuron system activity. Studies using TMS have demonstrated diminished modulation of motor cortical reactivity during both neutral action observation [30] and context-based action in schizophrenia [31]. Also, reduced MNS activity is related to poorer social cognition performance. In contrast, patients with mania demonstrate an elevated MNS response, perhaps reflecting disinhibition of the regulatory prefrontal brain regions. Higher MNS activity in mania was associated with higher manic symptom severity [32].

3.6 TMS as a tool for cortical mapping

TMS methods allow for the identification of a direct association between the studied site and the behavioral outcome in a temporary and non-invasive fashion, allowing for mapping of areas of cortex less accessible by previous techniques, hence providing a powerful tool to identify the brain-behavior relationship. A single TMS pulse or a short sequence of pulses has the potential to transiently disrupt ongoing cortical activity in the region being stimulated. This phenomenon has been termed a virtual lesion. TMS is an important tool in cognitive neuroscience and has changed the way we understand cognitive function [33]. TMS can create virtual lesions, thereby allowing us to obtain information about the contribution of a given cortical region to a specific behavior. For example, subjects asked to memorize and repeat a list of words would likely show increased activity in the prefrontal cortex using fMRI. This increased activity would provide an indirect association between the prefrontal cortex and the task. However, if stimuli from TMS over the prefrontal cortex were found to obstruct the ability to learn and recall the list, then researchers would have more convincing evidence to support the involvement of the prefrontal cortex in short-term memory. Pascual-Leone investigated the role of the dorsolateral prefrontal cortex (DLPFC) in implicit procedural learning. In this study, low-intensity rTMS was applied to the DLPFC, to the supplementary motor area, or directly to the ipsilateral hand used in testing. It was demonstrated that DLPFC stimulation markedly impaired implicit procedural learning, whereas stimulation of the other areas did not impair learning [34].

3.7 CombinatoryTMS approaches

TMS may also be paired with other investigative modalities to investigate connectivity between brain regions and can be used as a brain-mapping tool to complement information gained from functional magnetic resonance imaging (fMRI) and electroencephalography (EEG), thus improving both spatial and temporal accuracy of the biological signals derived.

DOI: http://dx.doi.org/10.5772/intechopen.103748

3.7.1 TMS-EEG-

TMS-evoked surface potentials from any cortical region can be recorded with scalp EEG electrodes and used to estimate the regional excitability of the extra-motor cortex [23]. This increases spatial benefits and also the very high temporal resolution of EEG makes it possible to detect differential effects of brain disturbance on TMS-induced responses. TMS-EEG recording is obtained using specialized amplifiers and electrode caps designed to minimize stimulus artifact. The high sampling frequency and other adjustments in the acquiring of signals permit recording cortical potentials induced with TMS and the spread of such oscillation to the different connected regions from the site of stimulation [35].

3.7.2 TMS- fMRI

Combining TMS and fMRI makes it possible to exploit both the good spatial resolution (can identify changes that occur in both cortical and subcortical structures) and the good temporal resolution of TMS. Such data can provide information on connectivity patterns. These patterns reflect the propagation of activity in the stimulated area to distal areas via a neural connection [36].

4. TMS safety

A general understanding of single-pulse stimulators is that they are safe. The high frequency of rTMS provides a much stronger effect on the brain and is unlikely, but can induce seizures. Other common side effects include nausea, arms jerking, transient headache, and facial pain caused by the activation of scalp and neck muscles. To help alleviate these problems, safe intensity limits using rTMS are suggested to help reduce the risk of discomfort. The general consensus of TMS is that it is safe; however, should remain mindful to minimize the risks.

Conflict of interest

The authors declare no conflict of interest.

Author details

Jitender Jakhar[1]*, Manish Sarkar[2] and Nand Kumar[3]

1 Fortis Hospital, Vasant Kunj, New Delhi, India

2 EHSAAS, New Delhi, India

3 All India Institute of Medical Sciences (AIIMS), Delhi, India

*Address all correspondence to: jitender.jak@gmail.com

References

[1] Barker AT, Jalinous R, Freeston IL. Non-invasive magnetic stimulation of human motor cortex. Lancet. 1985;**1**(8437):1106-1107. Available from: http://www.ncbi.nlm.nih.gov/pubmed/2860322

[2] Klomjai W, Katz R, Lackmy-Vallée A. Basic principles of transcranial magnetic stimulation (TMS) and repetitive TMS (rTMS). Annals of Physical and Rehabilitation Medicine [Internet]. 2015 58(4): 208-213. Available from: https://www.sciencedirect.com/science/article/pii/S1877065715000792

[3] Hornung JP, De Tribolet N. Distribution of GABA-containing neurons in human frontal cortex: A quantitative immunocytochemical study. Anatomy and Embryology. 1994;**189**(2):139-145. Available from: https://link.springer.com/article/10.1007/BF00185772

[4] Sahara S, Yanagawa Y, O'Leary DDM, Stevens CF. The fraction of cortical GABAergic neurons is constant from near the start of cortical neurogenesis to adulthood. The Journal of Neuroscience. 2012;**32**(14):4755

[5] Petty F, Sherman AD. Plasma GABA levels in psychiatric illness. Journal of Affective Disorders. 1984;**6**(2):131-138. Available from: http://www.ncbi.nlm.nih.gov/pubmed/6233345

[6] Schür RR, Draisma LWR, Wijnen JP, Boks MP, Koevoets MGJC, Joëls M, et al. Brain GABA levels across psychiatric disorders: A systematic literature review and meta-analysis of ^{1}H-MRS studies. Human Brain Mapping. 2016;**37**(9):3337-3352. DOI: 10.1002/hbm.23244

[7] De Haas SL, De Visser SJ, Van Der Post JP, Schoemaker RC, Van Dyck K, Murphy MG, et al. Pharmacodynamic and pharmacokinetic effects of MK-0343, a GABA (A) alpha2,3 subtype selective agonist, compared to lorazepam and placebo in healthy male volunteers. Journal of Psychopharmacology. 2008;**22**(1):24-32. Available from: https://pubmed.ncbi.nlm.nih.gov/18187530/

[8] Sanders SK, Shekhar A. Regulation of anxiety by GABAA receptors in the rat amygdala. Pharmacology, Biochemistry, and Behavior. 1995;**52**(4):701-706. Available from: https://pubmed.ncbi.nlm.nih.gov/8587908/

[9] Chen X, van Gerven J, Cohen A, Jacobs G. Human pharmacology of positive GABA-A subtype-selective receptor modulators for the treatment of anxiety. Acta Pharmacologica Sinica. 2018;**40**(5):571-582. Available from: https://www.nature.com/articles/s41401-018-0185-5

[10] Mombereau C, Kaupmann K, Froestl W, Sansig G, Van Der Putten H, Cryan JF. Genetic and pharmacological evidence of a role for GABA(B) receptors in the modulation of anxiety- and antidepressant-like behavior. Neuropsychopharmacology: Official publication of the American College of Neuropsychopharmacology. 2004;**29**(6):1050-1062. Available from: https://pubmed.ncbi.nlm.nih.gov/15039762/

[11] Croarkin PE, Levinson AJ, Daskalakis ZJ. Evidence for GABAergic inhibitory deficits in major depressive disorder. Neuroscience and Biobehavioral Reviews. 2011;**35**(3):818-825. DOI: 10.1016/j.neubiorev.2010.10.002

[12] Radhu N, Ravindran L, Levinson A, Daskalakis Z. Inhibition of the cortex

using transcranial magnetic stimulation in psychiatric populations: Current and future directions. Journal of Psychiatry & Neuroscience. 2012;**37**(6):369-378. Available from: http://jpn.ca/vol37-issue6/37-6-369/

[13] Farzan F. Single-Pulse Transcranial Magnetic Stimulation (TMS) Protocols and Outcome Measures. New York, NY: Humana Press; 2014. pp. 69-115. DOI: 10.1007/978-1-4939-0879-0_5

[14] Zewdie E, Kirton A. TMS basics: Single and paired pulse neurophysiology. In: Pediatric Brain Stimulation: Mapping and Modulating The Developing Brain. Elsevier Inc.; 2016. DOI: 10.1016B978-0-12-802001-2.00001-1

[15] Rotenberg A, Horvath JC, Pascual-Leone A. Transcranial Magnetic Stimulation. Neuromethods. 2014;**89**. DOI: 10.1007/978-1-4939-0879-0

[16] Hallett M. Transcranial magnetic stimulation: A primer. Neuron. 2007;**55**(2):187-199

[17] Epstein CM, Wassermann EM, Ziemann U. Oxford Handbook of Transcranial Stimulation. 2008:1-764. DOI: 10.1093/OXFORDHB/9780198568926.001.0001

[18] Streeter CC, Jensen JE, Perlmutter RM, Cabral HJ, Tian H, Terhune DB, et al. Yoga asana sessions increase brain GABA levels: A pilot study. The Journal of Alternative and Complementary Medicine. 2007;**13**(4):419-426. DOI: 10.1089/acm.2007.6338

[19] Radhu N, de Jesus DR, Ravindran LN, Zanjani A, Fitzgerald PB, Daskalakis ZJ. A meta-analysis of cortical inhibition and excitability using transcranial magnetic stimulation in psychiatric disorders. Clinical neurophysiology: Official journal of the International Federation of Clinical Neurophysiology. 2013;**124**(7):1309-1320. Available from: http://linkinghub.elsevier.com/retrieve/pii/S1388245713000564

[20] Kaskie RE, Ferrarelli F. Investigating the neurobiology of schizophrenia and other major psychiatric disorders with Transcranial Magnetic Stimulation. Schizophrenia Research. 2018;**192**:30-38. Available from: https://pubmed.ncbi.nlm.nih.gov/28478887/

[21] Daskalakis ZJ, Christensen BK, Fitzgerald PB, Roshan L, Chen R. The mechanisms of interhemispheric inhibition in the human motor cortex. The Journal of Physiology. 2002;**543**(Pt 1):317-326. Available from: http://www.ncbi.nlm.nih.gov/pubmed/12181302

[22] Koch G, Rothwell JC. TMS investigations into the task-dependent functional interplay between human posterior parietal and motor cortex. Behavioural Brain Research. 2009;**202**(2):147-152

[23] Koch G, Del Olmo MF, Cheeran B, Ruge D, Schippling S, Caltagirone C, et al. Focal stimulation of the posterior parietal cortex increases the excitability of the ipsilateral motor cortex. Journal of Neuroscience. 2007;**27**(25):6815-6822

[24] Classen J, Schnitzler A, Binkofski F, Werhahn KJ, Kim YS, Kessler KR, et al. The motor syndrome associated with exaggerated inhibition within the primary motor cortex of patients with hemiparetic. Brain. 1997;**120**(4):605-619. Available from: http://www.ncbi.nlm.nih.gov/pubmed/9153123

[25] Gothe J, Brandt SA, Irlbacher K, Roricht S, Sabel BA, Meyer B-U. Changes in visual cortex excitability in blind subjects as demonstrated by transcranial magnetic stimulation.

Brain. 2002;**125**(3):479-490 Available from: http://www.ncbi.nlm.nih.gov/pubmed/11872606

[26] Mehta UM, Thanki MV, Padmanabhan J, Pascual-Leone A, Keshavan MS. Motor cortical plasticity in schizophrenia: A meta-analysis of transcranial magnetic stimulation – Electromyography studies. Schizophrenia Research. 2019 May 1;**207**:37-47

[27] Classen J, Wolters A, Stefan K, Wycislo M, Sandbrink F, Schmidt A, et al. Paired associative stimulation. Supplements to Clinical Neurophysiology;**57**:563-569. Available from: http://www.ncbi.nlm.nih.gov/pubmed/16106657

[28] Huang YZ, Edwards MJ, Rounis E, Bhatia KP, Rothwell JC. Theta burst stimulation of the human motor cortex. Neuron. 2005;**45**(2):201-206

[29] Hall PA, Erickson KI, Lowe CJ, Sakib MN. Quantifying cortical resilience in experimental, clinical, and epidemiological studies: A conceptually grounded method using noninvasive brain stimulation. Psychosomatic Medicine. Apr 2020;**82**(3):281-286. DOI: 10.1097/PSY.0000000000000785. PMID: 32084068

[30] Mehta UM, Thirthalli J, Basavaraju R, Gangadhar BN, Pascual-Leone A. Reduced mirror neuron activity in schizophrenia and its association with theory of mind deficits: Evidence from a transcranial magnetic stimulation study. Schizophrenia Bulletin. 2014;**40**(5):1083-1094

[31] Bagewadi VI, Mehta UM, Naik SS, Govindaraj R, Varambally S, Arumugham SS, et al. Diminished modulation of motor cortical reactivity during context-based action observation in schizophrenia. Schizophrenia Research. 2019;**204**:222-229. DOI: 10.1016/j.schres.2018.07.043

[32] Basavaraju R, Mehta UM, Pascual-Leone A, Thirthalli J. Elevated mirror neuron system activity in bipolar mania: Evidence from a transcranial magnetic stimulation study. Bipolar Disorders. 2019;**21**:259-269

[33] Nevler N, Ash EL. TMS as a tool for examining cognitive processing. Current Neurology and Neuroscience Reports 15. 2015. DOI: 10.1007/S11910-015-0575-8

[34] Pascual-Leone A, Wassermann EM, Grafman J, Hallett M. The role of the dorsolateral prefrontal cortex in implicit procedural learning. Experimental Brain Research. 1996;**107**(3):479-485. Available from: https://pubmed.ncbi.nlm.nih.gov/8821387/

[35] Lioumis P, Kičić D, Savolainen P, Mäkelä JP, Kähkönen S. Reproducibility of TMS - evoked EEG responses. Human Brain Mapping. 2009;**30**(4):1387-1396

[36] Ruff CC, Driver J, Bestmann S. Combining TMS and fMRI: From 'virtual lesions' to functional-network accounts of cognition. Cortex. Oct 2009;**45**(9):1043-1049. DOI: 10.1016/j.cortex.2008.10.012. Epub 2008 Dec 6. PMID: 19166996; PMCID: PMC2726131

Chapter 11

Impact of Hypoxia on Astrocyte Induced Pathogenesis

Farwa Munir, Nida Islam, Muhammad Hassan Nasir, Zainab Anis, Shahar Bano, Shahzaib Naeem, Atif Amin Baig and Zaineb Sohail

Abstract

Astrocytes are the most abundant cells of the central nervous system. These cells are of diverse types based on their function and structure. Astrocyte activation is linked mainly with microbial infections, but long-term activation can lead to neurological impairment. Astrocytes play a significant role in neuro-inflammation by activating pro-inflammatory pathways. Activation of interleukins and cytokines causes neuroinflammation resulting in many neurodegenerative disorders such as stroke, growth of tumours, and Alzheimer's. Inflammation of the brain hinders neural circulation and compromises blood flow by affecting the blood–brain barrier. So the oxygen concentration is lowered, causing brain hypoxia. Hypoxia leads to the activation of nuclear factor kappa B (*NFkB*) and hypoxia-inducible factors (*HIF*), which aggravates the inflammatory state of the brain. Hypoxia evoked changes in the blood–brain barrier, further complicating astrocyte-induced pathogenesis.

Keywords: astrocyte, hypoxia, inflammation, neuroinflammation, brain

1. Introduction of astrocytes

In the nervous system, astrocytes are isotypes of neuroglia, also identified as astrocytic cells. They are Star-shaped; their countless progressions enclose synopses prepared by nerve cells. A particular astrocytic cell can simultaneously act together with the human being by two billion synapses. These specialized glial cells are more numerous than neurons by above fivefold. They closely tile the central nervous system (CNS) and apply for multiple important diverse roles in the energetic CNS. Astrocytes react to all methods of CNS offences through a procedure stated as reactive astrogliosis, which has developed a pathological mark of CNS fundamental abrasions. Two main subtypes of astrocytes are classified based on their structural and anatomical position. Those names are protoplasmic astrocytes and fibrous astrocytes. Protoplasmic astrocytes seem to spread equally in the interior of cortical grey matter, while the fibrous astrocytes are structured with white matter regions [1]. Astrocytes of the brain and spinal cord are very different in morphology and function. Brain cells

 IntechOpen

(astrocytes) are accountable for the homeostasis of ions and neurotransmitters in the synaptic cleft, native metabolic sustenance, and relief of sensitive oxygen species. Pathology of many nervous disorders containing neuropsychiatric and neurodegenerative syndromes is well-defined by loss of homeostatic role. Astrocytes play a significant role in the homeostasis of the central nervous system (CNS). Brain cells or astrocytes are extremely diverse cells that regulate the network, emergence and function, and homeostasis. Since it's involved in protective astrogliosis, it's become an essential component of neuropathology. Most neuropathology astroglial cells are impacted by degenerative alterations that inhibit their functional and neuroprotective capacities, allowing the pathology to proceed [2]. Astrocytes play an important role in data handling, and communicative mechanism proficiencies of brain circuits are unknown. Around all research studying the correlation among astrocyte cells' structure and function concentrates on its influences on nervous system activity and flexibility under functional and syndrome conditions. At synapses, a collective subject important to these outcomes is that astrocytes analyse, respond to, and control glutamate release and post-synaptic activity.

Removal and postponement of PAPs in reaction to glutamate improve post-synaptic responses, inhibit trans-synaptic activation, and prevent additional glutamate proclamation. Still, astrocyte operational flexibility exchange is not recognized upon declaration of other GABA, dopamine, somatostatin, serotonin, acetylcholine, etc. (neurotransmitters) [3]. The appearance of glial fibrillary acid protein (GFAP) has become a typical indicator for immunohistochemically astrocytic cells [4]. Research on transgenic mice showed that the appearance of GFAP isn't necessary for the usual form and role of the furthermost astrocytic cell in the energetic, nervous system of transgenic mice. Still, it's essential to develop reactive astrogliosis and glial scar development. Over and above, concerning the procedure of GFAP as an astrocyte indication, it's compulsory to pay attention that GFAP expression isn't limited to protoplasmic and fibrous astrocytes. In the interior of the nervous system, GFAP is too expressed by numerous cells that can be reflected as part of prolonged astroglial cells. On the outer side of the nervous system, GFAP is articulated extensively in countless nerves via a range of cell forms **Figures 1–4** [5].

2. Anatomical association

Astrocytes tile the whole CNS in a touching, non-covering, organized, and efficient way. Comparative individual astrocyte areas show up liable to exist in white matter. However, this has not yet been as broadly written about. Astrocytes are giant, intricate, and different from astrocytes in rodents. Astrocytes show controlled proliferation in intracellular calcium absorption [Ca++], which signifies a method of astrocyte excitability. An enormous suggestion is now obtainable that these delimited proliferations in astrocyte [Ca++] remain of purposeful importance in astrocyte and astrocyte-neuron intercellular communication. The advice is that calcium signalling permits astrocytes to show a direct role in synaptic transmission. It's worth noting that astrocytes do communicate with one another via gap junctions generated by connexins. Gap junctional partnering of astrocytes into multicellular systems may contribute to normal function and CNS disorders. The suggestion is that calcium signalling permits astrocytes to show a natural interest in synaptic transmission.

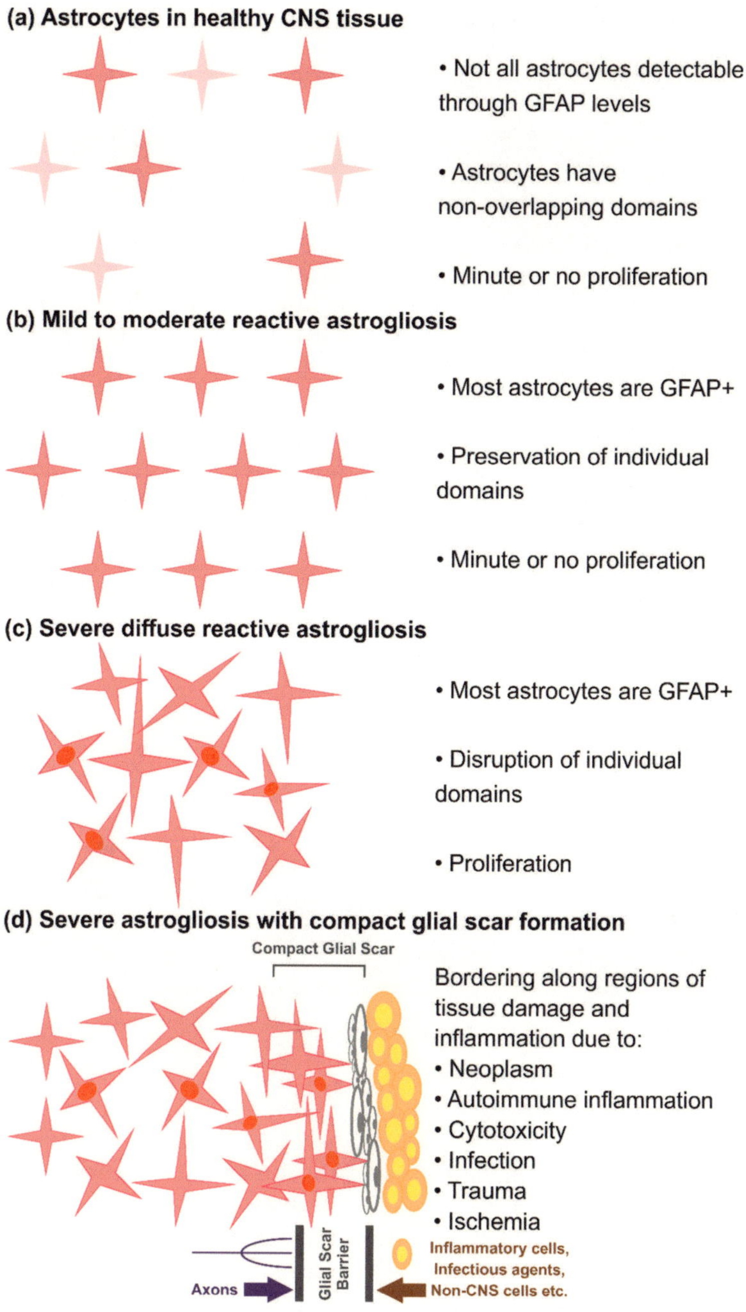

Figure 1.
Astrogliosis and GFAP articulation in astrocytes.

3. Astrocyte-induced pathogenesis

ASTROCYTIC cells are central homeostatic and protective cells of the nerves, and every kind of astrocyte plays an integral part in neuropathological changes. Hence, the decline in nerve cells or astrocytes causes a disease-permissive landscape and

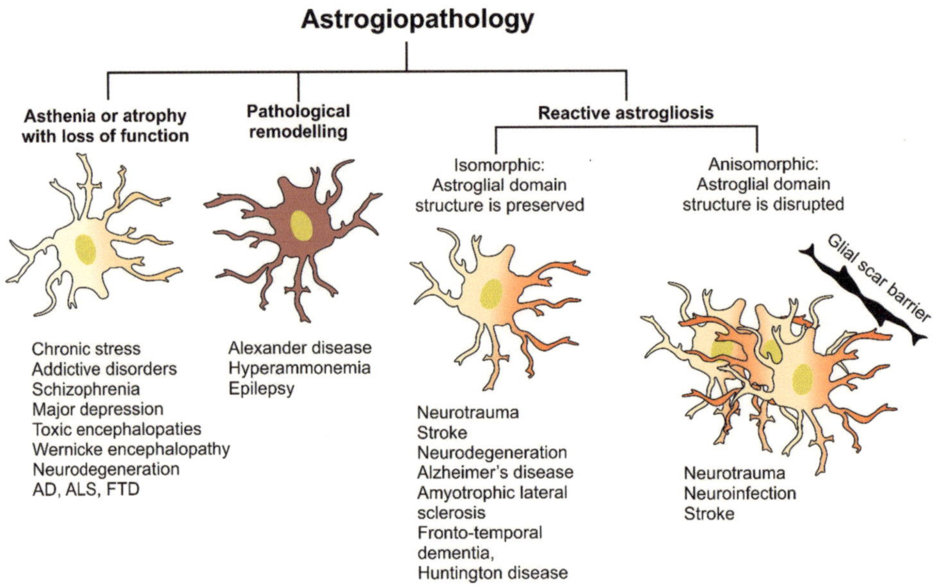

Figure 2.
Astroglial asthenia/atrophy and astrogliosis in neuropathology.

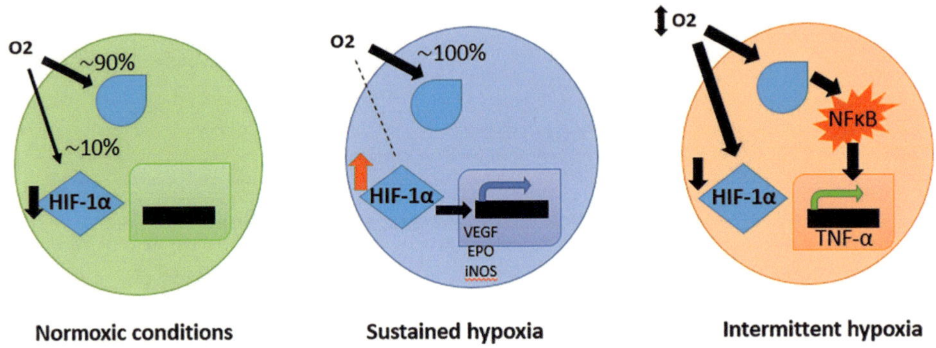

Figure 3.
Hypoxia and HIF-1α stabilization in brain.

triggers nerve cell malfunction, nerve cell death, and nerve cell deficiency. Glia cells are essential for sustaining nerve function, and nerves survive bodily procedures and pathology [6]. Most essential findings concerning astrocyte's functional significance depend on the dead animal model research. It's given a durable but incomplete base for a complete astrocyte role in physiopathology [7]. They are categorized into three kinds that are reactive astrogliosis, Astro-degeneration with astroglial atrophy, and pathological remodelling and loss of function of astrocytes. Altogether these pathological feedbacks proceed together. It's categorized on the base of neuroanatomical and severity. According to neuroanatomical, astrocytes are further distributed into isomorphic and isomorphic astrogliosis [8]. The isomorphic astrogliosis conserves astroglial defensive areas that are changeable.

In contrast, anisomorphic astrogliosis continues through the destruction of the defensive regions, cell relocation and territorial overlap, development of astroglial palisades, and eventually scar formation. While in severity, astrogliosis is categorized

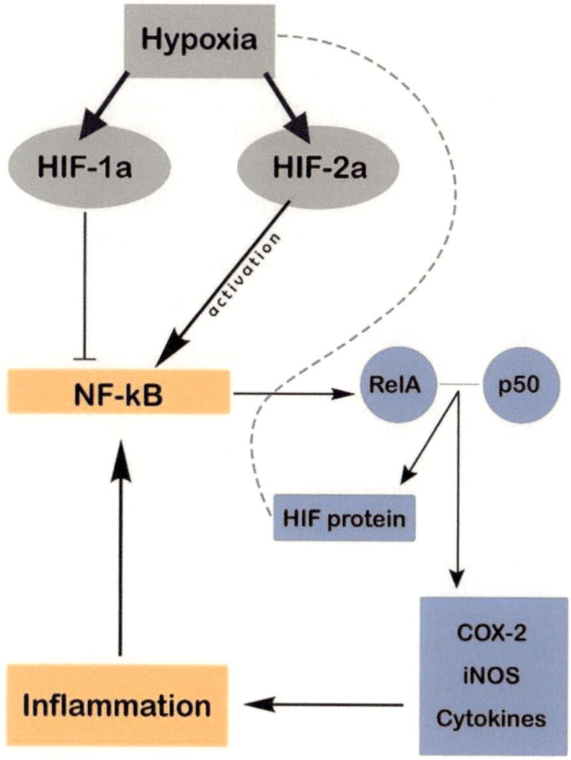

Figure 4.
Role of hypoxia in activation of inflammatory pathway.

into slight to adequate astrogliosis, severe diffuse astrogliosis and severe astrogliosis by dense scar development [9]. Astroglial atrophy is mainly noticeable in major psychiatric illnesses. For example, schizophrenia, a primary depressing condition, Wernicke–Korsakoff encephalopathy, and addictive disorders decrease the storing concentration of astrocytic cells. The conclusion is furthermost particularly accompanied by glutamate-glutamine shuttle and glutamate homeostasis; both are impaired in these conditions [10]. It promotes several leukodystrophies, especially Alexander disease, megalocephalic leukoencephalopathy with subcortical lumps or disappearing white matter disease, in which the astrocyte-pathy pledges destruction of the white matter [11]. It also describes mesial temporal lobe epilepsy, in which astrocytes obtain abnormal cell structure, decreases gap junction coupling, and decrease Kir4.1 channel expression; these alterations weaken K+ homeostasis, contributing to seizure start [12].

4. Pathology of astroglia in neurological diseases

Reactive astrogliosis and glial scar realization are noticeable structures of CNS trauma and are progressively involved by playing significant parts in the decisive prolonged medical result. Glial formation of scar and severe reactive astrogliosis at the location of the neurons trauma are glowing familiar to inhibit axonal regeneration and are extensively observed as harmful to medical results. Contagions of the CNS triggered by microbes, parasites, fungi, and viruses are categorized into inflammation of meninges, active tissues of the brain, or (pus-filled pocket of infected material in your brain) cerebral abscess. Not all germs can attack the nervous system. Somewhat, merely confirmed neurotropic parasites, fungi, bacteria, and viruses can enter obsessed by the cerebral and vertebral column. Utmost of the microbe are efficiently stopped by the cerebral obstacles [13].

The SAE (Sepsis-associated encephalopathy) explicitly uses a scientific disorder linked through the common cerebral disorders that progresses in sepsis to the lack of core contagion of the neural tissue. In the cerebral parenchyma, sepsis is frequently linked with the production of inflammations and tiny inflammation directly related to the SAE. Particularly in the initial phases, the disease associated with structural infection remains frequently linked through 'sickness behaviour' [14].

Dense metals are a source of extreme cephalic disorder with intellectual deficiencies, mainly targeting neuroglia. Heavy metals that are manganese, lead, aluminium or mercury primarily gather into an astrocytic cell by diverse plasma lemma transporters. Overall, it's also a down-regulating astroglial expression of glutamate transporters, reducing glutamate permission and activating excitotoxicity [15]. Aluminium toxic encephalopathy is demonstrated via mental losses, communication variations, seizures, and flapping wrist shake (asterixis) [16]. It disturbs the cerebrum due to liver encephalopathy, which is described by mental and developmental damage signs including misperception, amnesia, bad temper, and alterations in perception, such as fatigue and sleepiness. Cerebral swelling, unconsciousness, and death also occur in the severity of Hyperammonemia. In stroke, a blood vessel ruptures that restricts blood supply to the brain or parts of the brain due to a systemic decrease of vascular occlusion in blood supply, all-cause disruptions in blood flow. This is known as brain ischemia. As a result, cerebral ischemia could be either focal or global, the latter of which can lead to a stroke. In stroke, astrocytes serve as both neurotoxic and neuroprotective agents, complicating and diverse astrgliopathology [17]. In cases of Congenital Glutamine Deficiency with Glutamine, Synthetase Mutations newborns expire in a while after delivery. The prominent pathophysiological contrivance is related to the weakened capacity of astrocytes to yield glutamine, which disturbs excitatory and inhibitory conduction; furthermore, lacking glutamine synthetase cannot accurately decontaminate ammonium. Pyruvate carboxylase is principally articulated in neuroglia. Pyruvate carboxylase deficit is an autosomal recessive disease linked to a diminished metabolic rate. The warning sign comprises delay of cerebral growth, persistent seizures, and increased plasma acidity.

Aceruloplasminemia is a congenital condition of iron metabolism due to the deficiency of ceruloplasmin action. Most important abrasions describe this disorder as neuroglia, which disturbs their structure and consequences in the presence of frothy spheroid forms at the vascular end feet. Aceruloplasminemia is also linked with brain demise and the exterior of iron deposition [18].

Alexander disease is an exceptional, long-lasting, and ordinarily neurodegenerative severe condition. Its consequences on or after a dominant gain-of-function mutation of the gene encoding GFAP. It is sub-categorized into Type I and types II [19]. Autism Spectrum Disorders (ASD) are a few medically introverted conditions connected to cerebral inadequacies. Astrocytes are accountable for neuroprotection and detoxifying harmful bodies like receptive oxygen species. The principle procedure of ASDs is undoubtedly associated with brain network mutation and abnormal neurotransmission during the undeveloped turn of events [20]. In Down syndrome, the thickness of astrocytes is fundamentally diminished in the cortex with diminished capacity to uphold synaptogenesis and neuronal development appropriately. Astroglial asthenia, loss of homeostatic capacities, atrophy, and perhaps pathogenic remodelling are all related to schizophrenia, although reactive alterations are not. Epilepsy, mood disorders, and addictive disorders are linked with astrogliopathology [12].

5. Astrocyte up-regulation

Astrocyte is the essential part of the blood–brain barrier (BBB) that can be damaged by traumatic brain injury (TRI) or ischemia. Astrocytes provide the morphological and physiological link between neural networks and cerebral circulation. Astrocytes have the power to adjust blood flow to the brain to keep the brain parenchyma's PCO2 and PO2 steady [21]. The formation of ATP in the brainstem through local astrocytes aids respiration and counters hypoxia-induced respiratory network depression. Hypoxia-induced alterations in brain inflammation, neuroprotection, and blood–brain barrier permeability against ischemic injury appear to be mediated by astrocytes. Astrocytes play a critical function in neuronal function in everyday situations and pathological conditions when the supply of oxygen to the brain is disrupted [22].

The telomerase reverse transcriptase (TERT) gene is linked to cell injury and stress resistance. After hypoxia-ischemia, protein and TERT mRNA were increased in neurons after a few days but moved to astrocytes [23]. TERT overexpression decreased astrocyte multiplication by upregulating the cell-cycle regulatory protein p15. While neurons were cultured with precondition medium from astrocytes with TERT inhibition contrasted to neurotrophin-3 expression, TERT overexpression in astrocytes decreased, resulting in higher death [24]. In TERT-overexpressing brains with hypoxia-ischemia, neuronal damage and Ki67-positive astrocytes were also suppressed [25].

Matrix metalloproteinase (MMP)-9 is an endo-peptidase that plays a crucial role in Blood–Brain Barrier proteolysis post-trauma and leads to cell death with persistent convulsions [26]. Activation of mitogen-activated protein kinases (MAPKs) in astrocytes can be caused by thrombin, oxidative stress, tissue plasminogen activator, or tumour necrosis factor- α and includes stimulation of MMP-9. In astrocytes, albumin causes a rise in MMP-9 synthesis, which requires ROS formation and motivation of the MAPK pathway [27]. These results add albumin to signalling molecules that activate MMP-9 in astrocytes alongside thrombin. These findings connect albumin to MMP-9-mediated cellular mechanisms such as intracerebral haemorrhage, neuronal damage, dendritic remodelling, and epileptogenesis [28].

Stroke and Traumatic Brain Injury (TBI) are frequently linked with hypoxia, which causes glial initiation. Glial cells, particularly astrocytes, play an essential part in stress prevention and the homeostasis of the CNS by offering structural and metabolic stability [29]. Hypoxia can cause astrocyte homeostasis to be disrupted, resulting in cell enlargement. Wnt pathway suppression was the most substantially disrupted signalling pathway in the mechanism of hypothermia-induced responses in human astrocytes after oxidative stress activation and hypoxia [30]. Global suppression of Wnt signalling can be troublesome because of its essential role in controlling critical mechanisms associated with the functional regulation of immune and stem cells and its effect on post-mitotic neuronal and glial cells [31].

6. Brain hypoxia

Molecular oxygen (O2) is required for most organisms on the planet because it supports intracellular biogenesis and is utilized by several metabolic activities. As a result, low oxygen level (hypoxia) is a crucial stress factor that usually disrupts aerobic species' lives and is a common feature of pathologic conditions such as cardiovascular abnormalities, inflammation, wounds, bacterial infection, and cancer [32]. Since it is required throughout breathing, oxygen is essential in life. In oxidative phosphorylation, O2 acts as the ultimate electron acceptor, raising the possibility of reactive oxygen species formation (ROS). ROS interact with biological macromolecules, changing their metabolic or physical characteristics and causing cell death or malfunction [33].

To sustain a wide range of cellular functions to secure life, organism cells require sufficient oxygen. The oxygen concentration in the body and localized tissues fall (hypoxia) whenever the need for oxygen increases, resulting in a physiological crisis that compromises physiological processes and survivability. Organisms have formed an effective and quick oxygen sensing system called (hypoxia-inducible factors (*HIF*s) because of the importance of oxygen in metabolism, survival, and respiration [34].

Arterial and central chemoreceptors detect changes in the oxygen content in the external environment. The brainstem's medulla, beneath the respiratory centres, contains central chemoreceptors. The carotid and aortic bodies include arterial chemoreceptors [35]. The stimulation of arterial chemoreceptors promotes neurotransmission and alters the function of neprilysin (NEP), a neutral endopeptidase that changes the biological response to hypoxia by hydrolytic component P [36]. Neuroendocrine cells in neuroepithelial bodies also govern oxygen sensing by imposing chemosensitivity, critical for oxygen sensing in the early stages of life. These exciting chemoreceptors enhance sympathetic nervous activity (SNA) and the systemic and arterial pulmonary blood flow to receive enough oxygen. Therefore, in organism cells, the expression of many adapted genes is activated to improve the oxygen supply and enable anaerobic ATP production. Hypoxia-inducible factors regulate these hypoxic responses (*HIFs*) [37].

There is a difference between sustained and intermittent hypoxia. Mitochondrial respiration utilizes more than 90% oxygen in humans during every day physiological situations [38]. The remaining (10%) oxygen is used to degrade *HIF-1*. The mitochondria use practically all oxygen or eliminate free cytosolic oxygen during the response to sustained hypoxia, enabling *HIF-1* to stabilize quickly. *HIF-1* activation causes enhanced transcription of several genes, including EPO, inducible nitric oxide synthase (iNOS), and *VEGF* [39]. These variables contribute to the recovery after early hypoxic shocks by boosting oxygenation and tissue perfusion as part of the adapted strategy to hypoxia. A free oxygen deficit is not adequately generated in intermittent hypoxia to enable *HIF-1* stabilization. Due to oxidative stress, intermittent hypoxia can cause a delayed elevation in *HIF*-1, leading to the stimulation of *NFkB*-driven inflammation [40].

7. Role of HIF 1a in causing hypoxia in the brain

A transcription factor that binds to specific nuclear cofactors and transactivates several genes, causing a range of adaptive responses in response to low oxygen levels in the body, is called Hypoxia Inducible Factor *HIF* [41]. *HIF-1* alpha and beta subunits form an active heterodimer under hypoxic settings, driving the transcription of approximately 60 genes involved in cell survival, adaptability, anaerobic metabolism, cytokine generation, vascularization, immunological response, and tissue homeostasis [42]. The two isoforms of *HIFα*, *HIF1α,* regulate erythropoietin (EPO), whereas *HIF2α* regulates the heme-regulating gene (hemopoietin genes). Increased *HIF* signalling in the body can contribute to inflammation and tumour progression. *HIF2α* activation has been observed because it plays a fundamental role in inflammation. HIF-1α has neuroprotective properties, but it can potentially be neurotoxic. *HIF-1* is involved in forming the early brain and the proliferation of neural precursor cells. *HIF-1* is recognized as critical in hypoxic–ischemic brain damage under pathological conditions [43]. During hypoxia, *HIF-1* participates in the apoptotic process to increase the stability of the tumour suppressor protein p53, which has neurotoxic consequences.

The brain is the most vulnerable organ to hypoxia, resulting in coma, convulsions, cognitive impairment, other neurological impairments, and brain death if left untreated [44]. Cardiac arrest, asphyxia, or systemic metabolic abnormalities affecting the blood's oxygen content, systemic hypoxia, severe anaemia, and systemic hypotension can lead to hypoxic brain damage [45]. Hypoxia-induced autophagy is linked to the HIF-1 signalling pathway. According to studies, hypoxic preconditioning protection is lost in HIF-1α knockout mice exposed to neonatal hypoxia/ischemia [46]. HIF1α hydroxylation is prevented by blockage of prolyl- and asparaginyl-hydroxylases in hypoxic environments. Prolyl-hydroxylase inhibitors reduce HIF1α breakdown, resulting in fast *HIF1α* protein build-up [47]. Phosphorylation of the *HIF1α* protein causes it to dimerize with *HIF1α*. The *HIF1α/HIF1β* dimer interacts with p300/CBP, causing hypoxia response elements in *HIF* target genes to be activated [48].

PI3K (Phosphatidylinositol 3-kinase) and Akt (protein kinase B) signalling pathway is related to hypoxia-ischemia injury as it increases the phosphorylation of downstream molecules such as apoptosis-related family members, transcription factors, mammalian target of rapamycin (mTOR), and glycogen synthase kinase-3. Phosphatase and tensin homologue (PTEN) is a lipid phosphatase that inhibits the PI3K/Akt pathway by hydrolysing PIP-3 to PIP-2 and preventing downstream p-Akt. PI3K and its downstream effector. Akt is a member of a well-studied family of signal transduction enzymes that regulate cellular activation, inflammation, and apoptosis [49].

8. Role of hypoxia in activating inflammatory pathways such as interleukins, cytokines, *NFk-B1*

Chemokines and cytokines are low-molecular-weight proteins produced mostly by lymphocytes and macrophages. As neurotransmitters and hormones, they mediate intracellular and extracellular interactions in an autocrine, endocrine, and paracrine manner. They regulate various biological activities, including local and systemic anti- and pro-inflammation, chemotaxis, metabolism, cellular proliferation, and tissue repair, by adhering to certain cell surface receptors [50].

HIF-1α attenuates periapical inflammation and tissue destruction, resulting in downregulation of nuclear factor-kappa B (*NF-κB*) and gene expressions. These two substances also prevented macrophages from activating *NF-κB* and producing pro-inflammatory cytokines. Furthermore, stimulation of HIF-1 reduced lipopolysaccharide-stimulated macrophage differentiation into M1 cells, resulting in a higher ratio of M2 macrophages to M1 cells [51]. In another study, *HIF-2a* activation by pro-inflammatory cytokines increases iNOS expression and activity via the *NF-κB* pathway, resulting in nitric oxide (NO) production, which causes liver damage when generated in excess [52].

cAMP-mediated signalling pathways might be changed in the presence of HIF1A, causing inflammatory-like processes to worsen. Only in the presence of Ni-induced hypoxia-inducible factor 1 (*HIF1α*) does prostaglandin (*PGE2*) synergistically accelerate Ni-induced Interleukin (IL-8) production [53]. Elevated IL-13 expression can cause eosinophilia and pathologies such as excessive mucus production. IL-13 can activate genes in the hypoxia signalling pathway, producing CD73 (immunoinhibitory protein) on the cell surface [54].

9. Effect of hypoxia on astrocyte functioning

Astrocytes have a crucial role in maintaining the normal oxygen levels in the brain. If the PO2 goes less than 17MMHg, astrocytes robust the Calcium ions into the brain; they act as the source of ATP in the hypoxic state of the brain. Astrocytes also have the potential to sense an increase in PCO2 levels too [21]. This chemosensation helps the cells to provide the astroglial networks with ATPS that help spread Ca + 2 activation and excitation. It also increases breathing to maintain homeostasis. It suggests that ATP released helps to keep living in the face of the hypoxia-evoked depression of the respiratory network [55].

In the previous century, it has also become clear that astrocytes can protect neurons under hypoxia conditions. The potential process is similar to "hypoxic preconditioning," in which a temporary interval of moderate hypoxia protects neurons from subsequent ischemia episodes that are generally fatal [56]. Mild hypoxia synthesizes several protective astrocytic factors that help neurons survive. Hypoxia increases the production of specific proteins, such as connexin 43, which may promote ATP/adenosine transit towards the interstitial space [57]. Astrocytes release erythropoietin in reaction to hypoxia, which has a significant neuroprotective impact. The erythropoietin expression in astrocytes is increased once hypoxia-inducible factors are activated. HIF-1α and HIF-2α are two transcription factors. [58]. In an adult brain, astrocytes can alter and monitor synaptic functionality. It was believed that synaptic plasticity is solely based on neurons, but in recent research, it has been found that the glial network and astrocytes alter synaptic transmissions [59]. The activation of metabotropic receptors modulates synaptic alterations by astrocytes. It helps release glutamate, gliotransmitter ATP, and D-serine, which act on neurons [60]. As per Astrocyte-induced pathogenesis, astrocytes can cause adenosine accumulation that affects glial cells and cause sleep deprivation and cognitive impairment [61]. Experimental investigations have found that the astrocytes sense synaptic activity with the help of astrocytic calcium. Astrocytes elevate their Ca + levels to sense neural activity with the help of the Gq-coupled protein pathway [62].

10. Conclusion

Astrocytes play a significant role in the homeostasis of the central nervous system. Astrocytic cells are central homeostatic and protective cells of the nerves. The decline in nerve cells or astrocytes causes a disease-permissive landscape and triggers nerve cell malfunction, nerve cell death, and nerve cell deficiency. Astroglial atrophy is mainly noticeable in major psychiatric illnesses. Severe reactive astrogliosis and glial scar formation at the location of the neurons trauma are glowing familiar to inhibit axonal regeneration. More research into the processes of Astrocytes protection, particularly the substances they produce, will give crucial insights into how to protect the Blood–Brain Barrier throughout trauma and neurological condition.

Moreover, its expression timing in astrocytes is essential to determine the influence of hypoxia-induced signalling on stroke volume. Furthermore, in addition to the impacts of hypoxia-signalling in astrocytes on neuron viability, it seems necessary to consider how such alterations will affect astrocyte viability. We must fully comprehend how to lessen the harm caused by stroke if we can better define the various consequences of hypoxia signalling in astrocytes.

Funding

The research received no specific grant from funding agencies in the public, commercial, or not-for-profit sectors.

Competing interests

The authors have declared no competing interests.

Ethical approval

The study does not require any ethical approval.

Availability of data and materials

No data was generated in the current study.

Author details

Farwa Munir[1], Nida Islam[2], Muhammad Hassan Nasir[1], Zainab Anis[2], Shahar Bano[2], Shahzaib Naeem[3,4], Atif Amin Baig[1*] and Zaineb Sohail[3,4]

1 Universiti Sultan Zainal Abidin, Malaysia

2 School of Food and Agriculture sciences, University of Management and Technology, Lahore, Pakistan

3 Surriya Azeem Teaching Hospital, Pakistan

4 Department of Life Sciences, University of Management and Technology, Lahore, Pakistan

*Address all correspondence to: atifameen01@gmail.com

References

[1] Sofroniew MV, Vinters HV. Astrocytes: Biology and pathology. Acta Neuropathologica. 2010;**119**(1):7-35

[2] Verkhratsky A, Nedergaard M, Hertz L. Why are astrocytes important? Neurochemical Research. 2015;**40**(2):389-401

[3] Lawal O, Ulloa Severino FP, Eroglu C. The role of astrocyte structural plasticity in regulating neural circuit function and behavior. Glia. Aug 2022;**70**(8):1467-1483. DOI: 10.1002/glia.24191. Epub 2022 May 10. PMID: 35535566; PMCID: PMC9233050

[4] Herrmann JE, Imura T, Song B, Qi J, Ao Y, Nguyen TK, et al. STAT3 is a critical regulator of astrogliosis and scar formation after spinal cord injury. Journal of Neuroscience. 2008;**28**(28):7231-7243

[5] Puspitasari V, Gunawan PY, Wiradarma HD, Hartoyo V. Glial fibrillary acidic protein serum level as a predictor of clinical outcome in ischemic stroke. Open Access Macedonian Journal of Medical Sciences. 2019;**7**(9):1471

[6] Hashimi SR, Hashimi S. Anticholinergic Burden and Risk of Cognitive Impairment In Older Adults. Virginia Commonwealth University. 2020. DOI: 10.25772/CMQ2-4A23

[7] Kortebi I. Understanding Cortical Astrocyte Distribution through Perinatal Birth Dating [Doctoral dissertation]. University of Toronto Canada; 2021

[8] Pekny M, Wilhelmsson U, Pekna M. The dual role of astrocyte activation and reactive gliosis. Neuroscience Letters. 2014;**565**:30-38

[9] Sofroniew MV. Astrogliosis. Cold Spring Harbor Perspectives in Biology. 2015;**7**(2):a020420

[10] Verkhratsky A, Rodríguez JJ, Steardo L. Astrogliopathology: A central element of neuropsychiatric diseases? The Neuroscientist. 2014;**20**(6):576-588

[11] Lanciotti A, Brignone M, Bertini E, Petrucci T, Aloisi F, Ambrosini E. Astrocytes: Emerging stars in leukodystrophy pathogenesis. Translational Neuroscience. 2013;**4**(2):144-164

[12] Bedner P, Dupper A, Hüttmann K, Müller J, Herde MK, Dublin P, et al. Astrocyte uncoupling as a cause of human temporal lobe epilepsy. Brain. 2015;**138**(5):1208-1222

[13] Combes V, Guillemin GJ, Chan-Ling T, Hunt NH, Grau GE. The crossroads of neuroinflammation in infectious diseases: Endothelial cells and astrocytes. Trends in Parasitology. 2012;**28**(8):311-319

[14] Dantzer R, Kelley KW. Twenty years of research on cytokine-induced sickness behavior. Brain, Behavior, and Immunity. 2007;**21**(2):153-160

[15] Yin Z, Milatovic D, Aschner JL, Syversen T, Rocha JB, Souza DO, et al. Methylmercury induces oxidative injury, alterations in permeability and glutamine transport in cultured astrocytes. Brain Research. 2007;**1131**:1-10

[16] Struys-Ponsar C, Guillard O, de Aguilar PVDB. Effects of aluminum exposure on glutamate metabolism: A possible explanation for its toxicity. Experimental Neurology. 2000;**163**(1):157-164

[17] Zhao Y, Rempe DA. Targeting astrocytes for stroke therapy. Neurotherapeutics. 2010;7(4):439-451

[18] Oide T, Yoshida K, Kaneko K, Ohta M, Arima K. Iron overload and antioxidative role of perivascular astrocytes in aceruloplasminemia. Neuropathology and Applied Neurobiology. 2006;**32**(2):170-176

[19] Sanacora G, Banasr M. From pathophysiology to novel antidepressant drugs: glial contributions to the pathology and treatment of mood disorders. Biological Psychiatry. 2013;**73**(12):1172-1179

[20] Büttner A, Weis S. Neuropathological alterations in drug abusers. Forensic Science, Medicine, and Pathology. 2006;**2**(2):115-126

[21] Marina N, Kasymov V, Ackland GL, Kasparov S, Gourine AV. Astrocytes and Brain Hypoxia. Advances in Experimental Medicine and Biology. 2016;**903**:201-207. DOI: 10.1007/978-1-4899-7678-9_14. PMID: 27343098

[22] Ranaivo HR, Hodge JN, Choi N, Wainwright MS. Albumin induces upregulation of matrix metalloproteinase-9 in astrocytes via MAPK and reactive oxygen species-dependent pathways. Journal of Neuroinflammation. 2012;**9**(1):1-12

[23] Qu Yi, Duan Z, Zhao F, Wei D, Zhang J, Tang B, et al. Telomerase reverse transcriptase upregulation attenuates astrocyte proliferation and promotes neuronal survival in the hypoxic–ischemic rat brain. 2011;**42**:3542-3550. DOI: 10.1161/STROKEAHA.111.626325Stroke

[24] Wang Z, Xu J, Geng X, Zhang W. Analysis of DNA methylation status of the promoter of human telomerase reverse transcriptase in gastric carcinogenesis. Archives of Medical Research. Jan 2010;**41**(1):1-6. DOI: 10.1016/j.arcmed.2009.11.001. Epub 2010 Jan 29. PMID: 20430247

[25] Hamby ME, Sofroniew MV. Reactive astrocytes as therapeutic targets forCNS disorders. Neurotherapeutics. 2010;**7**:494-506

[26] Michaluk P, Wawrzyniak M, Alot P, Szczot M, Wyrembek P, Mercik K, et al. Influence of matrix metalloproteinase MMP-9 on dendritic spine morphology. Journal of Cell Science. 2011;**124**:3369-3380

[27] Barr TL, Latour LL, Lee KY, Schaewe TJ, Luby M, Chang GS, et al. Blood–brain barrier disruption in humans is independently associated with increased matrix metalloproteinase-9. Stroke. 2010;**41**:e123-e128

[28] Hill MD, Martin RH, Palesch YY, Tamariz D, Waldman BD, Ryckborst KJ, et al. The albumin in acute stroke part 1 trial: An exploratory efficacy analysis. Stroke. 2011;**42**(6):1621-1625

[29] Kou Z, VandeVord PJ. Traumatic white matter injury and glial activation: from basic science to clinics. Glia. 2014;**62**:1831-1855. DOI: 10.1002/glia.22690

[30] Blagodatski A, Poteryaev D, Katanaev VL. Targeting the Wnt pathways for therapies. Molecular Cell Therapy. 2014;**2**:28. DOI: 10.1186/2052-8426-2-28

[31] Munji RN, Choe Y, Li G, Siegenthaler JA, Pleasure SJ. Wnt signaling regulates neuronal differentiation of cortical intermediate progenitors. The Journal of Neuroscience. 2011;**31**:1676-1687. DOI: 10.1523/JNEUROSCI.5404-10.2011

[32] Lee P, Chandel NS, Simon MC. Cellular adaptation to hypoxia through hypoxia inducible factors and beyond. Nature Reviews Molecular Cell Biology. 2020;**21**(5):268-283

[33] Semenza GL. Hypoxia-inducible factors in physiology and medicine. Cell. 2012;**148**(3):399-408

[34] Yeo EJ. Hypoxia and aging. Experimental & Molecular Medicine. 2019;**51**(6):1-15

[35] Caravagna C, Seaborn T. Oxygen sensing in early life. Lung. 2016;**194**:715-722

[36] Jordan AS, McSharry DG, Malhotra A. Adult obstructive sleep apnoea. Lancet. 2014;**383**:736-747

[37] Antikainen H, Driscoll M, Haspel G, Dobrowolski R. TOR-mediated regulation of metabolism in aging. Aging Cell. 2017;**16**:1219-1233

[38] Taylor CT, Kent BD, Crinion SJ, McNicholas WT, Ryan S. Human adipocytes are highly sensitive to intermittent hypoxia induced NF-kappaB activity and subsequent inflammatory gene expression. Biochemical and Biophysical Research Communications. 2014;**447**:660-665

[39] Prabhakar NR, Semenza GL. Adaptive and maladaptive cardiorespiratory responses to continuous and intermittent hypoxia mediated by hypoxia-inducible factors 1 and 2. Physiological Reviews. 2012;**92**:967-1003

[40] Nanduri J et al. HIF-1alpha activation by intermittent hypoxia requires NADPH oxidase stimulation by xanthine oxidase. PLoS One. 2015;**10**:e0119762

[41] Shay JE, Celeste SM. Hypoxia-inducible factors: Crosstalk between inflammation and metabolism. Seminars in Cell & Developmental Biology. 2012;**23**(4):389-394

[42] Adams J, Difazio L, Rolandelli R, Luján J, Haskó G, Csóka B, et al. HIF-1: A key mediator in hypoxia (Review). Acta Physiologica Hungarica. 2009;**96**(1):19-28

[43] Fan X, Heijnen CJ, van der Kooij MA, Groenendaal F, van Bel F. The role and regulation of hypoxia-inducible factor-1α expression in brain development and neonatal hypoxic–ischemic brain injury. Brain Research Reviews. 2009;**62**(1):99-108

[44] Sharp FR, Bernaudin M. HIF1 and oxygen sensing in the brain. Nature Reviews Neuroscience. 2004;**5**(6):437-448

[45] Lacerte M, HSA, Mesfin FB. Hypoxic Brain Injury. StatPearls; 2022

[46] Lu N, Li X, Tan R, An J, Cai Z, Hu X, et al. HIF-1α/Beclin1-mediated autophagy is involved in neuroprotection induced by hypoxic preconditioning. Journal of Molecular Neuroscience. 2018;**66**(2):238-250

[47] Wang X, Cui L, Ji X. Cognitive impairment caused by hypoxia: From clinical evidences to molecular mechanisms. Metabolic Brain Disease. 2022;**37**(1):51-66

[48] Ciuffreda L, Falcone I, Incani UC, Del Curatolo A, Conciatori F, Matteoni S, et al. PTEN expression and function in adult cancer stem cells and prospects for therapeutic targeting. Advances in Biological Regulation. 2014;**56**:66-80

[49] Zhang Z, Yao L, Yang J, Wang Z, Du G. PI3K/Akt and HIF-1 signaling pathway in hypoxia-ischemia (Review). Molecular Medicine Reports. 2018;**18**(4):3547-3554

[50] Sarode P, Schaefer MB, Grimminger F, Seeger W, Savai R. Macrophage and tumor cell cross-talk is fundamental for lung tumor progression: We need to talk. Frontiers in Oncology. 2020:10

[51] Hirai K, Furusho H, Hirota K, Sasaki H. Activation of hypoxia-inducible factor 1 attenuates periapical inflammation and bone loss. International Journal of Oral Science. 2018;**10**(2):12

[52] Yoshida T, Okumura T, Matsuo Y, Okuyama T, Michiura T, Kaibori M, et al. activation of transcription factor HIF inhibits IL-1β-induced NO production in primary cultured rat hepatocytes. Nitric Oxide. 2022;**124**:1-14

[53] Brant KA, Fabisiak JP. Role of hypoxia-inducible factor 1, α subunit and cAMP-response element binding protein 1 in synergistic release of interleukin 8 by prostaglandin E2 and nickel in lung fibroblasts. American Journal of Respiratory Cell and Molecular Biology. 2013;**49**(1):105-113

[54] Khalil SM, Bernstein I, Kulaga H, Gour N, Rowan N, Lajoie S, et al., editors. Interleukin 13 (IL- alters hypoxia-associated genes and upregulates CD73). International Forum of Allergy & Rhinology; 2020: Wiley Online Library.

[55] Heurteaux C, Lauritzen I, Widmann C, Lazdunski M. Essential role of adenosine. Adenosine.

[56] Heurteaux C, Lauritzen I, Widmann C, Lazdunski M. Essential role of adenosine, adenosine A1 receptors, and ATP-sensitive K+ channels in cerebral ischemic preconditioning. Proceedings of the National Academy of Sciences. 1995;**92**(10):4666-4670

[57] Prass K, Scharff A, Ruscher K, Lowl D, Muselmann C, Victorov I, et al. Hypoxia-induced stroke tolerance in the mouse is mediated by erythropoietin. Stroke. 2003;**34**:1981-1986

[58] Semenza GL, Agani F, Booth G, Forsythe J, Iyer N, Jiang B-H. Structural and functional analysis of hypoxia-inducible factor 1. Kidney International. 1997;**51**(2):553-555. ISSN: 0085-2538. DOI: 10.1038/ki.1997.77

[59] Barker AJ, Ullian EM. Astrocytes and synaptic plasticity. The Neuroscientist. 2010;**16**(1):40-50

[60] Halassa MM, Haydon PG. Integrated brain circuits: Astrocytic networks modulate neuronal activity and behavior. Annual Review of Physiology. 2010;**72**:335

[61] Chung WS, Allen NJ, Eroglu C. Astrocytes control synapse formation, function, and elimination. Cold Spring Harbor Perspectives in Biology. 2015;**7**(9):a020370

[62] Bernardinelli Y, Muller D, Nikonenko I. Astrocyte-synapse structural plasticity. Neural Plasticity. 2014:2014

Chapter 12

Astrocytic Abnormalities in Schizophrenia

Kiarash Saleki, Mohammad Banazadeh, Banafshe Abadi, Zeynab Pirmoradi, Zahra Esmaili, Shiva Amiri, Ramtin Pourahmad, Kristi A. Kohlmeier and Mohammad Shabani

Abstract

Astrocytes are glial cells in the central nervous system (CNS), which contribute to CNS health and disease by participating in homeostatic, structural, and metabolic processes that play an essential role in facilitating synaptic transmission between neurons. Schizophrenia (SCZ) is a neuropsychiatric disorder associated with various positive and negative behaviors and interruption of executive function and cognition thought to be due partly to aberrations in signaling within neural networks. Recent research has demonstrated that astrocytes play a role in SCZ through various effects, including influencing immune system function, altering white matter, and mediating changes in neurotransmitters. Astrocytes are also known to play a role in inducing SCZ-associated changes in neuroplasticity, which includes alterations in synaptic strength and neurogenesis. Also, astrocyte abnormalities are linked to neurobehavioral impairments seen at the clinical level. The present chapter details general information on SCZ. It highlights the role of astrocytes in SCZ at molecular and behavioral levels, including neural changes seen in the disease, and the therapeutic implications of targeting astrocytes in SCZ.

Keywords: astrocytes, neurobehavioral disorders, schizophrenia, neurosciences, neuroimmunology

1. Introduction

Schizophrenia (SCZ) is a severe mental illness with typical onset during early adulthood, which confers a lifetime disability. The main signs of the disease comprise positive symptoms, such as delusions and hallucinations; negative symptoms, such as blunted emotions, social withdrawal, and apathy; and executive and cognitive neurobehavior interruption. Additionally, SCZ is associated with a greater suicide possibility and a shorter lifetime. The illness puts a substantial socioeconomic burden on caregivers, families, patients, and society [1].

While the mechanisms underlying SCZ remain unknown, dysfunctions in synaptic signaling are implicated. A synapse relying on chemical transmission is comprised of a neural presynaptic and neural postsynaptic component. However, an astrocyte is a vital member of the synapse leading to the term "tripartite synapse" and among other functions, plays a role in synaptic plasticity [2, 3]. Astrocytic processes are considered crucial members of the tripartite synapse because they enclose the pre- and postsynaptic components and are close to the synaptic cleft [4]. They can quickly reuptake excess glutamate produced by the presynaptic terminals because of their great expression of the high-affinity glutamate transporters EAAT1 and EAAT2, which limits excitatory transmission [5]. In turn, astrocytes provide a significant contribution to synaptic plasticity and transmission. Even though they are electrically inactive, they react to presynaptic activation by sending G-protein-mediated Ca^{2+} signals, which cause the release of "gliotransmitters" such as glutamate, ATP, and GABA, which influence local synaptic transmission [6–8]. The likelihood of inducing LTP is increased by astrocyte-derived glutamate-mediated of NMDA receptor activation on the postsynaptic sites [9]. Furthermore, astrocytic processes surrounding synapses undergo fast structural remodeling due to stimuli that cause synaptic long-term potentiation (LTP), altering their capacity to control synaptic transmission. When taken as a whole, the anatomical and functional data strongly suggest that astrocytes are essential active mediators of synaptic plasticity [10]. Recently, a fourth player at the synapse has been recognized as the important role of the extracellular matrix (ECM), especially in synaptic plasticity has emerged. Accordingly, synapses consist of four pre- and postsynaptic elements, glial processes, and an ECM [11, 12]. In SCZ, it has been shown that interactions between all synapse elements disrupt synaptic functions and alter plasticity [13], and a breakdown in communication between synaptic components and changes in synaptic plasticity is believed to cause SCZ. Given the clinical, genetic, and pathological diversity of SCZ, synaptic dysfunction in specific brain areas may represent a point of convergence, perhaps resulting from various unique molecular processes in different people [13].

The evidence supporting a role of astrocytes in altered synaptic signaling in SCZ is diverse and often includes changes seen in glutamate transmission. Transplanted Human Induced Pluripotent Stem Cell (hiPSC)-astrocyte progenitors from SCZ patients transformed into mature astrocytes in a mouse created behavioral alterations consistent with cognitive and olfactory changes seen in SCZ patients. Healthy neurons co-cultured with astrocytes from SCZ patients of both males and females showed a drastically heightened reaction to glutamate, suggesting modifications in gliotransmitter liberation and/or inadequate turnover of neurotransmitters [14]. Additionally, synaptic dysfunction, demyelination, and alterations in inflammation pathways were noted [14]. Dysregulated glia functions have been associated with endothelial cell stimulation and increased systemic inflammatory markers in brain pathology [15, 16].

Further evidence for a role of altered glutamate signaling in astrocytes playing a role in SCZ are findings that the astrocyte-derived N-methyl-D-aspartate (NMDA) receptor antagonist kynurenic acid (KYNA) is higher in SCZ patients [17, 18]. In addition, KYNA is not only produced in astrocytes but also in a diversity of cell types through activation of the kynurenine pathway (KP) resulting in tryptophan metabolism. As high concentrations of KYNA are associated with the pathophysiology of SCZ, enhanced knowledge of mechanisms leading to high KYNA production in patients with SCZ could aid in the design of novel diagnostics and therapeutics, which could focus on targeting astrocytes [19].

Astrocytes synthesize and release D-serine, which is a co-agonist of the NMDAR where it modulates synaptic activity. Reductions in D-serine release by astrocytes could play a role in SCZ by leading to inhibition of synaptic transmission and synaptic plasticity mediated by NMDA receptors. Hypofunctionality of NMDA receptors has been shown to be associated with behaviors reminiscent of SCZ, and further, a risk factor for SCZ is reduced functioning of NMDAR in cortical pyramidal neurons and interneurons. While astrocytes are not the sole source of D-serine, they do con¬tribute to the available pool, and their contribution can be local and region-specific. Accordingly, targeting astrocytic D-serine synthesis in SCZ represents a potential clinical strategy in order to reverse cortical hypofunctionality.

Neuroinflammation has been implicated in a role in SCZ, and astrocytes could play a role in this process [20, 21]. These comprise reduced astrocyte cellular characteristics and gene expression in chronic stress, anxiety, depression, and enhanced inflammation in SCZ [22]. Targeting changes in inflammatory markers in astrocytes might also represent a therapeutic strategy for SCZ patients.

In the present chapter, we review molecular aspects of astrocyte abnormality in SCZ, focusing on neuroplasticity in line with clinical features. We also summarize animal studies of the behavioral aspects of this topic. Finally, we propose therapeutic and diagnostic strategies focused on targeting astrocytes.

2. Overview of SCZ

2.1 Clinical presentation

The word SCZ originated by Eugen Bleuler in 1908, is derived from the Greek words 'schizo' (splitting) and 'phren' (mind) and defines a functional psychotic illness typified by the occurrence of hallucinations, delusional opinions, and disruptions of perception, thought, and behavior. Conventionally, symptoms have been separated into two major classifications: positive symptoms that comprise delusions, hallucinations, and formal thought disorders, and negative symptoms such as poverty of speech, anhedonia, and lack of motivation. SCZ diagnosis is clinical, solely done after acquiring a detailed psychiatric record and excluding further reasons for psychosis. Risk factors comprise the season of birth, severe maternal malnutrition, maternal influenza in pregnancy, family history, childhood trauma, social isolation, cannabis use, minority ethnicity, complications of giving birth, and urbanization [23–25].

SCZ is a developmental disorder and it is now widely accepted that astrocytes play an important role during both pre and postnatal development and continue their important role in development into adulthood by regulating establishment of neuronal circuits [26] and by regulating multiple homeostatic functions, such as buffering extracellular potassium or modulation of synaptic activity [27] and functional hyperemia [28], respectively. They offer metabolic support for synaptic activity and are also required for creating and maintaining synapses [29]. Changes in astrocytic numbers have been demonstrated to cause cognitive impairment, which is consistent with the essential roles of astrocytes in neural circuit functioning. In the astrocyte-specific toxin L-aminoadipate (L-AA) model [30] or in a transgenic mouse line with inducible and selective tetanus neurotoxin (TeNT) expression in astrocytes [31], mice exhibited deficits in set-shifting attention, working memory, reversal learning [30], recognition memory, and abnormal cortical gamma oscillations [31]. Similarly, decreased expression of the astrocytic glutamate transporter GLT-1 lowered prepulse suppression of the acoustic startle response [32], which is a well-established characteristic of SCZ. Changes in astrocytic cell density and morphology in the mouse prefrontal cortex (PFC) cause cognitive dysfunctions associated with the subcortical zone (SCZ) [30, 31].

2.2 Epidemiology

The prevalence of SCZ is around 0.6% and 1.9% in the U.S. population [33]. Based on insurance claims for the management of SCZ, the yearly prevalence in the U.S. is 5.1 per 1000 lives [34]. The illness prevalence appears identical in men and women, although the onset of symptoms occurs younger in men than in women [35]. Men tend to encounter their initial episode of SCZ in their early 20s, while women commonly encounter their initial episode in their late 20s or early 30s [36].

Study regarding a potential connection between geographical birthplace and the origination of SCZ has given inconsistent outcomes. A cooperative survey by the World Health Organization in 10 countries discovered that SCZ occurred with similar rates throughout the diverse geographically-stratified populations [37].

3. Molecular mechanisms

3.1 Neuroplasticity mechanisms

3.1.1 Synaptic plasticity and neurotransmitter alterations

3.1.1.1 Astrocyte function in synaptic plasticity and SCZ

Patients with SCZ exhibit profound cognitive impairment and negative symptoms resistant to current medication. Evidence supports the theory that these deficiencies are caused, at least in part, by changes in cortical synaptic plasticity, which is the ability of synapses to strengthen or weaken their activity. Targeting synaptic plasticity is a promising therapeutic approach to managing SCZ as synaptic transmission is a well-understood process, and the biochemical mechanisms behind short-term and long-term changes in synaptic strength are becoming even more evident as players at the molecular level are identified [38].

Long-term depression (LTD) and potentiation (LTP) of synaptic transmission are essential processes by which the brain changes the strength of synapses [39]. Many forms of synaptic plasticity depend on alterations in AMPA and NMDA receptors within the membrane, as these excitatory glutamate receptors are essential for synaptic transmission [38]. SCZ brains show reductions in synaptic functioning due to the loss of AMPA and NMDA receptors and loss of dendritic spines and synaptic markers. Loss of synapses and markers is consistent with microscopic analysis in patients showing reductions in brain volume, notably in the hippocampus, prefrontal and superior temporal cortices, and frontolimbic circuitry, which is accompanied with an increase in ventricular size [38].

3.1.1.2 Neurotransmitter hypotheses of SCZ

The role of dysfunctions in neurotransmitters, many of which are gliotransmitters, has been explored in SCZ and as many of these transmitters are gliotransmitters, alterations in astrocytic functioning have been suggested.

3.1.1.2.1 Glutamate

Glutamate is the principal excitatory neurotransmitter in the central nervous system (CNS) that initiates fast signal transmission. Its activity is critically involved in routine behaviors, including learning and memory, which mechanistically processes relying on synaptic plasticity. Dysfunctions in glutamate transmission are seen in SCZ [40]. Given the significance of astrocytes in altering glutamatergic transmission, it is probable that they are involved in the glutamate dysfunction seen in SCZ.

Activity of astrocytes regulates extracellular glutamate levels [41]. As high glutamate levels can be toxic, following impulse transmission, EAATs, immediately terminate glutamate transmission by removing glutamate from the tripartite synapse (**Figure 1**). In astrocytes, EAAT1 and EAAT2 are the most prominent EAATs [42]. Once transported into astrocytes by EAATs, glutamate is converted to glutamine-by-glutamine synthase. This newly created glutamine can then be transported back to

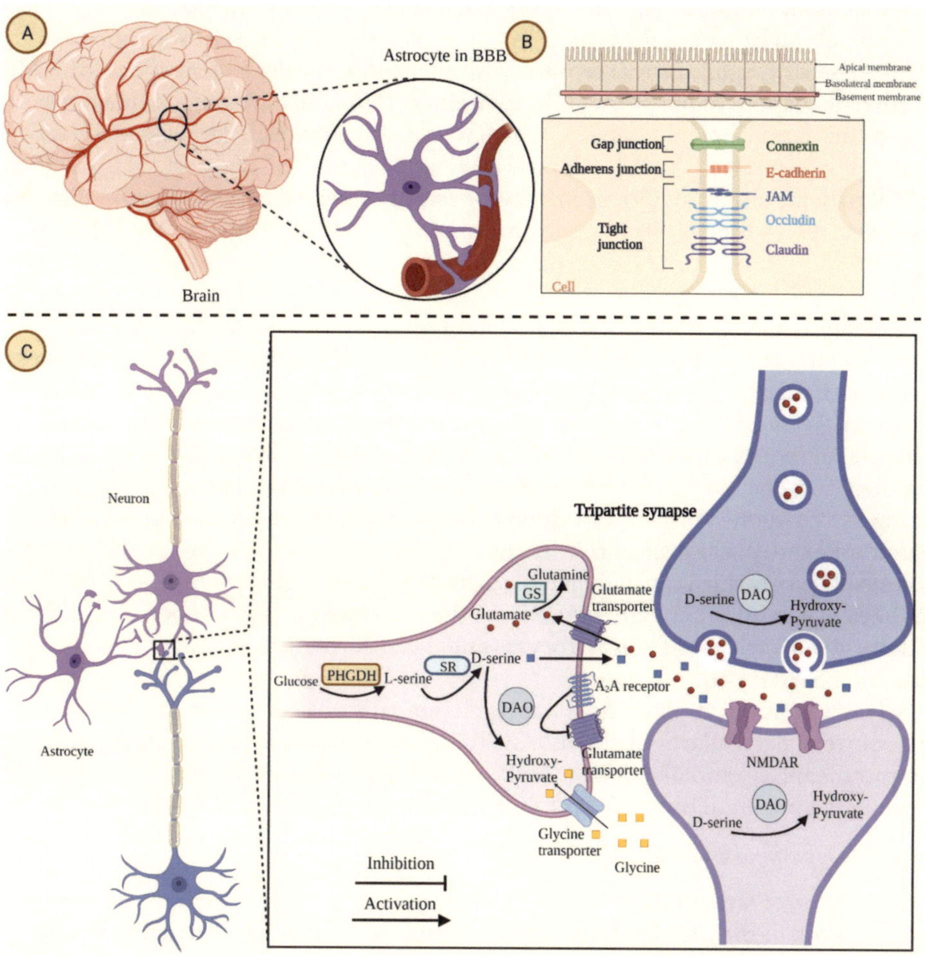

Figure 1.
A general depiction of the position of the astrocyte in the tripartite synapse and BBB. A) Astrocytes protect the brain from blood borne toxins and microbes by close opposition of their end feet on blood vasculature. B) Gap junctions play a crucial role in cellular communications between astrocytes. C) Astrocytes play a vital role in neuronal communication via participation in the tripartite synapse which is shown with important glial and neuronal intracellular and extracellular pathways. Created with BioRender.com. Abbreviations: BBB: Blood brain barrier, D: Dextro, DAO: D-amino acid oxidase, GS: Glutamine synthetase, L: Levo, NMDAR: N-methyl D-aspartate receptor, PHGDH: 3-phosphoglycerate dehydrogenase, and SR: Serine racemase.

presynaptic neurons in a process known as glutamate-glutamine cycling, where it is converted into glutamate [43].

The glutamate theory of SCZ is based in part on findings that NMDA antagonists like ketamine and phencyclidine induce an SCZ-like psychosis exhibiting positive symptoms [40]. Further evidence of a role played by the NMDA receptor in SCZ is that disruptions in NMDA receptors in interneurons result in the lack of inhibitory signals to glutamate neurons, mainly in the prefrontal cortex, which may be associated with negative symptoms of SCZ [40]. The endogenous NMDAR antagonist KYNA can mimic SCZ-like symptoms similar to other exogenous NMDAR antagonists such as phencyclidine and ketamine. Studies of the role it could play in SCZ have been conducted [19, 44]. At low concentrations, KYNA acts as an antagonist at the strychnine-insensitive glycine-binding site of NMDARs. However, higher concentrations also block the glutamate-binding site of NMDARs [19]. Furthermore, KYNA has modest antagonistic effects on kainate and amino-3-hydroxy-5-methyl-4-isoxazolepropionic acid (AMPA)-sensitive glutamate receptors, with concentration-dependent effects on AMPA receptor-mediated actions [19, 44]. In addition, KYNA is an endogenous antagonist of α7 nicotinic acetylcholine receptors, which would also be expected to reduce synaptic excitability. Increased KYNA levels have been noted in the CSF and cortical brain areas in SCZ patients due to alterations in enzyme activity/expression in the kynurenine pathway (KP), which transfers tryptophan metabolism to KYNA synthesis [19, 44].

3.1.1.2.2 Glycine

Glycine, a nonessential amino acid that plays a critical role in inhibitory and excitatory neurotransmission is upregulated in SCZ. SCZ patients have higher glycine levels, particularly in the parietal and occipital cortex [45]. Astrocytes are a major source of glycine. While thought of as a classic inhibitory amino acid as it interacts with the inhibitory canonical strychnine glycine receptor (GlyR), which is a chloride channel [46], glycine also functions as a potentiator of excitatory transmission by acting as a co-agonist of NMDAR. Interestingly, it is glycine's actions at the NMDAR which appear to play a crucial role in the neurodevelopmental phases of SCZ pathogenesis [47]. D-serine is an endogenous ligand at the NMDAR glycine B-site [48] and has been proposed to activate NMDARs in the amygdala, however, upon high afferent activity glycine released from astrocytes enhanced NMDAR activity, which affected induction of LTP [49].

Glycine concentration in the synaptic cleft is carefully controlled by glycine transporters, particularly GlyT1 and GlyT2, located in astrocytes, which regulate neurotransmitter reuptake. Accordingly, targeting the astrocytic glycine transporters represents a potential treatment for SCZ [50].

3.1.1.2.3 Dopamine and adenosine systems

Alteration of striatal D2 dopamine receptors leads to positive symptoms, whereas the alteration of the prefrontal cortex D1 dopamine receptor leads to negative and cognitive symptoms. Dopamine transmission within the striatum is also impaired [51]. The number of synapses in the lateral part of the ventral tegmental area and the substantia nigra is reduced in SCZ. Among other effects, loss of synapses would lead to loss of NMDA receptors, which would result in reductions of activity of these striatal dopamine-containing cells, resulting in alterations in dopamine release at terminals. This is supported by findings that there is a deficiency of dopaminergic activation in the prefrontal cortex in SCZ. This could also result from decreased communication between the striatum and the prefrontal cortex, due in part to NMDAR dysfunction that has been shown in both prefrontal cortex and the striatum in postmortem SCZ brain [52].

Adenosine has two known types of receptors: A1 and A2. A1 receptors block the release of neurotransmitters including glutamate. Activating A2A-receptors causes glutamate release, which activates NMDA receptors and inhibits A1-receptors [53]. A2A receptors are found in dopamine-rich locations, such as the prefrontal cortex and the striatum, and their activation causes reduced dopaminergic innervation [53]. Hypofunction of A2A receptors in the striatum leads to hyperfunction of D2 receptors, which are implicated in disorders linked to neuroinflammatory processes, as well as to triggering immunological responses and heightening dopaminergic neurons' vulnerability to neurotoxic injury. Striatal astrocytes express A2A-D2 receptor heterodimers. While D2 receptor activation decreases presynaptic glutamate release, stimulation of A2A receptor reverses this action [53]. Dysfunction of the striatal astrocytic A2A receptor, which causes damage to the D2 receptor and disrupts glutamate homeostasis, is believed to be linked to SCZ [53, 54].

3.1.1.2.4 GABA

Postmortem investigations have extensively identified changes in various GABA-related markers in SCZ patients. Some studies show the reduction of inhibitory GABAergic neurotransmission across several brain regions affected by SCZ. These anomalies may cause difficulties in emotional functioning and cognitive control. Furthermore, one clinical investigation found reduced GABA concentrations in CSF samples from patients with first-episode psychosis compared to healthy volunteers, which were linked to total and general positive and negative syndrome scale ratings, disease severity, and poor performance on attention testing [55].

Astrocytes react to GABA through various pathways, including GABA receptors and transporters. GABA-activated astrocytes may then influence local neuronal activity by releasing gliotransmitters such as glutamate and ATP. Furthermore, astrocyte activation through various inputs can influence GABAergic neurotransmission. The complexity of communication within the brain is enhanced by reciprocal signaling between GABAergic neurons and astrocytes, and our improved understanding of this complexity could lead to new treatments for brain disorders [55].

3.1.1.2.5 Endocannabinoid systems

The endocannabinoid system (ECS), which consists of two well-characterized receptors and enzymes responsible for their production and degradation, is engaged in various physiological and pathological processes of the CNS [19].

There are two types of cannabinoid receptors (CBRs), cannabinoid receptor type 1 (CB1R) and cannabinoid receptor type 2 (CB2R), belonging to the family of Gi/o protein-coupled receptors (GPCRs). Therefore, activating them inhibits cAMP production, activates mitogen-activated protein kinases, and presynaptically inhibits several neurotransmitters involved in SCZ through presynaptic mechanisms [56, 57].

CB1Rs are involved in regulation of mood or emotion, antinociception, energy balance, immunological function, and endocrine activities [58]. CB2Rs, on the other hand, are expressed mainly in immunological and hematopoietic cells. CB2Rs have a protective function by reducing inflammation-induced pain via cytokine modulation and immune cell migration [58, 59].

Several alterations in the ECS have been reported in SCZ patients, including changes in CB1R availability, density, and/or mRNA expression, differences in levels of endocannabinoid in specific brain regions and CSF have been noted in SCZ patients. Endogenous CBR ligands are lipid-derived hydrophobic molecules, the best researched of which are N-arachidonoylethanolamine (AEA) and 2-arachidonoyl glycerol (2-AG). The fatty acid amide hydrolase (FAAH) enzyme quickly metabolizes AEA, whereas the monoacylglycerol lipase (MAGL) enzyme hydrolyzes 2-AG. Blocking the FAAH enzyme in order to heighten effects of AEA has been proposed as a potential therapy for SCZ. Furthermore, phytocannabinoids produced from plants, such as 9-tetrahydrocannabinol (9THC), the main psychoactive component of cannabis, and non-psychoactive cannabidiol (CBD) have intriguingly been suggested as potential treatments for SCZ due to actions involving glutamate release from astrocytes [60, 61]. This is supported by findings that CB1R is located on internal PFC astrocytes and reduces the negative symptoms of SCZ by reducing glutamate release [62]. On the other hand, exogenously applied cannabinoids can induce SCZ-like symptoms in adolescents if exposure is frequent and early in life [62].

3.1.2 Myelination and white matter

Astrocytes play a significant role in the repair and recovery of neurons, which includes a role in the repair of damage to white matter through their regulation of oligodendrocytes. White matter damage has been linked to developing various demyelinating conditions, including SCZ.

White matter astrocytes differ from those in gray matter in terms of development, morphology, residence, protein synthesis, and the role they play in supporting neighboring cells. During demyelination and remyelination, the functions of astrocytes are dynamic and are modified in response to specific stimuli or reactive processes, leading to vastly different biological outcomes. The effect of astrocytes on oligodendrocytes and various cellular subtypes in the oligodendrocyte lineage includes serving as an energy supply, a modulator of immune system function, a mediator of inflammation processes, a resource for trophic factors, and a regulator of iron. Features of astrocytes that lead to their neuroprotective properties include anti-oxidative properties, stabilization of glutamate homeostasis, and production of growth factors.

Ultrastructural evaluations showed induced microglia near dystrophic and apoptotic oligodendroglia, demyelinating and demyelinating axons, and swollen and vacuolated astroglia in cases with SCZ, in contrast to healthy subjects [15]. Theoretically, targeting astrocyte function represents a rational approach to repair injured myelin white matter-associated diseases such as SCZ [63] and cell therapy using stem cells or progenitor cell-derived astroglia has been recommended for patients with neuropsychiatric disorders associated with white matter degeneration or synaptic loss, [64]. Unfortunately, re-myelinating strategies have to date proved inadequate, which could stem from an unsuitable microenvironment. When taken together, although white matter alterations are implicated in SCZ, the astrocyte-specific alterations in this disease need to be explored further and those alterations need to take into account the vast array of cell types that astrocytes interact with before we can use astrocytic cell therapy for management of SCZ.

3.1.3 Adult neurogenesis

Adult neurogenesis could play a role in the pathological mechanisms of SCZ, but also could perhaps be exploited therapeutically. In the process of adult hippocampal neurogenesis, neural stem cells (NSCs) in the dentate gyrus (DG) transform into neurons, which is a process that continues throughout life [65]. Natural proliferation and maturation of NSCs in DG contribute to emotional behaviors and cognitive function, and disturbances to this process can cause neuropsychiatric disorders such as anxiety, mood disorders, and memory and learning disorders [66–70]. Transformation of NSCs into neurons is regulated by adult astrocytes in the hippocampus [71]. Following transformation, astrocytes contribute to the maturation and integration of neurons in the hippocampal circuit. Inhibiting the exocytosis of astrocytes leads to a decrease in dendritic spine count and dendritic branching, as well as disrupted dendrite survival and maturation of the new neurons [72]. Astrocytes control neurogenesis by releasing a variety of factors. D-Serine, BDNF, fibroblast growth factor 2 (FGF-2), glial cell line-derived neurotrophic factor (GDNF), and VEGF are examples of astrocyte-derived factors that may promote neurogenesis [73].

3.1.3.1 D-serine

Astrocytic D-serine is believed to play a role in neurogenesis. Obstruction of vesicle proliferation in astrocytes led to a decrease in dendritic formation in adult-born granule cells (abGCs) associated with a reduction in extracellular D-serine [72]. Further, mutations leading to decreases in D-serine result in changes reminiscent of SCZ. The DISC1 (disrupted-In-SCZ 1) gene is located on chromosome 1q42 and the product binds to and stabilizes serine racemase (SR), resulting in an increase in the conversion of L-Serine to D-Serine in astrocytes. A mutation in the DISC1 gene was first detected and linked to SCZ in a Scottish family with several major mental disorders, and now this gene mutation is recognized as a major risk factor for disorders involving abnormalities of neural development [74]. As a dominant-negative molecular tool, DISC1 mutants truncated at the C end of the full-length protein are used to elucidate the role of DISC1 in astrocytes [75]. In the mouse, dominant-negative C-terminus-truncated human DISC1 resulted in dendritic dysplasia and reduced neurogenesis [65]. Further, dominant-negative effects resulted in SR degradation, decreases in NMDA neurotransmission, and reductions in neurogenesis [65]. Further, this mutation appears to affect the neurogenesis of DG neurons more than other brain regions, such as the subventricular zone [76]. Therefore, it is tempting to speculate that D-serine depletion plays a role in SCZ.

3.1.3.2 BDNF and FGF2

Another astrocyte secretory factor that is involved in adult neurogenesis is BDNF. Astrocytes regulate the removal and recycling of pro-BDNF [77]. Pro-BDNF neuronal precursors are taken up by astrocytes through endocytosis and then converted to adult BDNF. Overexpression of BDNF-containing hippocampal astrocytes promotes the survival and maturation of abGCs and was associated with an anxiolytic/antidepressant-like effect in mice [78]. The process of neurogenesis is also supported by FGF-2, which is mainly produced by astrocytes. In addition, astrocyte-derived FGF-2 facilitates abGC differentiation and function [73].

3.1.3.3 Lactate and VEGF

Astrocytes store large amounts of glycogen as a source of energy for cells [79]. Astrocytes break down glycogen into lactate that is released from cells by a monocarboxylate transporter 4 (MCT4). The neurotransmitter MCT2 takes lactate up as an energy source for neurons. This pathway is called the astrocyte-neuron lactate shuttle [80, 81].

Increasing lactate concentrations in the blood can raise brain VEGF levels by binding to the hydroxycarboxylic acid 1 receptor, which presents on vascular endothelial cells [82]. Then the VEGF stimulates angiogenesis and neurogenesis. Intraperitoneal administration of lactate to mice increases abGC viability via MCT2 [83], however, further research is needed to understand how increased lactate mediates this action on neurogenesis.

3.2 Immune mechanisms

3.2.1 Innate immunity

Individuals affected by neuropsychiatric conditions usually present with hyperinflammation-associated dysfunctions in the peripheral blood, such as raised concentrations of inflammatory cytokines or chemokines, augmented quantities of circulating monocytes and neutrophils, along with a higher reactivity of astrocytes, microglia, as well as brain endothelial cells to different proinflammatory signals [84].

Due to the prominent role of inflammation in neuropsychiatric conditions, it is important to understand the role of inflammasomes. These essential multimeric complexes regulate inflammation by mediating the secretion of cytokines. Specifically, inflammasome pathways can be categorized into canonical and noncanonical signaling, which depend on caspase-1 and caspase-4/5 activation in humans, respectively [85]. Caspase-1 activation helps the maturation of IL-1β and IL-18, two vital inflammatory cytokines. Two neuroimaging studies also showed a link between carrier status for a functional single nucleotide polymorphism (SNP) in the IL-1β gene and aberrant white and gray matter proportions in SCZ patients, unlike healthy patients individuals [15]. Researchers also found that orbitofrontal white matter neuronal compactness was enhanced in SCZ cases with high transcription concentrations of proinflammatory cytokines compared to those with low concentrations [15]. In another study, the animals presented SCZ-like neurobehavior at 45 postnatal days linked with the increase of NLRP3 inflammasome expression and IL-1β levels on 7, 14, and 45 postnatal days [86]. This study shows maternal immune activation (MIA) may be associated with an SCZ-like neurobehavior. This neurobehavior can be induced to a neuroinflammatory profile in the brain. This evidence may base future studies on the relationship between neuroinflammation and psychiatric disorders [87].

Extracellular adenosine triphosphate (ATP) and the metabolite adenosine are key mediators of the immune response. Excess shedding of ATP into the intercellular area from induced/injured nerve cells or abnormal purinergic signaling induces the nod-like receptor proteins (NLRPs) of inflammasomes in astrocytes/microglia [84]. A new review article details the involvement of aberrant purinergic signaling in the pathogenesis of SCZ, in addition to other neuropsychiatric disorders such as major depressive disorder, bipolar disorder, autism, anxiety, and attention-deficit/hyperactivity disorders (ADHDs) [88]. Each of these conditions could be attributed to abnormalities in signaling from P1 and P2 receptors along with enzymes processing the metabolism of purinergic mediators.

In addition, SCZ patients are often treated with antipsychotic medications that cause brain volume reduction and astrocyte expiry in a process called pyroptosis or proinflammatory cellular expiry [89]. This process involves the formation of inflammatory bodies and enhanced production of complexes such as NLRP3, parallel with the induction of caspases and gasdermin D (GSDMD). These components are strongly linked to innate immunity, hyperinflammation, and cell damage/expiry.

The same study found the main effect of antipsychotic treatments on astrocyte pyroptotic pathways and the molecular processes that could be exerted through inflammasome pathways [89]. In this experiment, 72-h therapy with olanzapine, quetiapine, risperidone, or haloperidol strongly attenuated the astrocytes' viability. 24-h therapy by olanzapine, quetiapine, risperidone, or haloperidol dose dependently augmented the protein synthesis of astrocytic NLRP3, NLRP6, caspase-1, caspase-4, and GSDMD. Co-administration with a histamine H1 receptor agonist, 2-(3-trifluoromethylphenyl) histamine (FMPH), attenuated the raised synthesis of NLRP3, caspase-1, and GSDMD activated via olanzapine, quetiapine, risperidone, or haloperidol [89]. Moreover, olanzapine, quetiapine, risperidone, or haloperidol treatment-induced pore formation in the astrocyte membranes was suppressed via FMPH co-treatment [89].

When taken together, astrocytic inflammasomes and hyperinflammation are implicated in SCZ, and activation of astrocyte pyroptotic pathways could be due to antipsychotic-activated astrocyte expiry. Further, H1 receptor activation could be a robust therapeutic approach to inhibit antipsychotic-activated astrocyte pyroptosis and hyperinflammation [90]. However, more studies should assess astrocyte-specific (or other CNS cell-specific) inflammasomes and consider the noncanonical pathways to understand more fully how these innate complexes contribute to hyperinflammatory cytokine secretion and neuronal/glial damage. Also, various toll-like receptors (TLKs) are involved in SCZ. These receptors are essential as they bridge innate and adaptive immunity and interact with inflammasomes. More studies are needed to decipher the role of astrocyte-specific TLRs in SCZ patients.

3.2.2 Adaptive immunity

Adaptive immunity, which arises following activity of the innate immune system, occupies a pivotal function in neurodevelopment [91]. A key component of the adaptive immune response is specialized immune cells known as T lymphocytes (T cells). The ability of T cells to penetrate the brain, stimulate microglia, and cause neuroinflammation is widely established, and these activities have been shown to disrupt several brain functions and cause progressive neuro alterations [92, 93]. While a role of inflammatory states has been noted in SCZ [94, 95], the impact of the adaptive immune system, particularly T lymphocyte cells, on the essential characteristics and severity of SCZ is unknown; however, epidemiological, immunological, and gene expression research do suggest a degree of dysfunction of T cells-related processes in SCZ [95–98]. Higher concentrations of T cells in the hippocampus and an elevated number of activated lymphocytes in the CSF have been noted in SCZ patients suggesting blood-brain barrier disruption and T-cell infiltration [99]. Multiple genome-wide association studies (GWAS) have noted genetic variations in SCZ patients of CD28 and CTLA-4 genes, which code for regulatory molecules of the adaptive immune system suggesting that these proteins modulate T-cell activity and are associated with T-cell functions, including antigen processing and cell adhesion, this provides further evidence that the adaptive immune system plays a role in SCZ [100, 101].

SCZ is linked to modestly elevated blood cytokines, which are thought to be the result/stimulus for the activation of microglia [102]. Overlap between activated microglia, proinflammatory cytokines, and translocator protein (TSPO) leads to the basis of using TSPO-PET imaging to monitor neurodegeneration. In SCZ patients, findings of lower levels of TSPO in frontal and subcortical regions, including caudate, putamen, and thalamus, lead to the speculation that suppression of microglial inflammation results in reduced TSPO binding [103]. However, discrepancies that exist in this literature and utility of this technique in SCZ have been debated given differences in degeneration seen in SCZ when compared to other neurodegenerative diseases [104, 105]. The discrepancy between experiments that show an increase, decrease, or no change in TSPO in the subcortical regions of schizophrenic patients has prompted several follow-up investigations employing first and second-generation tracers with mixed results [106, 107].

Postmortem immunohistochemistry analysis, genetic association studies, and transcriptome investigations indicate increased astrocyte activity in SCZ [29].

The expression of marker gene profiles of various cortical cell types was investigated in a prior study, which gave compelling evidence of an increase in astroglial gene expression in SCZ [108].

Further studies strengthened these results and suggested that increased astroglial gene expression followed by a decrease in microglial gene expression is a primary cause of disease rather than a side effect [109, 110].

Astrocytes play a significant role in microglial excitation and function via TGF-β. TGF-β regulates microglial previous studies discovered that TGF-β regulates microglial activation and activity, and astrocytes are crucial actors in this process [111–113]; A critical process that actively suppresses the inflammatory TSPO-expressing phenotype of microglia via elimination of the TGF-β receptor type 2 from adult microglia could be pertinent to the SCZ patient debate [111].

3.3 Other mechanisms

3.3.1 Gap junction

Connexins make gap junction channels that facilitate the transmission of intercellular calcium currents. Most brain gap junctions are situated among glial cells and gap junctions between astrocytes and astrocytic processes are known as reflexive gap junctions and together an astroglial system is created [114]. Gap junctions allow two-sided interchange of molecules, ions, nutrients, etc. The gating of brain gap junction channels is modulated dynamically by alterations in the number of cell connections, conductance properties, and subunit configurations. Astrocyte signaling happens primarily via intercellular calcium currents in response to neuronal activity and/or through flux in the endoplasmic reticulum.

Gap junctions comprise hemichannels (connexons) that attach through their extracytoplasmic processes. Each hemichannel is an oligomer of six connexin proteins (Cx). An astrocytic syncytium gap junction is comprised of four connexins, Cx32, Cx26, Cx43, and Cx45, which can create homotypic (*i.e.,* gap junction channels created by hemichannels of the identical type) or heterotypic gap junction channels (*i.e.,* created by hemichannels of various types) [115].

Loss of gap junction function in astrocytes has been hypothesized to play a role in SCZ [116]. Use of computational modeling of astrocyte gap junction activity also supported the conclusion that a loss of astrocytic gap junctions would alter activity in the neural network which could play a role in neuronal-glial changes seen in SCZ [117]. Finally, reduced gap junctions between astrocytes were found to concentrate signaling among the most connected astrocytes, which would be expected to impact communication at the tripartite synapse [118] that could lead to cognitive deficits.

3.3.2 Extracellular matrix system (ECM)

Neurons and glial synthesize the ECM, which impacts the maturing and maintenance of synapses. Moreover, at the hyaluronan level, the ECM divides the exterior of neurons, restricting the exterior movement of integrated membrane proteins. In the pathogenesis of SCZ, failures in ECM development have been suggested. The ECM is comprised in part of chondroitin sulfates. Patients diagnosed with SCZ demonstrated a massive rise in chondroitin sulfate proteoglycan (CSPG) - positive glia in the entorhinal cortex and deep amygdala, whereas the density of GFAP – positive cells was not altered. As CSPGs mediate adult synaptic plasticity, higher levels in SCZ could play a role in alterations in this process. Further, modification of the ECM could result in differences in distribution of receptors and transmitters. For example, decreased expression of Reelin, a component of the ECM primarily expressed in adult GABA neurons, was observed in patients with SCZ in caudate nucleus, cerebellum, hippocampus, and prefrontal and temporal cortices. A decrease in Reelin is typically associated with a reduction in glutamic acid decarboxylase expression, suggesting a robust functional correlation between Reelin expression and GABAergic neurotransmission. Consequently, the variations in ECM could be expected to have an impact on release of inhibitory neurotransmitter [119].

3.3.3 Epigenetics

A connection between SCZ risk and epigenetic pathways has been suggested based on association studies. In a study of the promotor hypermethylation status of the glutathione S-transferase TI (GSTT1) and glutathione S-transferase P1 (GSTP1) genes in a Tunisian SCZ population, a significant relationship between SCZ and the GSTT1 and GSTP1 active genotype was noted [120]. A similar relationship was noted in another study [121].

The single nucleotide polymorphisms rs1043618 and rs2075799 in *HSPA1A* (heat shock protein family A "HSP70" member 1A) have been linked to SCZ [122, 123], which is interesting as HSP70 facilitates astrocytic neuroinflammation [124]. No epigenetic processes have been defined that might modulate the expression of HSP70 in SCZ.

Valproic acid (VPA) and other HDAC inhibitors are molecules that facilitate chromatin remodeling to modify gene transcription selectively, such as MS-275, trichostatin A, sodium butyrate, and apicidin appear to enhance H3K4Me3 and H3K4Me2 levels at the *HSP70* astrocytic promoter. H3K4me3 and H3K4me2 are related to transcriptionally active chromatin areas. Curiously, VPA provoked stimulation of the astrocytic *HSP70* promoter through employing histone acetyltransferase p300 in astrocytes of the rat cortex [125].

Moreover, there are data for enhanced *HSPA1A* expression and additional genes associated with immune function, such as the proinflammatory mediators *IFITM2*, *IFITM1*, and *IFITM3* in postmortem dorsolateral prefrontal cortex (dlPFC) sections from SCZ [120]. A considerable link was noted between first-episode psychosis in Greek schizophrenic patients and *the HSPA8* variant (rs1136141) [126].

Regulator of G-protein–signaling 4 (RGS4) is a GTPase-triggering protein that regulates G-protein-coupled receptor signaling, regulating receptor-facilitated synaptic neural signaling [127]. A variant of the *RGS4* polymorphism (rs951436) leads to declines in the structural volume of the white matter [128]. Downregulation of *RGS4* transcripts in the dlPFC of SCZ patients has been reported that suggest that this gene could be a candidate gene for SCZ [129].

Type of OMICS	Findings	Reference
Genomics	Quantitative-PCR determination of Human Induced Pluripotent Stem Cells (hiPSC)-obtained Neural progenitor cells (NPCs), neurons, NGN2 neurons, and astrocytes revealed a baseline expression of all five SCZ risk genes, TOAK2, NRXN1, SNAP91, KCTD13, and CLCN3.	[133]
Proteomics	In comparison with the controls, astrocytes acutely treated with 20 μM MK-801 demonstrated 11 differentially expressed proteins and those acutely treated with 50 μM demonstrated 30. The expressed proteins most often regulate cell development and/or protection (28%) or energy pathways (24%). Additionally, 10 of these proteins (33.3%) were also observed differentially expressed in specific areas of SCZ human brain samples.	[134]
Proteomics	A whole of 124 PFC proteins were observed to be substantially differentially expressed among the isolated-rearing (IR) group and the social-rearing (SR) group, the most notable of them were Annexin A2 (ANXA2), glial fibrillary acidic protein (GFAP), and vimentin (VIM), three astrocyte biomarkers. Additional Western blot tests proved that the GFAP, ANXA2 and VIM levels were improved considerably in IR rats. Adolescent social isolation promoted SCZ-like behaviors and considerably different expression of 124 PFC proteins in mature rats, particularly ANXA2, GFAP, and VIM, that proposes that astrocyte development could be implicated in the neuronal process of SCZ.	[135]
Transcriptomics	Four differentially expressed miRNAs (miR-206, miR-127-5p, miR-337-3p, miR-1185-1-3p) were found in SCZ astrocytes that demonstrated significantly lesser basic expression relative to controls.	[136]
Transcriptomics	There was seen an ongoing detection of an enhancement in the expression of cortical astrocytes. No alterations in astrocyte expression were detected in subcortical areas.	[108]

Table 1.
An overview of studies evaluating OMICS approaches for astrocytic abnormalities in SCZ.

A different study demonstrated that *RSG4* expression of the lengthiest variant, *RGS4*-1, was reduced in schizophrenic patients' dlPFC [130]. The methylation of *RGS4* regulatory region CpG islands in the *postmortem* dlPFC sections was evaluated in SCZ and results showed that the reduced *RGS4*-1 mRNA expression levels were not related to hypermethylation of the 5′ region CpG islands. Moreover, the more deficient *RGS4*-1 expression was associated with the 5′ regulatory regions SNPs rs2661347, rs951436, rs10917670, and rs2661319 [131].

Research by Vrajová *et al*. assessed the potential epigenetic process of *RGS4* expression via silenced *RGS4* gene utilizing siRNA targeted toward human *RGS4* and analyzed the impact of neuroblastoma cell lines differential expression. They noted that downregulated *RGS4* mRNA alters the expression of 67 genes comprising crucial transcription factors, such as brain-derived neurotrophic factor (BDNF) and DISC1, related to the pathology of SCZ [132]. Additionally, several OMICS approaches have evaluated the interaction among astrocytes and SCZ (**Table 1**).

4. Behavioral outcomes

Development of preclinical models of astrocyte dysfunction in SCZ is quickly expanding. We will summarize the existing animal models, which have relied on altering key functions of astrocytes that have been discovered to be aberrant in SCZ. We will focus on genetic models. However, nongenetic preparations will be discussed in the case of modifying the astrocytes to trigger SCZ-like behavioral phenotypes. For further details, the readers are referred to the comprehensive review paper by Xia *et al.* [137].

4.1 Models with structural change

To reduce glia in cortical areas in order to model changes seen in SCZ patients, an astrocyte-exclusive toxin, L-alpha-aminoadipic acid (L-AAA), was administered in the prefrontal cortex (PFC) of adult rats. L-AAA triggered anhedonia in sucrose preference evaluation, anxiety, and helplessness in forced swim test (FST). Of note, these effects were not seen following ibotenate-triggered neurotoxic lesion of the PFC, highlighting the specificity of astrocyte deterioration in inducing the affective neurobehaviors. The toxin also influenced attentional set-shifting, working memory, and reversal learning. The influences of L-AAA seem to support the role of astrocytes in behavioral disorders due to dysfunction of the medial PFC. The limitation of utilizing this toxin for neurobehavioral exploration is progressive neuronal death and dendritic degeneration in the surviving nerve cells in L-AAA-treated animals [138, 139].

4.2 Models related to glutamate signaling

Glutamate uptake is one of the major functions of astrocytes [140]. GLT-1 actions are heightened in the PFC of SCZ [141]. To model this scenario, the antibiotic ceftriaxone that specifically augments GLT-1 production and activity was administered to rodents. In this model, evaluation of the inhibition of prepulse inhibition (PPI) of the acoustic startle reflex, which is a biomarker of SCZ was conducted. Ceftriaxone-triggered GLT-1 up-regulation was linked to attenuated PPI that was reversed via dihydrokainate (DHK), a GLT-1 blocker [142]. Intriguingly, ceftriaxone's PPI suppressive influences were further augmented by administering only one dosage of phencyclidine (PCP) [143]. However, the NMDAR antagonist, MK-801, exhibited conflicting results regarding ameliorating sensorimotor, cognitive, and memory performance in environmental enrichment (EE) models [144–146].

Other models relate to D-serine, purine, trophic factors, and extracellular matrix proteins [137].

4.3 Models with hyperinflammation

Intraperitoneal injection in rodents of kynurenine (100 mg/kg) resulting in a 37-fold CNS enhancement likely leading to immune system activation was associated with a higher rate of error in the radial arm maze, although no alteration was found in locomotor function or the latency to obtain food reward [147]. When kynurenine was applied on postnatal days 7–10 and rats were evaluated as adults' social neurobehavior and locomotor activity was attenuated; however, no effects on attention or fear conditioning were detected [148].

While the dosages of kynurenine utilized in these studies were high, the outcomes suggest that neuroinflammation can affect cognitive function, which could involve astrocytes [137], and that early life activation of the processes of neuroinflammation can have long-lived consequences. Further evidence of this is provided by data showing that infection by *Toxoplasma gondii* (*T. gondii*) is a predisposing factor for

SCZ [149]. *T. gondii* induces the synthesis of KYNA, potentially in astrocytes [150]. *T. gondii* involvement may play a role in the disease by enhancing KYNA synthesis within astrocytes. While *T. gondii*-infected animals have exhibited SCZ-like neurobehaviors in multiple studies, the contribution of KYNA or other astrocyte-related elements needs further study because the parasite exerts a plethora of direct and indirect influences on the brain [151].

Short-term one-week exposure of mice to cuprizone resulted in flawed working memory and augmented responses to methamphetamine and phencyclidine. These cognitive behaviors were associated with perturbation of astrocytes/microglia and enhancement of IL-6 in GFAP+ cells as well as an increase in other proinflammatory biomarkers, which have been observed in SCZ [152].

The production of many proinflammatory elements is mediated through the transcription of nuclear factor-kappa B (NF-κB) [153]. Several genetic, biomarker, and postmortem data indicate the role of NF-κB in SCZ. In transgenic rodents in which NF-κB function was specifically blocked in astrocytes while there were no differences in overall health outcomes, locomotor function, sensorimotor actions, or anxiety, female GFAP-IκBα-dn rodents exhibited mild deficiencies in the terminal stage of the non-cued type of the Barnes maze, as corroborated via the increased latency to the first correct nose poke and reduced time length in the portion of the maze already enclosing the goal box [154].

5. Therapeutic and diagnostic approaches for SCZ: Focus on astrocytes

5.1 Therapeutic targets for SCZ

Astrocytes are involved in numerous critical physiological processes in the brain, which directly or indirectly contribute to the pathogenesis of SCZ, including receptor trafficking, development and maturation of synapses, synaptic glutamate metabolism, regulation of CNS homeostasis, maintenance of integrity of the blood-brain barrier (BBB), nutrient provision to neural tissues, and regulation of neurogenesis. Based on data showing astrocyte involvement in the pathology of SCZ, astrocytes should be considered as therapeutic targets for treating this disease. In SCZ, abnormalities in neurons and neurotransmitters mainly result from the malfunction of astrocytes. Hence, correct functioning of astrocytes is required for the processes of synaptic activity and synaptic plasticity within neural networks, which is activity necessary for normal cognitive functions [155]. Therefore, targeting pathways associated with astrocytes' abnormal function may help ameliorate SCZ complications.

As hypofunctionality of NMDA receptors is believed to be one of the leading causes of SCZ, modifying NMDA receptors' function can be considered a therapeutic strategy for SCZ treatment. Targeting astrocytic glutamate reuptake presents a viable strategy for increasing glutamate sufficient to restore NMDA functionality [156]. Clozapine decreases glutamate reuptake through downregulation of glutamate transporter (GLT1) in astrocytes resulting in ameliorating hypofunctionality of NMDA receptors [157].

In addition to glutamate, several small molecules that can affect NMDA receptor functioning are linked to astrocyte activity. In preliminary trials, the NMDA receptor co-agonists glycine and D-serine have been applied as well as D-cycloserine [158]. An encouraging effect of high-dose D-serine administration was seen in SCZ patients [159]. Unfortunately, while small studies showed promising results, more extensive trials indicated no significant difference between intervention and placebo groups when D-serine was exogenously applied [160].

An alternative approach to enhance D-serine levels in the synaptic space is to inhibit its metabolism in astrocytes. D-serine is metabolized by D-amino acid oxidase (DAAO) in astrocytes; thus, DAAO inhibitors may enhance NMDA activity by increasing endogenous D-serine concentrations. However, none of the identified human DAAO inhibitors have been approved for use in SCZ patients. Low bioavailability, high clearance rate, and inability to cross the BBB are considered the primary restrictions of these inhibitors [161].

Strategies focused on suppressing glycine reuptake in order to increase glycine's extracellular concentration in the synaptic space where it might be able to enhance NMDA functionality have been implemented, and inhibiting glycine transporter 1 (GlyT1) in astrocytes has been one approach considered for management of some SCZ symptoms. Among various GlyT1 inhibitors that have been trialed, only bitopertin has reached phase III clinical trials [155]. However, lack of efficacy has led to the discontinuation of its development as an antipsychotic [162].

In addition to glycine and D-serine, Kynurenic acid (KYNA), a metabolite of tryptophan degradation in astrocytes, can influence the function of NMDA receptors. KYNA has a preferential affinity for the NMDA receptor and can inhibit NMDA activity. According to the direct relationship between KYNA concentration and cognitive impairments in SCZ, interventions that lower brain KYNA levels may be clinically beneficial. Regrettably, at the current time, it is not feasible to target degradative enzymes or reuptake sites to enhance the removal of excess KYNA from its effector site in the brain. Moreover, exploiting the ability of depolarization events or cellular energy scarcity to reduce cerebral KYNA production is not possible. Pharmacological kynurenine aminotransferase (KAT) inhibitors are the most effective strategy to reduce KYNA production in the brain. The practicality of this approach is supported by findings that the nonspecific aminotransferase inhibitor aminooxyacetic acid readily prevents cerebral KYNA neosynthesis *in vivo*. In this regard, KAT II is the preferential target to suppress KYNA synthesis in the brain due to its high specificity toward kynurenine [158]. According to previous preclinical studies, the administration of selective KAT II inhibitors could successfully reduce extracellular KYNA levels in various rat brain regions. When taken together, considering KYNA's inhibitory effects on several neurotransmitters with a critical role in cognitive processes, any therapeutic agent or intervention that decreases KYNA levels or otherwise hinders KYNA function in the brain may lead to cognitive enhancement in SCZ or other psychiatric disorders, and the data suggest that KAT II inhibitors or pharmacological agents that weaken the function of KYNA at its receptor(s) have a high potential to be used for cognitive deficits in SCZ [158].

Astrocytes can also be a target to repair synaptic functions by moderating their effects on glycogen/lactate metabolism, as glucose uptake into astrocytes is reduced in SCZ due to a decrease in glycolysis and decreased lactate production. The decrease in lactate could have a profound effect on reductions in neurogenesis. Therefore, modifying glycogen/lactate metabolism sufficient to compensate for reduction of lactate could facilitate lactate-mediated neurogenesis, and lead to improvement of behavioral deficits in SCZ patients [163].

Reducing inflammation resulting from astrocytes can also be considered a therapeutic approach for SCZ and this approach has been shown to improve SCZ symptoms. For instance, minocycline, an antibiotic with anti-inflammatory effects, induced improvements in some SCZ patients. COX2 inhibitors, which are non-steroid anti-inflammatory drugs, have been shown to improve SCZ symptoms.

Lending support to the effectiveness of this strategy, many antipsychotic drugs exhibit anti-inflammatory effects, which could be important in their therapeutic efficacy. Given the link between inflammation and SCZ, a clear understanding of the cytokines involved in SCZ and the role played by astrocytes in linking inflammation and SCZ could lead to therapeutic strategies [156].

5.2 Diagnostic approach for SCZ

Postmortem studies identified significant changes in astrocyte density and morphology, as well as deregulated expression of several common astrocyte markers, including glial fibrillary acidic proteins (GFAP), aquaporin 4 (AQ-4), S100, glutaminase, thrombospondin (TSB-1), and excitatory amino acid transporter 2 (EAAT2) [22, 164–166]. When taken together, while data are suggestive of a role of altered astrocytic function in SCZ, the findings do differ, with some studies indicating a drop in marker levels and the number of astroglial cells compared to controls and others a rise. Although dysregulation in developing astroglial cells may have profound effects on the formation and maturation of neuronal networks, few studies have examined the status of astroglial cells during postnatal brain development, instead focusing on the postmortem examination of adult brain tissues [36]. Due to confounding factors associated with the use of postmortem tissues, differences in the brain regions evaluated, variety in the severity of the disease, and disparities in pharmacological treatments, it remains to be determined what the contribution of these markers to the disease is and if they play a role, at which developmental stage their role is most important. In light of the profound changes in astrocytic morphology and function, monitoring of alterations in astrocytes has been considered a diagnostic approach in SCZ. However, it is difficult to know what can easily be monitored from tissue non-invasively extracted in patients, which reflects astrocytic status. At the present time, identification of peripheral biomarkers that reflect neuropathological changes in SCZ has received a great deal of interest and in this arena, exosomes have been a focus of study as they are relatively easy to detect and have been proposed to be involved in psychiatric disorders [167]. Intriguingly, it is possible to identify the parent cell from which exosomes source.

Exosomes are nano-sized extracellular vesicles containing nucleic acids, proteins, lipids, and other bioactive substances secreted by cells into the surrounding body fluids, which regulate cellular communications in addition to neuroplasticity [168], trafficking of microRNA (miRNA) [169], and neuroinflammation [170, 171]. They can cross the BBB and be assayed peripherally, Exosomes derived from astrocytes would be expected to exhibit changes across the progression of SCZ. As proof of concept that exosomes can be detected and traced back to their parent cell, a high concentration of exosomal GFAP, resulting from astrogliosis was detected in plasma obtained from SCZ patients [172]. Thus, astrocytes-derived exosomes have the potential to be used for SCZ diagnosis and assessment of disease progression. However, further studies are needed to clarify to what extent circulating exosomes can serve as novel peripheral biomarkers of SCZ.

6. Conclusions

Astrocytic changes have been linked to SCZ at the neurobehavioral, structural, functional, and molecular levels. ECM, gap junctions, and epigenetics are also

Highlights
1. SCZ is a debilitating disorder with an estimated prevalence of 0.6% to 1.9% in the US population.
2. Astrocytic abnormalities end in cognitive disturbances such as memory, learning, and attention, and also abnormal cortical gamma oscillations in SCZ.
3. Various neurotransmitters such as glutamate, glycine, dopamine, adenosine, GABA, and the endocannabinoid system are implicated in astrocytic abnormalities in SCZ
4. Astrocyte-meditated remyelination is impaired in SCZ.
5. Neurogenesis-regulating molecules including D-serine, BDNF, FGF2, Lactate, and VEGF apparently fail in modulating neurogenesis via astrocytes in SCZ.
6. Innate (TLRs and Inflammasomes) and adaptive (T lymphocytes) immune responses exacerbate astroglial mediated abnormalities in SCZ.
7. The synaptic microenvironment (ECM, and Gap junctions) is highly altered in astrocyte-neuron communications in SCZ.
8. Epigenetic studies highlight a derailed cascade of regulatory molecular pathways.
9. Animal models of SCZ also demonstrate astrocytic abnormalities.
10. Astrocytes show promise as therapeutic and diagnostic targets in SCZ.

Table 2.
A summary of important points from this chapter.

implicated in the astrocytic abnormalities associated with SCZ. Various neurotransmitter systems that are regulated by astrocytes including GABA, glutamate, and adenosine are involved in SCZ. Also, different types of neuroplasticity governed by astrocytes are altered in SCZ. Moreover, hyperinflammation that is in part regulated by astrocytic inflammasomes (e.g., NLRPs) is present in SCZ patients and is affected by pharmacotherapy with antipsychotics. Clinical behavioral deficits in animal models are also related to aberrancies in astrocytes. When taken together, the plethora of studies that indicate a link between astrocytic dysfunction and SCZ should warrant future research to explore the role played by astroglial cells in SCZ to bridge the clinical and molecular findings and pave the path for developing future therapeutics that correct, or exploit, astrocyte functions in SCZ (**Table 2**).

Acknowledgements

We thank the USERN MUBabol Office and Kerman Neuroscience Research center for helping with the preparation of this manuscript. No funding was received for the preparation of this manuscript.

Conflict of interest

The authors declare no conflict of interest.

Author details

Kiarash Saleki[1,2†], Mohammad Banazadeh[3†], Banafshe Abadi[4,5], Zeynab Pirmoradi[6], Zahra Esmaili[6], Shiva Amiri[7,8], Ramtin Pourahmad[9], Kristi A. Kohlmeier[10] and Mohammad Shabani[6*]

1 Student Research Committee, Babol University of Medical Sciences, Babol, Iran

2 USERN Office, Babol University of Medical Sciences, Babol, Iran

3 Pharmaceutical Sciences and Cosmetic Products Research Center, Kerman University of Medical Sciences, Kerman, Iran

4 Herbal and Traditional Medicines Research Center, Kerman University of Medical Sciences, Kerman, Iran

5 Brain Cancer Research Core (BCRC), Universal Scientific Education and Research Network (USERN), Kerman, Iran

6 Kerman Neuroscience Research Center, Neuropharmacology Institute, Kerman University of Medical Sciences, Kerman, Iran

7 Pharmaceutics Research Center, Institute of Neuropharmacology, Kerman University of Medical Sciences, Kerman, Iran

8 Faculty of Pharmacy, Department of Toxicology and Pharmacology, Kerman University of Medical Sciences, Kerman, Iran

9 School of medicine, Tehran University of Medical Sciences, Tehran, Iran

10 Faculty of Health Sciences, Department of Drug Design and Pharmacology, University of Copenhagen, Copenhagen, Denmark

*Address all correspondence to: shabanimoh@yahoo.com and shabani@kmu.ac.ir

† Co-first authorship and contributed equally to the present work

References

[1] Tayeb HO et al. Pharmacotherapy of schizophrenia: Toward a metabolomic-based approach. CNS Spectrums. 2019;**24**(3):281-286

[2] Genoud C et al. Plasticity of astrocytic coverage and glutamate transporter expression in adult mouse cortex. PLoS Biology. 2006;**4**(11):e343

[3] Ventura R, Harris K. Three-dimensional relationships between hippocampal synapses and astrocytes. Journal of Neuroscience. 1999;**19**(16):6897-6906

[4] Witcher MR, Kirov SA, Harris KMJG. Plasticity of perisynaptic astroglia during synaptogenesis in the mature rat hippocampus. Glia. 2007;**55**(1):13-23

[5] Danbolt N. Glutamate Uptake. Progress in Neurobiology. 2001;**65**(1): 1-105

[6] Araque A et al. Gliotransmitters travel in time and space. Neuron. 2014;**81**(4):728-739

[7] Harada K, Kamiya T, Tsuboi T. Gliotransmitter release from astrocytes: Functional, developmental, and pathological implications in the brain. Frontiers in Neuroscience. 2016;**9**:499

[8] Rivera A, Vanzulli I, Butt A. A central role for ATP signalling in glial interactions in the CNS. Current Drug Targets. 2016;**17**(16):1829-1833

[9] Perea G, Araque AJS. Astrocytes potentiate transmitter release at single hippocampal synapses. Science. 2007;**317**(5841):1083-1086

[10] Perez-Alvarez A et al. Structural and functional plasticity of astrocyte processes and dendritic spine interactions. Journal of Neuroscience. 2014;**34**(38):12738-12744

[11] Dityatev A et al. Extracellular matrix and synaptic functions. Cell Communication in Nervous and Immune System. 2006;**43**:69-97

[12] Song I, Dityatev A. Crosstalk between glia, extracellular matrix and neurons. Brain Research Bulletin. 2018;**136**:101-108

[13] Chelini G et al. The tetrapartite synapse: A key concept in the pathophysiology of schizophrenia. European Psychiatry. 2018;**50**:60-69

[14] Koskuvi M et al. Patient iPSC-astrocytes show transcriptional and functional dysregulation in schizophrenia. bioRxiv. 2020

[15] Najjar S, Pearlman DM. Neuroinflammation and white matter pathology in schizophrenia: Systematic review. Schizophrenia Research. 2015;**161**(1):102-112

[16] Dieset I et al. Association between altered brain morphology and elevated peripheral endothelial markers — Implications for psychotic disorders. Schizophrenia Research. 2015;**161**(2):222-228

[17] Takahashi N, Sakurai T. Roles of glial cells in schizophrenia: Possible targets for therapeutic approaches. Neurobiology of Disease. 2013;**53**:49-60

[18] Cohen SM et al. The impact of NMDA receptor hypofunction on GABAergic neurons in the pathophysiology of schizophrenia. Schizophrenia Research. 2015;**167**(1-3):98-107

[19] Plitman E et al. Kynurenic acid in schizophrenia: A systematic review and meta-analysis. Schizophrenia Bulletin. 2017;**43**(4):764-777

[20] Khandaker GM et al. Inflammation and immunity in schizophrenia: Implications for pathophysiology and treatment. The Lancet Psychiatry. 2015;**2**(3):258-270

[21] Miyaoka T et al. Remission of psychosis in treatment-resistant schizophrenia following bone marrow transplantation: A case report. Frontiers in Psychiatry. 2017;**8**:174

[22] Kim R et al. Astroglial correlates of neuropsychiatric disease: From astrocytopathy to astrogliosis. Progress in Neuro-Psychopharmacology and Biological Psychiatry. 2018;**87**:126-146

[23] Davis J et al. A review of vulnerability and risks for schizophrenia: Beyond the two hit hypothesis. Neuroscience and Biobehavioral Reviews. 2016;**65**:185-194

[24] Messias EL, Chen CY, Eaton WW. Epidemiology of schizophrenia: Review of findings and myths. The Psychiatric Clinics of North America. 2007;**30**(3):323-338

[25] Banazadeh M et al. Evaluating the effect of black myrobalan on cognitive, positive, and negative symptoms in patients with chronic schizophrenia: A randomized, double-blind, placebo-controlled trial. Phytotherapy Research. 2022;**36**(1):543-550

[26] Eroglu C, Barres BA. Regulation of synaptic connectivity by glia. Nature. 2010;**468**(7321):223-231

[27] Hu W et al. The glutamate hypothesis of schizophrenia: Evidence from human brain tissue studies. Annals of the New York Academy of Sciences. 2015;**1338**(1):38-57

[28] Verkhratsky A, Nedergaard M. Physiology of astroglia. Physiological Reviews. 2018;**98**(1):239-389

[29] Chung WS, Allen NJ, Eroglu C. Astrocytes control synapse formation, function, and elimination. Cold Spring Harbor Perspectives in Biology. 2015;**7**(9):a020370

[30] Lima A et al. Astrocyte pathology in the prefrontal cortex impairs the cognitive function of rats. Molecular Psychiatry. 2014;**19**(7):834-841

[31] Lee HS et al. Astrocytes contribute to gamma oscillations and recognition memory. Proceedings of the National Academy of Sciences. 2014;**111**(32):E3343-E3352

[32] Bellesi M et al. GLT-1 upregulation impairs prepulse inhibition of the startle reflex in adult rats. Glia. 2009;**57**(7):703-713

[33] van Os J, Kapur S. Schizophrenia. Lancet. 2009;**374**(9690):635-645

[34] Wu EQ et al. Annual prevalence of diagnosed schizophrenia in the USA: A claims data analysis approach. Psychological Medicine. 2006;**36**(11):1535-1540

[35] Dipiro JT et al. Pharmacotherapy: A Pathophysiologic Approach. Vol. 4. Connecticut: Appleton and Lange; 2014. pp. 141-142

[36] American Psychiatric Association, D. and A.P. Association. Diagnostic and Statistical Manual of Mental Disorders: DSM-5. Vol. 5. Washington, DC: American Psychiatric Association; 2013

[37] Sartorius N et al. Early manifestations and first-contact

incidence of schizophrenia in different cultures. A preliminary report on the initial evaluation phase of the WHO collaborative study on determinants of outcome of severe mental disorders. Psychological Medicine. 1986;**16**(4):909-928

[38] Blanco-Suárez E, Caldwell ALM, Allen NJ. Role of astrocyte-synapse interactions in CNS disorders. The Journal of Physiology. 2017;**595**(6):1903-1916

[39] Malenka RC, Bear MF. LTP and LTD: An embarrassment of riches. Neuron. 2004;**44**(1):5-21

[40] Moghaddam B, Javitt D. From revolution to evolution: The glutamate hypothesis of schizophrenia and its implication for treatment. Neuropsychopharmacology. 2012;**37**(1):4-15

[41] Jia M et al. Taming glutamate excitotoxicity: Strategic pathway modulation for neuroprotection. CNS Drugs. 2015;**29**(2):153-162

[42] Pajarillo E et al. The role of astrocytic glutamate transporters GLT-1 and GLAST in neurological disorders: Potential targets for neurotherapeutics. Neuropharmacology. 2019;**161**:107559-107559

[43] Andersen JV et al. Glutamate metabolism and recycling at the excitatory synapse in health and neurodegeneration. Neuropharmacology. 2021;**196**:108719

[44] Zádor F et al. Kynurenines and the endocannabinoid system in schizophrenia: Common points and potential interactions. Molecules (Basel, Switzerland). 2019;**24**(20):3709

[45] Pei J-C et al. Directly and indirectly targeting the glycine modulatory site to modulate NMDA receptor function to address unmet medical needs of patients with schizophrenia. Frontiers in Psychiatry. 2021;**12**:742058-742058

[46] de Bartolomeis A et al. Glycine signaling in the framework of dopamine-glutamate interaction and postsynaptic density. Implications for treatment-resistant schizophrenia. Frontiers in Psychiatry. 2020;**11**:369-369

[47] Yoshikawa M et al. Expression of the mRNA and protein of serine racemase in primary cultures of rat neurons. European Journal of Pharmacology. 2006;**548**(1-3):74-76

[48] Chatterton JE et al. Excitatory glycine receptors containing the NR3 family of NMDA receptor subunits. Nature. 2002;**415**(6873):793-798

[49] Li Y et al. Identity of endogenous NMDAR glycine site agonist in amygdala is determined by synaptic activity level. Nature Communications. 2013;**4**:1760-1760

[50] Betz H et al. Glycine transporters: Essential regulators of synaptic transmission. Biochemical Society Transactions. 2006;**34**(Pt 1):55-58

[51] Juárez Olguín H et al. The role of dopamine and its dysfunction as a consequence of oxidative stress. Oxidative Medicine and Cellular Longevity. 2016;**2016**:9730467-9730467

[52] Errico F et al. Decreased levels of D-aspartate and NMDA in the prefrontal cortex and striatum of patients with schizophrenia. Journal of Psychiatric Research. 2013;**47**(10):1432-1437

[53] Guidolin D et al. Chapter eight—Adenosine A2A-dopamine D2 receptor-receptor interaction in neurons and astrocytes: Evidence and perspectives.

In: Giraldo J, Ciruela F, editors. Progress in Molecular Biology and Translational Science. London, United Kingdom: Academic Press; 2020. pp. 247-277

[54] Tarasov VV et al. Alterations of astrocytes in the context of schizophrenic dementia. Frontiers in Pharmacology. 2020;**10**:1612-1612

[55] Losi G, Mariotti L, Carmignoto G. GABAergic interneuron to astrocyte signalling: A neglected form of cell communication in the brain. Philosophical Transactions of the Royal Society of London. Series B, Biological Sciences. 2014;**369**(1654):20130609

[56] Matsuda LA et al. Structure of a cannabinoid receptor and functional expression of the cloned cDNA. Nature. 1990;**346**(6284):561-564

[57] Di Marzo V, Bifulco M, De Petrocellis L. The endocannabinoid system and its therapeutic exploitation. Nature Reviews. Drug Discovery. 2004;**3**(9):771-784

[58] Maccarrone M et al. Endocannabinoid signaling at the periphery: 50 years after THC. Trends in Pharmacological Sciences. 2015;**36**(5):277-296

[59] Hashimotodani Y, Ohno-Shosaku T, Kano M. Endocannabinoids and synaptic function in the CNS. The Neuroscientist. 2007;**13**(2):127-137

[60] Devane WA et al. Isolation and structure of a brain constituent that binds to the cannabinoid receptor. Science. 1992;**258**(5090):1946-1949

[61] Mechoulam R et al. Identification of an endogenous 2-monoglyceride, present in canine gut, that binds to cannabinoid receptors. Biochemical Pharmacology. 1995;**50**(1):83-90

[62] Secci ME et al. Astrocytic mechanisms involving Kynurenic acid control Δ(9)-tetrahydrocannabinol-induced increases in glutamate release in brain reward-processing areas. Molecular Neurobiology. 2019;**56**(5):3563-3575

[63] Li J et al. Astrocytes in oligodendrocyte lineage development and white matter pathology. Frontiers in Cellular Neuroscience. 2016;**10**:119-119

[64] Li J et al. Astrocytes in oligodendrocyte lineage development and white matter pathology. Frontiers in Cellular Neuroscience. 2016;**10**:119-125

[65] Terrillion CE et al. DISC1 in astrocytes influences adult neurogenesis and hippocampus-dependent behaviors in mice. Neuropsychopharmacology. 2017;**42**(11):2242-2251

[66] Christian KM, Song H, Ming G-L. Functions and dysfunctions of adult hippocampal neurogenesis. Annual Review of Neuroscience. 2014;**37**:243-262

[67] Clelland C et al. A functional role for adult hippocampal neurogenesis in spatial pattern separation. Science. 2009;**325**(5937):210-213

[68] Revest J et al. Adult hippocampal neurogenesis is involved in anxiety-related behaviors. Molecular Psychiatry. 2009;**14**(10):959-967

[69] Snyder JS et al. Adult hippocampal neurogenesis buffers stress responses and depressive behaviour. Nature. 2011;**476**(7361):458-461

[70] Zhou M et al. mTOR inhibition ameliorates cognitive and affective deficits caused by Disc1 knockdown in adult-born dentate granule neurons. Neuron. 2013;**77**(4):647-654

[71] Song H, Stevens CF, Gage FH. Astroglia induce neurogenesis from adult neural stem cells. Nature. 2002;**417**(6884):39-44

[72] Sultan S et al. Synaptic integration of adult-born hippocampal neurons is locally controlled by astrocytes. Neuron. 2015;**88**(5):957-972

[73] Araki T, Ikegaya Y, Koyama R. The effects of microglia-and astrocyte-derived factors on neurogenesis in health and disease. European Journal of Neuroscience. 2021;**54**(5):5880-5901

[74] Millar JK et al. Disruption of two novel genes by a translocation co-segregating with schizophrenia. Human Molecular Genetics. 2000;**9**(9):1415-1423

[75] Ma TM et al. Pathogenic disruption of DISC1-serine racemase binding elicits schizophrenia-like behavior via D-serine depletion. Molecular Psychiatry. 2013;**18**(5):557-567

[76] Bernstein H-G et al. Reduced density of DISC1 expressing astrocytes in the dentate gyrus but not in the subventricular zone in schizophrenia. Neuropsychopharmacology. 2018;**43**(3):457-458

[77] Bergami M et al. Uptake and recycling of pro-BDNF for transmitter-induced secretion by cortical astrocytes. The Journal of Cell Biology. 2008;**183**(2):213-221

[78] Quesseveur G et al. BDNF overexpression in mouse hippocampal astrocytes promotes local neurogenesis and elicits anxiolytic-like activities. Translational Psychiatry. 2013;**3**(4):e253-e253

[79] Brown AM, Ransom BR. Astrocyte glycogen and brain energy metabolism. Glia. 2007;**55**(12):1263-1271

[80] Suzuki A et al. Astrocyte-neuron lactate transport is required for long-term memory formation. Cell. 2011;**144**(5):810-823

[81] Zhou J et al. Lactate potentiates angiogenesis and neurogenesis in experimental intracerebral hemorrhage. Experimental & Molecular Medicine. 2018;**50**(7):1-12

[82] Morland C et al. Exercise induces cerebral VEGF and angiogenesis via the lactate receptor HCAR1. Nature Communications. 2017;**8**(1):1-9

[83] Lev-Vachnish Y et al. L-lactate promotes adult hippocampal neurogenesis. Frontiers in Neuroscience. 2019;**13**:403

[84] Ratajczak MZ et al. ATP-Nlrp3 Inflammasome-complement Cascade Axis in sterile brain inflammation in psychiatric patients and its impact on stem cell trafficking. Stem Cell Reviews and Reports. 2019;**15**(4):497-505

[85] de Zoete MR et al. Inflammasomes. Cold Spring Harbor Perspectives in Biology. 2014;**6**(12):a016287

[86] Ventura L et al. Involvement of NLRP3 inflammasome in schizophrenia-like behaviour in young animals after maternal immune activation. Acta Neuropsychiatr. 2020;**32**(6):321-327

[87] Ventura L et al. Involvement of NLRP3 inflammasome in schizophrenia-like behaviour in young animals after maternal immune activation. Acta Neuropsychiatrica. 2020;**32**(6):321-327

[88] Pasquini S et al. Adenosine receptors in neuropsychiatric disorders: Fine regulators of neurotransmission and potential therapeutic targets. International Journal of Molecular Sciences. 2022;**23**(3):1219

[89] He M et al. NLRP3/Caspase-1-mediated Pyroptosis of astrocytes induced by antipsychotics is inhibited by a histamine H1 receptor-selective agonist. Frontiers in Aging Neuroscience. 2022;**14**:847561

[90] He M et al. NLRP3/Caspase-1-mediated Pyroptosis of astrocytes induced by antipsychotics is inhibited by a histamine H1 receptor-selective agonist. Frontiers in aging. Neuroscience. 2022;**14**:847561

[91] Filiano AJ, Gadani SP, Kipnis J. Interactions of innate and adaptive immunity in brain development and function. Brain Research. 2015;**1617**:18-27

[92] Amor S et al. Inflammation in neurodegenerative diseases. Immunology. 2010;**129**(2):154-169

[93] González H, Pacheco R. T-cell-mediated regulation of neuroinflammation involved in neurodegenerative diseases. Journal of Neuroinflammation. 2014;**11**:201

[94] Najjar S et al. Neuroinflammation and psychiatric illness. Journal of Neuroinflammation. 2013;**10**(1):816

[95] Kirkpatrick B, Miller BJ. Inflammation and schizophrenia. Schizophrenia Bulletin. 2013;**39**(6):1174-1179

[96] Nyland H, Naess A, Lunde H. Lymphocyte subpopulations in peripheral blood from schizophrenic patients. Acta Psychiatrica Scandinavica. 1980;**61**(4):313-318

[97] Smith RS. A comprehensive macrophage-T-lymphocyte theory of schizophrenia. Medical Hypotheses. 1992;**39**(3):248-257

[98] Matloubi H et al. Decreased T cell response to mitogen and increased anti-cytoplasmic antibody in drug-free schizophrenic patients. Iranian Journal of Immunology. 2007;**4**(1):32-37

[99] Busse S et al. Different distribution patterns of lymphocytes and microglia in the hippocampus of patients with residual versus paranoid schizophrenia: Further evidence for disease course-related immune alterations? Brain, Behavior, and Immunity. 2012;**26**(8):1273-1279

[100] Aberg KA et al. A comprehensive family-based replication study of schizophrenia genes. JAMA Psychiatry. 2013;**70**(6):573-581

[101] Frydecka D et al. The role of genetic variations of immune system regulatory molecules CD28 and CTLA-4 in schizophrenia. Psychiatry Research. 2013;**208**(2):197-198

[102] Pillinger T et al. A meta-analysis of immune parameters, variability, and assessment of modal distribution in psychosis and test of the immune subgroup hypothesis. Schizophrenia Bulletin. 2019;**45**(5):1120-1133

[103] Corsi-Zuelli F, Deakin B. Impaired regulatory T cell control of astroglial overdrive and microglial pruning in schizophrenia. Neuroscience and Biobehavioral Reviews. 2021;**125**:637-653

[104] Schain M, Kreisl WC. Neuroinflammation in neurodegenerative disorders-a review. Current Neurology and Neuroscience Reports. 2017;**17**(3):25

[105] Enache D, Pariante CM, Mondelli V. Markers of central inflammation in major depressive disorder: A systematic review and meta-analysis of studies examining cerebrospinal fluid, positron emission tomography and post-mortem brain

tissue. Brain, Behavior, and Immunity. 2019;**81**:24-40

[106] Marques TR et al. Neuroinflammation in schizophrenia: Meta-analysis of in vivo microglial imaging studies. Psychological Medicine. 2019;**49**(13):2186-2196

[107] Plavén-Sigray P et al. Meta-analysis of the glial marker TSPO in psychosis revisited: Reconciling inconclusive findings of patient-control differences. Biological Psychiatry. 2021;**89**(3):e5-e8

[108] Toker L et al. Transcriptomic evidence for alterations in astrocytes and Parvalbumin interneurons in subjects with bipolar disorder and schizophrenia. Biological Psychiatry. 2018;**84**(11):787-796

[109] Gandal MJ et al. Transcriptome-wide isoform-level dysregulation in ASD, schizophrenia, and bipolar disorder. Science. 2018;**362**(6420):eaat8127

[110] González-Peñas J et al. Enrichment of rare genetic variants in astrocyte gene enriched co-expression modules altered in postmortem brain samples of schizophrenia. Neurobiology of Disease. 2019;**121**:305-314

[111] Zöller T et al. Silencing of TGFβ signalling in microglia results in impaired homeostasis. Nature Communications. 2018;**9**(1):4011

[112] Schilling T et al. Astrocyte-released cytokines induce ramification and outward K+ channel expression in microglia via distinct signalling pathways. The European Journal of Neuroscience. 2001;**14**(3):463-473

[113] Travis MA, Sheppard D. TGF-β activation and function in immunity. Annual Review of Immunology. 2014;**32**:51-82

[114] Miterauer B, Baer W. Disorders of human consciousness in the tripartite synapses. Medical Hypotheses. 2020;**136**:109523

[115] Mitterauer BJ. Pathophysiology of schizophrenia based on impaired glial-neuronal interactions. Open journal of medical. Psychology. 2014;**03**(02):15

[116] Mitterauer B. Loss of function of glial gap junctions may cause severe cognitive impairments in schizophrenia. Medical Hypotheses. 2009;**73**(3):393-397

[117] Genocchi B, Lenk K, Hyttinen J. Influence of Astrocytic Gap Junction Coupling on in Silico Neuronal Network Activity. Berlin: Springer;

[118] Lenk K et al. Larger connection radius increases hub astrocyte number in a 3-D neuron–astrocyte network model. IEEE Transactions on Molecular, Biological and Multi-Scale Communications. 2021;7(2):83-88

[119] Tarasov VV et al. Alterations of astrocytes in the context of schizophrenic dementia. Frontiers in Pharmacology. 2019;**10**:1612

[120] Arion D et al. Molecular evidence for increased expression of genes related to immune and chaperone function in the prefrontal cortex in schizophrenia. Biological Psychiatry. 2007;**62**(7):711-721

[121] Kordi-Tamandani DM et al. Association of glutathione s-transferase gene methylation with risk of schizophrenia in an Iranian population. Pharmacology. 2014;**94**(3-4):179-182

[122] Kim JJ et al. Association analysis of heat shock protein 70 gene polymorphisms in schizophrenia. European Archives of Psychiatry

and Clinical Neuroscience. 2008;**258**(4):239-244

[123] Kowalczyk M et al. Heat shock protein 70 gene polymorphisms are associated with paranoid schizophrenia in the polish population. Cell Stress & Chaperones. 2014;**19**(2):205-215

[124] Yu WW et al. Heat shock protein 70 suppresses neuroinflammation induced by α-synuclein in astrocytes. Molecular and Cellular Neurosciences. 2018;**86**:58-64

[125] Marinova Z et al. Histone deacetylase inhibition alters histone methylation associated with heat shock protein 70 promoter modifications in astrocytes and neurons. Neuropharmacology. 2011;**60**(7-8):1109-1115

[126] Bozidis P et al. HSP70 polymorphisms in first psychotic episode drug-naïve schizophrenic patients. Life Sciences. 2014;**100**(2):133-137

[127] Gerber KJ, Squires KE, Hepler JR. Roles for regulator of G protein signaling proteins in synaptic signaling and plasticity. Molecular Pharmacology. 2016;**89**(2):273-286

[128] Buckholtz JW et al. Allelic variation in RGS4 impacts functional and structural connectivity in the human brain. The Journal of Neuroscience. 2007;**27**(7):1584-1593

[129] Liu YL et al. Evaluation of RGS4 as a candidate gene for schizophrenia. American Journal of Medical Genetics. Part B, Neuropsychiatric Genetics. 2006;**141b**(4):418-420

[130] Mirnics K et al. Disease-specific changes in regulator of G-protein signaling 4 (RGS4) expression in schizophrenia. Molecular Psychiatry. 2001;**6**(3):293-301

[131] Ding L et al. Expression of the longest RGS4 splice variant in the prefrontal cortex is associated with single nucleotide polymorphisms in schizophrenia patients. Frontiers in Psychiatry. 2016;**7**:26

[132] Vrajová M et al. The effects of siRNA-mediated RGS4 gene silencing on the whole genome transcription profile: Implications for schizophrenia. Neuro Endocrinology Letters. 2011;**32**(3):246-252

[133] Ho SM et al. Evaluating synthetic activation and repression of neuropsychiatric-related genes in hiPSC-derived NPCs, neurons, and astrocytes. Stem Cell Reports. 2017;**9**(2):615-628

[134] Martins-de-Souza D, Lebar M, Turck CW. Proteome analyses of cultured astrocytes treated with MK-801 and clozapine: Similarities with schizophrenia. European Archives of Psychiatry and Clinical Neuroscience. 2011;**261**(3):217-228

[135] Sun L et al. Adolescent social isolation affects schizophrenia-like behavior and astrocyte biomarkers in the PFC of adult rats. Behavioural Brain Research. 2017;**333**:258-266

[136] Akkouh IA et al. Transcriptome analysis reveals disparate expression of inflammation-related miRNAs and their gene targets in iPSC-astrocytes from people with schizophrenia. Brain, Behavior, and Immunity. 2021;**94**:235-244

[137] Xia M et al. Behavioral sequelae of astrocyte dysfunction: Focus on animal models of schizophrenia. Schizophrenia Research. 2016;**176**(1):72-82

[138] Banasr M, Duman RS. Glial loss in the prefrontal cortex is sufficient to induce depressive-like behaviors. Biological Psychiatry. 2008;**64**(10):863-870

[139] Lima A et al. Astrocyte pathology in the prefrontal cortex impairs the cognitive function of rats. Molecular Psychiatry. 2014;**19**(7):834-841

[140] Danbolt NC. Glutamate uptake. Progress in Neurobiology. 2001;**65**(1):1-105

[141] Matute C et al. Increased expression of the astrocytic glutamate transporter GLT-1 in the prefrontal cortex of schizophrenics. Glia. 2005;**49**(3):451-455

[142] Bellesi M et al. GLT-1 upregulation impairs prepulse inhibition of the startle reflex in adult rats. Glia. 2009;**57**(7):703-713

[143] Melone M, Bellesi M, Conti F. Synaptic localization of GLT-1a in the rat somatic sensory cortex. Glia. 2009;**57**(1):108-117

[144] Nozari M et al. Postnatal MK-801 treatment of female rats impairs acquisition of working memory, but not reference memory in an eight-arm radial maze; no beneficial effects of enriched environment. Psychopharmacology. 2015;**232**(14):2541-2550

[145] Nozari M et al. Sex-specific restoration of MK-801-induced sensorimotor gating deficit by environmental enrichment. Neuroscience. 2015;**299**:28-34

[146] Nozari M et al. Enriched environment prevents cognitive and motor deficits associated with postnatal MK-801 treatment. Psychopharmacology. 2014;**231**(22):4361-4370

[147] Chess AC et al. Elevations of endogenous kynurenic acid produce spatial working memory deficits. Schizophrenia Bulletin. 2007;**33**(3):797-804

[148] Iaccarino HF et al. The effect of transient increases in kynurenic acid and quinolinic acid levels early in life on behavior in adulthood: Implications for schizophrenia. Schizophrenia Research. 2013;**150**(2-3):392-397

[149] Yolken R, Dickerson F, Fuller Torrey E. Toxoplasma and schizophrenia. Parasite Immunology. 2009;**31**(11):706-715

[150] Notarangelo F et al. Evaluation of kynurenine pathway metabolism in toxoplasma gondii-infected mice: Implications for schizophrenia. Schizophrenia Research. 2014;**152**(1):261-267

[151] Kannan G, Pletnikov MV. Toxoplasma gondii and cognitive deficits in schizophrenia: An animal model perspective. Schizophrenia Bulletin. 2012;**38**(6):1155-1161

[152] Tezuka T et al. Cuprizone short-term exposure: Astrocytic IL-6 activation and behavioral changes relevant to psychosis. Neurobiology of Disease. 2013;**59**:63-68

[153] Petegnief V et al. Neuronal injury-induced expression and release of apolipoprotein E in mixed neuron/glia co-cultures: Nuclear factor κB inhibitors reduce basal and lesion-induced secretion of apolipoprotein E. Neuroscience. 2001;**104**(1):223-234

[154] Brambilla R et al. Inhibition of astroglial nuclear factor κB reduces

inflammation and improves functional recovery after spinal cord injury. The Journal of Experimental Medicine. 2005;**202**(1):145-156

[155] Singer P, Dubroqua S, Yee KB. Inhibition of glycine transporter 1: The yellow brick road to new schizophrenia therapy? Current Pharmaceutical Design. 2015;**21**(26):3771-3787

[156] Takahashi N, Sakurai T. Roles of glial cells in schizophrenia: Possible targets for therapeutic approaches. Neurobiology of Disease. 2013;**53**:49-60

[157] Vallejo-Illarramendi A et al. Clozapine reduces GLT-1 expression and glutamate uptake in astrocyte cultures. Glia. 2005;**50**(3):276-279

[158] Pocivavsek A et al. Chapter 25—Astrocytes as pharmacological targets in the treatment of schizophrenia: Focus on Kynurenic acid. In: Pletnikov MV, Waddington JL, editors. Handbook of Behavioral Neuroscience. Amsterdam, Netherlands: Elsevier; 2016. pp. 423-443

[159] Kantrowitz JT et al. High dose D-serine in the treatment of schizophrenia. Schizophrenia Research. 2010;**121**(1):125-130

[160] Girgis RR et al. The past and future of novel, non-dopamine-2 receptor therapeutics for schizophrenia: A critical and comprehensive review. Journal of Psychiatric Research. 2019;**108**:57-83

[161] Egerton A et al. Glutamate in schizophrenia: Neurodevelopmental perspectives and drug development. Schizophrenia Research. 2020;**223**: 59-70

[162] Kantrowitz JT et al. Neurophysiological effects of Bitopertin in schizophrenia. Journal of Clinical Psychopharmacology. 2017;**37**(4):447-451

[163] Powell CL, Davidson AR, Brown AM. Universal glia to Neurone lactate transfer in the nervous system: Physiological functions and pathological consequences. Biosensors. 2020;**10**(11):183

[164] Katsel P et al. Astrocyte and glutamate markers in the superficial, deep, and white matter layers of the anterior cingulate gyrus in schizophrenia. Neuropsychopharmacology. 2011;**36**(6):1171-1177

[165] Tarasov VV et al. Alterations of astrocytes in the context of schizophrenic dementia. Frontiers in Pharmacology. 2020;**10**:1612

[166] Trépanier MO et al. Postmortem evidence of cerebral inflammation in schizophrenia: A systematic review. Molecular Psychiatry. 2016;**21**(8):1009-1026

[167] Lee EE et al. Plasma levels of neuron- and astrocyte-derived Exosomal amyloid Beta1-42, amyloid Beta1-40, and phosphorylated tau levels in schizophrenia patients and non-psychiatric comparison subjects: Relationships with cognitive functioning and psychopathology. Frontiers. Psychiatry. 2021;**11**:532624

[168] Tavakolizadeh J et al. MicroRNAs and exosomes in depression: Potential diagnostic biomarkers. Journal of Cellular Biochemistry. 2018;**119**(5):3783-3797

[169] Banigan MG et al. Differential expression of exosomal microRNAs in prefrontal cortices of schizophrenia and bipolar disorder patients. PLoS One. 2013;**8**(1):e48814

[170] Saeedi S et al. The emerging role of exosomes in mental disorders. Translational Psychiatry. 2019;**9**(1):122

[171] Zhang Y et al. Exosomes: Biogenesis, biologic function and clinical potential. Cell & Bioscience. 2019;**9**(1):19

[172] Ranganathan M et al. Analysis of circulating exosomes reveals a peripheral signature of astrocytic pathology in schizophrenia. The World Journal of Biological Psychiatry. 2022;**23**(1):33-45